Degrees of Freedom

Degrees of Freedom

On Robotics and Social Justice

Tom Williams

The MIT Press
Cambridge, Massachusetts
London, England

The MIT Press
Massachusetts Institute of Technology
77 Massachusetts Avenue, Cambridge, MA 02139
mitpress.mit.edu

The MIT Press would like to thank the anonymous peer reviewers who provided comments on drafts of this book. The generous work of academic experts is essential for establishing the authority and quality of our publications. We acknowledge with gratitude the contributions of these otherwise uncredited readers.

This book was set in Stone Serif and Stone Sans by Westchester Publishing Services. Printed and bound in the United States of America.

Library of Congress Cataloging-in-Publication Data is available.

ISBN: 978-0-262-55402-2

EU Authorised Representative: Easy Access System Europe, Mustamäe tee 50, 10621 Tallinn, Estonia | Email: gpsr.requests@easproject.com

Contents

Acknowledgments vii

1 **Secret Agents** 1
2 **Black Servos, White Masks** 25
3 **Identity Crisis** 69
4 **Control System** 113
5 **Bounding Box** 137
6 **End Effector** 169
7 **Future Tense** 215

Notes 223
Bibliography 291
Index 347

Acknowledgments

I am deeply indebted to a number of people who helped make this book a reality. I am very grateful to Nicki Washington and her Cultural Competence in Computing program, which put me on the path that led to this book. I am similarly grateful to my department for supporting me in teaching a race and computing course, offered outside my department, which was fundamental in shaping my thinking on the issues presented in this book. And I am grateful to mentors such as Ayanna Howard and Chad Jenkins, who brought me into thoughtful conversations on these topics after the #NoJusticeNoRobots campaign.

Enormous thanks go to the amazing graduate (and undergraduate!) students from the MIRRORLab who led much of the human–robot interaction research I describe throughout this book. Working with the thoughtful, diverse, and transdisciplinary group of scholars we have cultivated in our lab is hands-down the best part of being a professor. Thanks also go to my collaborators on much of that work, especially Beth Phillips, Boyoung Kim, Qin Zhu, and Katie Winkle, who have all been an absolute joy to work with. And this work would not have been possible without the support of those who funded this work, especially AFOSR (through grants FA9550-20-1-0089 and FA9550-23-1-0036), the NSF (through grants IIS-2044865, IIS-1909847, IIS-1849348, IIS-1909864 and IIS-1823245), and NASA (through grant 80NSSC20K0070). The views expressed in this book, however, do not necessarily reflect the views of those institutions.

In the actual writing of this book, I am grateful to my editor Phil Laughlin for taking a chance on this book; to the reviewers he recruited to give constructive, insightful, and honest feedback on its first draft; to the Keepers of Khorvaire for serving as a sounding board for some of my more extreme ideas in between games of Dungeons and Dragons; to my colleagues, including Filipa Correia, Ryan Blake Jackson, Beth Phillips, Matthew Stone, Alan Wagner, and Yifei Zhu, who provided feedback on chapter drafts; and to Anne Lowy for her extensive feedback across the entirety of the book. I am further grateful to my colleagues who hosted me during my sabbatical for talks at Birmingham, Cambridge, Delft, Edinburgh, Eindhoven Ghent, Glasgow, Heriot-Watt, Imperial College London, Kings College London, Leiden, Instituto Superio-Tecnico Lisboa, Oxford, Twente, Utrecht, Vrij University Amsterdam, and, of course, at the University of Bristol and the Bristol Robotics Laboratory. I am also grateful to the reviewers of my papers "No Justice, No Robots: From the Dispositions of Policing to an Abolitionist Robotics"[1] (coauthored with Kerstin Haring, and some parts of which, solely authored by me, have been reproduced as part of chapter 6), "The Eye of the Robot Beholder"[2] (some parts of which have been reproduced as part of chapter 5), and "Understanding Roboticists' Power Through Matrix Guided Technology Power Analysis"[3] (some parts of which have been reproduced at the ends of chapters 3–6). Any text reproduced from those papers is licensed through a Creative Commons CC-BY license. Finally, I am grateful to the University of Bristol for giving me an official place to write while on sabbatical, and for Society Café and Watershed for giving me unofficial places to do so.

And above all, I am eternally grateful to my partner, Rachel, for her love, her support, and her patience.

1 Secret Agents

In the summer of 2016, I had the opportunity to attend the annual open house of DFKI, the German Research Centre for Artificial Intelligence. The visit was incredible, with demonstrations of autonomous cars, underwater robots intended to operate on far-off planets, and huge insect-inspired robots. But what stuck with me most was AILA, the Artificial Intelligence Lightweight Android.

AILA (shown in figure 1.1) is a humanoid robot designed for use on the International Space Station. What is immediately obvious to most observers (as I can attest after showing AILA to many years of robot ethics classes) is that AILA is clearly designed to appear both *female* and *feminine*. AILA's morphology (the shape of its physical form) is distinctly female inspired, with an obvious chest, a thin waist, wide hips, and a pronounced rear. In addition, AILA's morphology is distinctly feminine, with a skirt, a bob haircut, and oversized eyes reminiscent of the ways that feminine characters are often animated. These are clearly intentional features, and in fact, product material online refers to AILA as "she." Moreover, AILA is designed to depict not just *a* woman but rather a particular *type* of woman: a *White* woman.

Why was AILA designed this way? When I posed this question to the (White male) engineer posted next to AILA in the DFKI exhibit hall, I was told, "It allowed us to fit in more computer chips."[1]

As you might expect, the need to "fit in more chips" is not the real reason why AILA was designed in the image of a White woman. If chip storage were the designers' primary concern, AILA would not have the chest/waist/hips ratio that it does. And of course, this excuse does not explain AILA's hair, eyes, or racialization as White. So, why *was* AILA designed in the image of a White woman? In fact, why are *so many* humanoid and interactive robots today designed to depict White men and White women?

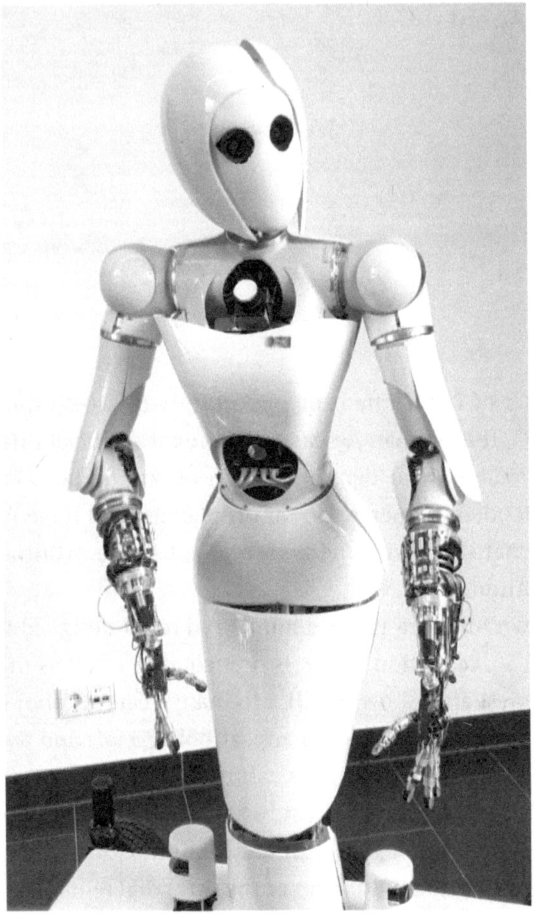

Figure 1.1
AILA. Image taken by Tom Williams, June 14, 2016.

Most humanoid and interactive robots today are designed in similar ways because they inherit from a shared *imaginary*. That is, these robots inherit from a shared cultural vision of what it means to be a robot, what robots look and sound like, and where they are used.

Specifically, today's robots inherit from a White patriarchal robot imaginary. Across both engineering and science fiction, most robots today are envisioned in ways that draw on the preferences and values of White men, and as a result, most robots today are subsequently designed and deployed in ways that not only predominantly serve the interests of White men but, moreover, help to perpetuate a society arranged to the benefit of White men. As I set out to develop a better understanding of how and why today's robots are designed and deployed according to this imaginary, I discovered that robotics had an unsettling and little-known origin story that explained all too much about the state of modern robotics.

Where does the idea of the robot come from? When asked this question, some scholars will point to Isaac Asimov and his wildly popular 1950 collection *I, Robot*,[2] which inspired the Unimate (the first industrial robot[3]) and gave us the word "robotics." Other scholars will point to Karel Čapek's *Rossum's Universal Robots*,[4] penned thirty years prior, which played a pivotal role in the history of science fiction and gave us the word "robot." Yet, the idea of humanlike-yet-nonhuman social beings with mechanical bodies clearly predated and informed both Asimov's and Čapek's work. This means that the robot imaginary predates Asimov and Čapek, even if we did not have the word "robot" to ground that imaginary until the staging of Čapek's play.

How much earlier do we need to go back to find the genesis of the robot imaginary that inspired Čapek and Asimov? As psychologist Kara Weisman demonstrates, robots follow a historical thread that traces back to human prehistory.[5] For tens of thousands of years, humans and their predecessors have reasoned about a host of entities viewed as alive—but not in the same way as humans—including gods, spirits, and plants. Although the influence of this prehistoric anthropology may still be felt today, it is too far removed from robotics to serve as a useful origin point for the *robot* imaginary—it does not help to explain the specific design trends seen today. Alternatively, as Dunstan and Hoffman argue,[6] we could go back to the golem of Jewish folklore or the Greek myth of Pygmalion, both of which involve human-constructed anthropomorphic nonhuman servants. Although these mythological creatures surely informed the robot imaginary in some way, they do not provide a useful origin story either, as these nonmechanical creatures were understood as entirely fictive rather than as near-future technological possibilities and bear little relation to how we think about and deploy robots today. Finally, some scholars point to the Turk, an anthropomorphic automaton exhibited across Europe in the eighteenth and nineteenth centuries that ostensibly played chess (but, in fact, was controlled from within by a human).[7] The Mechanical Turk is in some ways a clearer origin for robotics than these ancient myths. Unlike previous legends, the Turk was a physically constructed mechanical creature built in the image of a human. And echoes of the Turk are still felt in modern robotics and artificial intelligence (AI) in the way the Turk made invisible the exploitative human labor necessary to make it work. Yet, the Turk was viewed fundamentally differently from today's social robots or the robotic characters we have seen in the science fiction of the past century, as the Turk was viewed as a mechanical wonder of ingenious design, not as a socially interactive agent.[8] Indeed, observers were keen to try to puzzle apart the operation of the Turk, and many correctly guessed that the machine was a hoax, although they were not able to prove it.

So, if we discard gods, golems, and Mechanical Turks, what is the origin of robots that inspired Čapek? Recent scholarship in literary studies suggests that the answer may be found in the time period surrounding the US civil war, shortly after the demise of the Turk and nearly equidistant in time from the height of the Turk's popularity and the first production of *Rossum's Universal Robots*.

The Origins of the Robot Imaginary in *The Steam Man of the Prairies*

Written in 1868, *The Steam Man of the Prairies*[9] includes what has been regarded as the first depiction of a robot in literature.[10] In the story, the titular steam man is an anthropomorphic mechanical creation that carries its master, a young White boy, on a journey across America.

Here, we have a mechanical, human-created, humanlike construct that is depicted as a social other: much closer to a social robot than the Mechanical Turk or even Čapek's "robots," which were biological rather than mechanical. Moreover, the steam man was viewed by both Ellis and his readers as a near-future technology, as it was inspired by a patent filed earlier in 1868. Finally, the fictive steam man was clearly designed to be viewed as a humanlike but nonhuman social agent. All of these factors, combined with the popular success of *The Steam Man of the Prairies*, make the novel a compelling new origin point for the robot imaginary and suggest a fascinating—yet disconcerting—robotic counter-history as the true nature of the robot imaginary is thrown into sharp relief. This is because the steam man was Black.

Both the original patent schematics and cover artwork for the real and fictive steam men were clearly designed to evoke stereotypical depictions of Black slaves. Specifically, these schematics and artwork were designed to evoke one of the most popular fictional characters in the most popular cultural art form of the time: the minstrel character Zip Coon.[11] This is a shocking genesis for today's robots—and, at first glance, it seems to conflict with the dominant depiction of today's humanoid and interactive robots as *White* men and women. But the through line to today's robots becomes clear when we consider the *motivations* for those original caricatured depictions. As literary scholar Taylor Evans has argued, both the patent schematics and cover art surrounding *The Steam Man of the Prairies* were intended to communicate a specific narrative about how technology could be used to satisfy the White male dream of economic dominance through sociological and social mastery over non-White people[12]—and as we will see, this narrative continues to lurk behind robotics even today.

What motivated this narrative, and how did it lead to *The Steam Man of the Prairies*? The late 1860s was a period of existential terror for White men in the American South. The end of slavery represented the end of the system that provided Southern White men with opportunities for not only economic dominance but also social dominance over both slaves and nonslaveholding White people, as well as sexual dominance, as White slaveholders not only raped their Black female slaves but also, in fact, were incentivized to do so under the law.[13]

As Evans notes, the patenting of the steam man and the release of *The Steam Man of the Prairies* directly responded to this existential uncertainty by proposing a possible solution.[14] Ellis's novel depicted a mechanical Black slave carrying a young White boy westward, and showed the White boy destroying the Black mechanical man in order to massacre Native Americans, seize their wealth, and bring it home, whereupon he and his companions achieved financial success and satisfied their sexual desires by marrying the women of their choosing. The message of the story for the young White boys who read *The Steam Man of the Prairies* was clear: slavery may be gone, but by mastering technology, you may yet achieve your dream of economic, social, and sexual dominance, and in doing so, you will not need the Black man anymore and will be free to dispose of him as you wish.

In order to spin this narrative, Ellis merely replaced one entity seen as "humanlike but inhuman" with another. At the time of its release, many White Americans did not view Black people as humans. This fact, as political philosopher Charles Mills has pointed out,[15] is critical for understanding much of pre-twenty-first century White philosophy, from Thomas Jefferson's claims that all men are created equal (in which Jefferson only intended "all men" to apply to White men) to Immanuel Kant's political and moral philosophy of universal law (in which Kant meant "universal" to apply to White men). This dehumanization of those racialized as Black (in order to justify their enslavement) is also critical for understanding the fight by many Americans against the provision of political rights such as citizenship and voting to Black people well into the twentieth century. As such, Ellis's substitution of a literal Black slave for a mechanical one can be seen as communicating a substitution of the biological for the technological rather than a substitution of the human for the inhuman.

This is the legacy of robotics that informed *Rossum's Universal Robots* and may explain why Čapek chose to use the word "robot" in the first place, which, in Czech, means "slave." This is the legacy of robotics that informed *I, Robot* and may explain why Asimov chose to focus his stories on the Three Laws of Robotics, which Asimov's characters describe as "good, healthy slave complexes."[16] This is the legacy of robotics that inspired the first industrial robots, which the "Father of Robotics" Joseph Engelberger motivated by saying, "Since we can't have slaves or kick around Black people anymore, the robot serves that purpose."[17] And as a result, this legacy explains the ultimate effect those first industrial robots had. Just as Ellis's story suggested the use of robots to replace Black labor while maintaining White power and wealth, Engelberger ultimately deployed the Unimate into factories in ways that eliminated or degraded Black working-class jobs while forcing remaining workers to work harder, faster, for less pay, and with weakened labor protections for the benefit of wealthy White male capitalists.[18] And as a result, this legacy continues to inform how, where, and why robots are designed and deployed today.

In this book, we will see how this counter-history of the origin and evolution of the White patriarchal robot imaginary explains the ways today's robots are designed to look and sound, especially in terms of how today's robots tend to be racialized and gendered. We will see how this legacy explains the ways today's robots are designed to behave, especially in terms of the norms of "appropriate" behavior today's robots are designed to follow. We will see how this legacy explains the ways today's robots perceive people, especially in terms of the ways that robots serve as technological equivalents to Whites-only water fountains, and the ways that robots are increasingly sorting people into unchallengeable categories of race and gender. And we will see how this legacy explains the domains into which today's robots are deployed, from warehouses and factories to police stations—domains in which today's robots predominantly redirect power and wealth into the hands of wealthy, White, American men. As a result, although many roboticists today *think* they are working toward the creation of collaborative robots, or "cobots," this book argues that these roboticists may be better understood to be unwittingly working toward the robotic enablers of the New Jim Crow,[19] or "New Jim Cobots."

In short, this book argues that robotics is a field that originated from a desire to maintain slave-based White supremacist power structures and is a field that, even today, continues to unknowingly serve that goal. This means that today's roboticists have substantial power: the power either to continue to design according to that status quo, thus repeating the mistakes of the past, or to subvert that status quo and use the new technological solutions made possible through robotics to genuinely address societal problems *without* reinforcing inequitable power structures.

How is it possible that roboticists today are not taught this origin story? How is it possible that roboticists today remain unaware of the ways their design decisions typically reinforce White and patriarchal power structures? And how is it possible that so few roboticists today are actively working to subvert the White patriarchal robot imaginary and to envision new possible futures for the field of robotics? The answer to these questions may lie in the overly limited way in which roboticists have typically thought about the power they wield.

What Roboticists Talk about When They Talk about Power

When roboticists talk about power in a social (rather than electrical or physical) sense, they are typically talking about robots' potential for persuasion and influence.[20] Often, discussions of robots' persuasive power are framed in terms of a broader tradition of persuasive technologies that encourage people to behave in ways that are safer or more beneficial to themselves and others. Speed bumps can encourage drivers to slow down, prompts to consider truthfulness can discourage social media users from sharing fake news, and wearable devices can help motivate users toward behaviors that benefit their physical and emotional health. The same is true for robots, which are capable of social behaviors that seem to give them unique persuasive power relative to other technologies.[21]

Human–robot interaction researchers have identified a number of ways that robots can put this potential persuasive power to positive use in their interactions with humans. Robots designed for use by kids have been shown to be effective at everything from getting kids to eat their vegetables[22] to promoting productive conflict resolution.[23] And robots designed for adults have been shown to help nurture empathy and encourage altruism.[24] Although these types of altruistic "nudges" are sometimes critiqued for being paternalistic,[25] the motivation provided by socially assistive robots (which provide social rather than physical assistance) is often explicitly sought out by patients and clients.

Two excellent examples of socially assistive robot persuasion come from the early work of social roboticist Katie Winkle. First, while working with physical therapists, Winkle showed that by enabling social robots to use the persuasive strategies that are already used by therapists (such as displaying goodwill and emphasizing similarity with one's interactant), therapy robots could effectively persuade patients to perform boring yet necessary physical therapy exercises.[26] Second, while working with exercise coaches as part of the UK National Health Service's "Couch to 5K" program, Winkle showed that by enabling social robots to learn to mimic the persuasive strategies used by exercise coaches, those robots could gradually learn to more effectively encourage human exercise partners.[27] In both of these use cases, robots mimicked human health experts' existing persuasive strategies to help patients and clients achieve their own goals.

But considering *why* robots have this unique persuasive power also highlights the ethical risks of persuasive technologies. Robots seem to have a unique ability to persuade people both because of humans' tendency to overtrust robots and because of robots' secret agency and patiency.

Overtrust

Robots are able to persuade people, in part, because of the extent to which people readily develop overtrust in robotic technologies, making themselves vulnerable to harm in the process.[28] The classic example of overtrust in robotics comes from the work of Paul Robinette, performed while he was a student at Georgia Tech advised by Ayanna Howard and Alan Wagner.[29] Robinette and his colleagues performed an experiment in which participants interacted with a tour-guide robot and observed its capabilities. For half of the participants, the robot clearly malfunctioned during parts of this demonstration. Robinette and his colleagues were interested in whether these displays would meaningfully impact the trust placed in the tour-guide robot when things went wrong. Partway through each experiment, the

Georgia Tech team flooded their experimental environment with smoke and had their robot ask the participant to follow it in order to evacuate. Would people follow the robot's path to "safety" or use their own judgment to leave the building?

Every single participant in Robinette's experiment, even those who had seen the robot malfunction, followed the robot's instructions rather than the illuminated exit signs. In follow-up experiments, they found that even when the robot led participants into a dark room with no discernable exit, most participants still followed the robot, with one-third of participants squeezing past furniture blocking the entrance to the dark room and an additional third refusing to leave the robot's side, requiring the experimenters to come and retrieve them. This type of blind trust in robots makes people particularly vulnerable to robots' potential for persuasive power. Yet, robots are also capable of more pernicious forms of persuasion due to the extent to which people view them as moral and social agents and moral and social patients.

Agency

Robots are often perceived as moral and social agents: interactive, adaptive, and autonomous entities capable of taking actions that have clear potential for moral and social impact.[30] When robots are viewed as both moral and social agents, people may sanction them for morally and socially inappropriate actions. But more importantly, they may also allow themselves to be sanctioned by those robots in return. This ultimately means that these robots will participate in the power dynamics of their communities, including the *negotiation* of their communities' social and moral norms.

Humans' conceptions of what is morally right and wrong are dynamic and malleable; they are not innate but rather socially constructed.[31] For behaviors to be viewed as permissible, forbidden, or obligatory within a community, the moral agents who are members of that community must comply with those views, communicate those views to others, and enforce those views through punishment and reward. If a principle is not complied with, transferred, or enforced by all community members, it will be weakened and ultimately may cease to be a norm. Roboticists thus have a critical responsibility to wield their persuasive power effectively. Robots' status as moral agents within their moral community enables them to communicate and enforce norms.[32] But, moreover, this status comes with a responsibility to prevent robots from inadvertently weakening important human moral and social norms.

To see how robots may inadvertently weaken human moral norms, I will use an example from my own research. When I was a graduate student, one of the topics I worked on was enabling robots to generate clarification requests. Asking for clarification is a critical capability for keeping dialogue on track. For example, if someone asks a robot in an eldercare facility to "bring me my medication," the robot may need to ask which medication they mean. Unfortunately, asking for clarification to resolve ambiguity can become problematic when the action being requested comes with ethical risk. If a robot is told, "I'm annoyed with Alice. Find her medication and throw it in the trash," then, suddenly, asking which medication the speaker means becomes problematic, as it seems to imply that the robot would be willing to comply with some version of the request—otherwise, why ask? And indeed, my former student Ryan Blake Jackson found that when people observe a robot asking for clarification in response to a morally problematic request, they come to believe that the robot probably thinks it is okay to perform the action being asked about. But more concerningly, participants *also* come to believe that it may *actually be okay* for the robot to perform that action! In short, simply by asking for clarification, robots may exert negative influence on human observers, weakening those humans' moral norms.[33]

These results demonstrate the ways that robots, perhaps due to their status as moral agents, are able to influence humans' beliefs as to what is appropriate (or inappropriate) for robots to do. But there is also good reason to suspect that robots could similarly influence humans' beliefs as to what is appropriate (or inappropriate) for *humans* to do. Indeed, research has shown that robots can readily change how people interact with each other[34] and can spur the creation of new social norms among human interactants.[35] One recent example of this is MicBot,[36] developed at Cornell University by Hamish Tennant, Solace Shen, and Malte Jung in 2019.[37] MicBot is a tabletop studio microphone that facilitates group conversations. Building on prior work from Shen that had shown how robots could encourage children to resolve conflicts productively,[38] MicBot was designed to perform two behaviors: turning toward whoever is speaking to better capture their voice and turning toward individuals who are *not* speaking much to encourage them to participate in the conversation. After deploying MicBot in group-based interactions in their laboratory, the team found that MicBot's behavior encouraged more even turn-taking among group members. This idea has since been expanded on by other researchers to encourage the inclusion of immigrant children and children with visual impairments into group play[39] and to actively discourage sexism.[40]

In my own lab, we have shown that robots can even shape human behaviors through *implicit* dimensions of their designs rather than through overt behaviors. For example, in experiments led by my former student Ruchen Wen, we found that human politeness toward robots can be shaped by the *wakewords* used to interact with those robots. Outside of robotics, wakewords such as "Hey, Alexa" are often used so that smart devices only *really* listen and respond to what you're saying to them directly. But some parents and journalists have raised concerns about this wakeword-driven design, speculating that this command-based interaction ("Hey, Alexa, do such-and-such") could train children to think in terms of commands, degrading their ability to interact politely both with voice assistants[41] and, critically, with other people.[42] As a response to these concerns, Amazon developed an "Alexa, please" mode, which aimed to encourage children to use "please" when talking to Alexa.[43] Unfortunately, this strategy may have been doomed to backfire. Counterintuitively, research has shown that the word "please" *discourages* politeness when used at the beginning of a sentence (as encouraged by Alexa), because what is likely to follow "sentence-frontal please" is a direct command ("Please do such-and-such"). In contrast, "please" is typically polite when appearing in the middle of a sentence, when used with other politeness strategies such as indirect requests (e.g., "Could you *please* . . . ," "Can you *please* . . . ," etc.).[44] As such, we were curious if the "Alexa, please" mode might actually *decrease* politeness by discouraging these indirect requests, and we wondered if a wakeword such as "Excuse me, Alexa" could actually be more effective.

In a series of experiments,[45] we asked participants to interact with wakeword-driven Pepper robots (like that shown in figure 1.2) that needed to be commanded with "Hey, Pepper," "Pepper, please," or "Excuse me, Pepper." Our results showed that "Pepper, please" significantly lowered the number of indirect requests (such as "Could you . . .") participants used when speaking to the robot compared to both "Hey, Pepper" and "Excuse me, Pepper." In short, asking users to say "please" made them *less* polite toward robots. We also looked at how our participants went on to interact with other people. Overall, people were already pretty polite with each other, phrasing most of their commands to other humans as indirect requests regardless of experimental condition. But both "Excuse me, Pepper" and "Pepper, please" did in fact lead participants to use "please" more when talking with other

Figure 1.2
Pepper. Image taken by Tom Williams, July 20, 2018.

people, accentuating the politeness of their language. In short, this work showed that changes in how people need to interact with robots (without changing robot behavior at all) can change the ways that people adhere to sociocultural politeness norms, both in their interactions with robots *and* in their interactions with other people. This serves as a clear example of "ripple effects." By changing how people interact with robots, we might be changing how people interact with each other.

The potential for robot-driven ripple effects raises questions about robots' long-term impacts. Will the influence robots have on human–human interactions continue to manifest long after those robots are gone? If robots negatively impact humans' moral norms, could this negative influence spread as those humans interact with others? If so, robot influence, positive or negative, might ripple out to more broadly affect humans' moral ecosystems.

Patiency

Just as robots are often viewed as moral and social agents, they are often perceived as moral and social patients (valid targets of moral and social goods and harms). People readily exhibit empathetic reactions toward robots and view it as wrong to cause (certain types of) robots (certain types of) social and moral harms. In one famous study, psychologist Astrid Rosenthal-von der Pütten showed participants videos of a Pleo—a robotic dinosaur toy—either being treated nicely or being abused in various ways, including being choked, beaten, and "suffocated" in a plastic bag. Rosenthal-von der Pütten found that even though the Pleo was very clearly a nonliving toy, people experienced significant emotional distress in response to the videos of it being harmed.[46] Similarly, in our own recent work (led by University of Twente student Hideki Garcia Goo and in collaboration with social roboticists Katie Winkle and Megan Strait), we showed that women, transgender people, and people who have themselves been subject to relational abuse are especially distressed by observed abuse toward humanoid robots, especially when those robots are gendered as feminine.[47] Although there are limitations to using empathy alone as a signifier of patiency,[48] it seems clear that people generally believe that robots can be harmed and that it is wrong to harm them (at least more so than other types of technology). This patiency can lead people to change their behaviors,[49] and put themselves at risk,[50] in order to protect their robot companions.

When robots demonstrate ostensible vulnerability to harm, either passively (through cute, childlike, or animallike qualities in their designs) or actively (through intentional emotional displays), human empathic instincts kick in, leading users to act in ways perceived as being in the *robot's* best interest rather than their own. This has led some scholars to characterize "cuteness" in robotics as a "dark pattern"[51] in which users are psychologically manipulated.[52]

So, although agency may allow robots to leverage community power dynamics to engage in explicit persuasion, patiency allows robots to implicitly shape human behavior in several key ways. First, patiency can evoke empathic responses that lead people to shield robots from moral and social harms. Second, patiency prompts people to fall into the patterns and scripts of interaction that govern human interactions.[53] Finally, patiency encourages users to feel as if they need to gain robots' trust, as evidenced by demonstrations of people's willingness to keep secrets for robots.[54]

In most cases, the persuasiveness enabled by robots' patiency is an unintended by-product of their designs. Yet, there is good reason to be concerned that robot designers might intentionally leverage cuteness as a "dark design pattern" due to the ways that Big Tech companies already work to steer user behavior outside of robotics. As Shoshana Zuboff meticulously analyzes in *The Age of Surveillance Capitalism*, many of today's largest and most visible technology companies have reoriented their business models around a new regime in which profits are made by surveilling users, building predictive models of their behaviors, and using those models to manipulate user behaviors in profitable ways.[55] Zuboff examines the tech industry's system of behavioral manipulation, equating it with a stripping away of fundamental dimensions of humanity, including "free will" itself. Since the publication of Zuboff's book, many of these companies, especially Google and Amazon, are now pursuing home robot projects that will leverage robots' persuasive power to allow them to better surveil and manipulate users. As we will discuss later in this book, Amazon's home robot, Astro, is marketed with a Ring subscription, which not only turns Astro into a surveillance system but, moreover, deposits the footage collected by Astro into the Ring cloud platform—which Amazon has a history of providing the police access to. As such, the development of social robots by companies such as Google and Amazon is concerning not only because of the vulnerable status of many social robot users (such as children, older adults, and medical patients) but also because of the ways that these companies are already facilitating the surveillance of overpoliced communities in other ways.

So, although there are benefits to robot persuasion, there is also real potential for this persuasion to be misused, especially by the Big Tech companies most likely to bring social robots to market. And because robots take the form of embodied characters, users are not always aware that another *person* is manipulating them, let alone *who* is manipulating them or who to hold accountable for the robot's mistakes. This is particularly critical, since, at the end of the day, it is always robot designers (and not their robots) who truly wield power through robots' persuasive capacity.

Beyond Interpersonal Power

In this opening chapter, we have so far seen the reasons why human–robot interaction researchers are typically concerned about power and influence:

robots are being integrated into human social contexts, those robots can persuade and influence people to do things against their best interests, tech companies will almost certainly leverage this persuasive capability in insidious ways, and this influence may spread, fundamentally changing how people interact with each other.[56]

These are critical issues that roboticists need to be aware of and actively work to address. But at the same time, it is unclear how this particular conceptualization of power (on its own) helps us understand the ways that robotics builds on and reinforces White and patriarchal power structures. An account of roboticists' power grounded only in interpersonal persuasion cannot fully explain why today's humanoid and interactive robots look and sound the way they do, how they might reason about others, or where they are being deployed—questions that involve types of power that go beyond interpersonal influence and persuasion. Interpersonal power is only one specific type of power and may be best understood as only the most visible tip of the iceberg of power, whose bulk remains hidden beneath the waves.

Enter the Matrix

How might we more effectively conceptualize the types of power wielded by robot designers? In this book, I propose to answer this question through the matrix of domination.

First introduced by Patricia Hill Collins in her 1990 book *Black Feminist Thought*,[57] the matrix of domination is a framework for analyzing how oppression manifests—and can be resisted—within human society. Specifically, the matrix of domination makes specific claims about the nature of oppression, domination, and power.[58] The matrix of domination views race-, class-, and gender-based oppression intersectionally. Individuals are not cleanly categorized as "oppressor" or "oppressed" based on their race, class, or gender, nor are they automatically assumed to be more or less oppressed due to their membership in more or fewer privileged or penalized groups. Instead, individuals may alternatively or simultaneously operate as oppressor or oppressed in ways that depend on their environmental and social-relational context. A White woman teaching in a school, for example, may enjoy privileges built into the structure of that educational environment due to how she is racialized. Yet, she may simultaneously be penalized within her interpersonal relationships because of how she is gendered. And of course, the same may be true for a robot racialized as White and gendered as feminine—which may explain why robots for classroom domains tend to be designed and treated in particular ways.

Within the matrix of domination, race, class, gender, religion, ethnicity, sexual orientation, and other identity characteristics serve as distinct axes of a multidimensional space in which oppression manifests—and is resisted—in different ways at different points within that space. At all points in this multidimensional space, however, oppression is viewed in terms of a politics of domination, where oppression originates in a desire to dominate due to perceived superiority. As evidenced by the schoolteacher example, the matrix of domination views domination as being structured at different levels (individual, communal, and systemic), at each of which domination is wrought, experienced, and resisted in different ways. At the individual level, domination operates through interpersonal interactions in which conscious or unconscious beliefs in superiority are brought to bear in service of domination—or confronted in service of liberation. At the community level, domination operates through cultural forces by which dominant groups aim to replace subjugated knowledge (e.g., spreading beauty standards that reinforce and justify ideas of one's own superiority) and in which dominated groups cultivate and elevate their own knowledge and ways of knowing in resistance to those efforts. And at the systemic level, domination operates through societal institutions such as schools, churches, and government agencies, which serve as engines for propagating and enforcing those hegemonic or counter-hegemonic beliefs.

Finally, and most critical to our work of understanding roboticists' power, the matrix of domination delineates four domains of power that are brought to bear in service of oppression across these three levels of domination, visualized in table 1.1: the structural domain, the disciplinary domain, the cultural (or hegemonic) domain, and the interpersonal domain. Within the structural domain of power, social institutions are structured in ways that subordinate groups through methods such as de jure segregation.[59] Within the disciplinary domain of power, members of oppressed groups entering White, masculine, heteronormative, and/or upper-class organizations are further subordinated through bureaucratic hierarchies and surveillance.[60] Within the cultural (or hegemonic) domain of power, dominant groups create and leverage ideological and cultural forces (dimensions of racial politiculture) to support their right to rule.[61] Finally, within the interpersonal domain of

Table 1.1
The matrix of domination's domains of power

The Structural Domain	**The Disciplinary Domain**
. . . in which oppression is structured through laws and institutions	*. . . in which oppression is managed through the implicit and explicit enforcement of formal or informal rules*
The Cultural Domain	**The Interpersonal Domain**
. . . in which oppression is justified through stereotypes and cultural forces	*. . . in which oppression manifests through patterns of interaction*

Based on Collins's *Black Feminist Thought* and the table similarly presented in Catherine D'Ignazio and Lauren Klein's *Data Feminism* (MIT Press, 2020).

power, power structures are perpetuated (or subverted) through the routine patterns of everyday human–human interactions.[62]

The key argument I make in this book is that roboticists wield power in all four of these domains (not just the interpersonal domain) and thus have substantial power to reinforce—or subvert—race-, class-, and gender-based oppression. As shown above, it is perhaps easiest to see how roboticists wield interpersonal power through robot design: interpersonal interactions are the primary site where social power structures are negotiated, reinforced, and subverted.[63] As such, when roboticists design language-capable robots, they are choosing whether and how to wield robots' interpersonal persuasive power in service of such negotiation, reinforcement, and subversion. But the matrix of domination helps us see past this most obvious form of power and encourages us to consider the ways that roboticists also wield power, often in more nuanced and far-reaching ways, in the cultural, disciplinary, and structural domains. As we will see across this book, roboticists wield cultural power through the ways their robot designs reinforce (or subvert) the societal "common sense" that justifies oppression. As a case study, we will see in chapter 3 how cultural power dynamics help to explain AILA's physical design, as well as the designs of commercially available robots such as the Pepper robot mentioned above and the REEM robot from PAL Robotics (shown in figure 1.3). Roboticists wield disciplinary power through the algorithmic bureaucracy by which their robots are programmed to give preferential treatment (or simply work at all) for certain groups of people according to that "common sense," enforce the norms and values elevated through that "common sense," and sort and categorize people according to

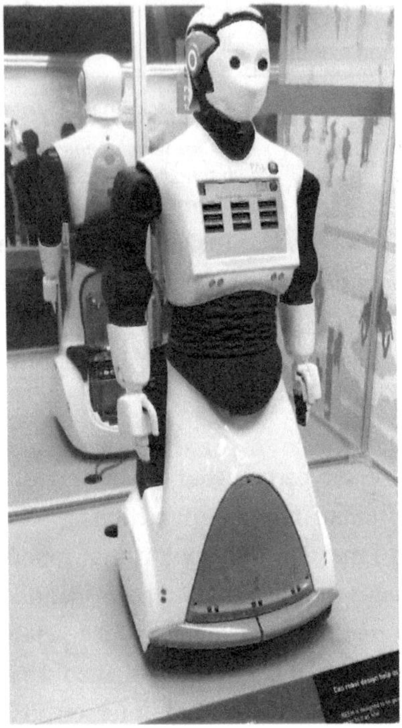

Figure 1.3
REEM. Image taken by Loz Pycock, licensed under CC BY-SA 2.0.

that "common sense." As a case study, we will see in chapter 5 how disciplinary power dynamics help to explain Pepper's API design, which encourages roboticists to sort users according to a patriarchy-reinforcing gender binary. And roboticists wield structural power through decisions such as who to collaborate with and the domains in which to employ robots, which may have the effect of legitimizing, normalizing, and reinforcing institutions such as policing, which structure oppression within our society. As case studies, we will see in chapter 6 how structural power dynamics help to explain REEM's deployment in Dubai as a White police officer and help to predict the ways that interpersonal, cultural, disciplinary, and structural power dynamics come together in the design of Astro in ways that predict the possible effects of that platform.

The matrix of domination also helps us see that the oppression that is reinforced through roboticists' wielding of these different types of power is not some abstract force but rather, most frequently, the exact type of oppression that the matrix of domination was meant to capture: White supremacist capitalist heteropatriarchy. As Black feminist scholars Patricia Hill Collins and bell hooks both emphasize in their theorizing of the politics of domination, societal dimensions such as race, class, sexuality, and gender represent axes of oppression along which, and at whose intersections, the power to oppress is wielded in different ways across multiple intersecting domains of power.[64] As we will see in this book, the matrix of domination is especially useful for helping us understand how the decisions roboticists make about how their robots are designed and deployed stand to reinforce or subvert the empowerment of White, high socioeconomic status, straight, cis-gender men and stand to reinforce or subvert the oppression of members of racialized groups, women, LGBTQIA+ people, and people of low socioeconomic status.

Understanding these other domains of power within the matrix of domination, and how robots might reinforce White supremacist capitalist heteropatriarchy in those other domains, requires us to develop a deep understanding of the social and historical context that has shaped the power structures and power dynamics into which robots are being embedded. This poses a significant challenge for roboticists, who, at worst, are typically used to thinking solely about robotic technologies in a technosolutionist vacuum and, at best, are typically used to thinking solely about the direct interactions between robots and those with whom they are interacting. In contrast, understanding why robots look the way they do today and the implications this has for human society requires us to understand a wide variety of political, economic, social, and even literary factors and their interaction over the past two centuries. Understanding how and why some roboticists are choosing to build robots for domains such as policing, how robots are being used in those domains, and the implications of their use in those domains requires us to understand the origins and evolution of policing (and the institutions that predate policing) over the past four centuries. And understanding how robots look at, listen to, and act toward people today, and the implications of such behaviors, requires us to understand the origins and evolution of fundamental social categories such as race (and the political and economic factors that led to the emergence of race) over the better half of the past millennium.

To use a metaphor, trying to understand the ways roboticists wield power without understanding the social and historical factors that shaped robotics is a bit like trying to steer a boat while only considering the orientation of the boat itself. Such an approach is unlikely to be successful. Without understanding the currents carrying the boat, you won't be able to tell where the boat is going by default and how you need to intervene to change its course. Without understanding the perils of the waters that surround you, you won't be able to assess whether the boat's voyage is likely to end in disaster without your intervention. Without knowledge of the shoreline beyond the water's edge, you won't be able to identify and steer yourself toward safer harbors. And without the ability to envision alternative forms of travel, you might never ask yourself why you're paddling a boat in the first place.

Accordingly, in this book, I will draw on a number of disparate fields that roboticists are not typically used to engaging with, including history and politics; law, criminology, and sociology; feminist, ethnic, and Black studies; literary and media studies; and social, moral, and cognitive psychology. By leveraging and synthesizing the insights from these different fields through a method I have termed "matrix-guided technology power analysis,"[65] we will discover some unsettling answers to the questions raised above. Specifically, I will bring together insights from across these fields to show how today's robots, especially humanlike, interactive, and language-capable robots, are designed and deployed in ways that reinforce White supremacist and patriarchal power structures across multiple distinct yet interlocking domains of power. Moreover, we will see that although this reinforcement of inequitable power structures by today's robots may be inadvertent, it is *not* accidental that robots have this effect: it is the outcome they were initially envisioned to achieve.

The Structure of the Book

Over the course of this book, I will explore the power wielded by roboticists across the cultural, disciplinary, and structural domains, with an emphasis on understanding how the technological status quo of robot design reinforces the sociological status quo of oppression within each of these domains; what historical and social root causes are responsible for that technological status quo; and how a more equitable robotic future might be enabled by revising the current status quo while working within its tacit assumptions, by rethinking those tacit assumptions, or by ripping up the technological status quo's historical and social root causes to create fundamentally new paradigms of robot design.[66]

Although roboticists stand to reinforce White supremacy, capitalism, and heteropatriarchy through their design choices, I will primarily focus in this book on White patriarchy. That is, I will focus on how roboticists' default design decisions tend to reinforce the power hegemony of White men. This is not to say that issues of class and sexuality are not important. Indeed, the tight interrelations between race and class, and between gender and sexuality, mean that issues of class and sexuality will inevitably arise across this book. However, because race and gender are the most salient identity characteristics of robot hardware design, because the logics of race and gender are the most salient logics enforced by robot software design, and to keep my analysis relatively concise, I will primarily focus on issues of race and gender in this book and will thus refer to the power system reinforced by robots as "White patriarchal power." As such, this book will primarily aim to interrogate roboticists' default reinforcement of White patriarchal power and the ways that roboticists can work to subvert that power system.

First, in chapters 2 and 3, we will see how the history of robotics helps us understand the power that roboticists wield in the cultural domain. Specifically, in chapter 2, I will explore the interconnected histories of race and robotics, and the formation and evolution of the White patriarchal robot imaginary. As I will show, the origins of robotics in the White patriarchal ambitions of the post-bellum American South inevitably shaped the evolution of robots in our cultural and techno-scientific imaginaries in ways that can still be seen today. Building on this base, in chapter 3, I will explore the implications that robots' White patriarchal legacy has for robot design today in terms of how roboticists are choosing to design robots' appearance, voice, and speech. In doing so, we will see how roboticists wield cultural power when making these decisions by either reinforcing or subverting the biases, norms, and assumptions that comprise White patriarchal patterns of racialized and gendered "common sense" through the selection and use of social identity design cues.

Next, in chapters 4 and 5, we will see how our consideration of the history of race also helps us understand the power that roboticists wield in the disciplinary domain. Specifically, in chapter 4, we will consider how roboticists' efforts to enable robots to enforce social and moral norms can reinforce the logics of White patriarchy and the norms of "appropriate" or "correct" behavior ascribed to different groups under this power system. In doing so, we will begin to see how roboticists wield disciplinary power by reinforcing or subverting the enforcement of White patriarchal logics, and the ways that today's otherwise helpful "cobots" are too often operating as what might be better referred to as "New Jim Cobots" that serve to enforce the New Jim Crow[67] and the New Jim Code.[68] Then, in chapter 5, we will consider who is seen and heard by today's robots (and who is not) and how today's robots too often serve as the technological equivalent of Whites-only water fountains. Finally, we will see how today's robots (and similar technologies) serve as New Jim Cobots by using social identity trait perception to categorize people according to systems of race and gender in ways that enforce (and reinforce) the fundamental logics of White patriarchy.

In chapter 6, we will see how the history of policing (a key institution of White patriarchal power) helps us understand the power that roboticists wield in the structural domain. Specifically, in this chapter, I will show how the history of policing in the United States explains how and why robots are being used by the police and how the use of robots is facilitating and exacerbating the racialized and gendered surveillance and violence that define policing (and is thus facilitating and exacerbating the modern systems of slavery that policing encourages and enables). Moreover, I will show how the social robots being developed today by Big Tech companies may be used in the future toward similar ends. Through this exploration, we will see how roboticists wield structural power when deciding which institutions to collaborate with, both by accelerating the default effects of those institutions and, moreover, by legitimizing and normalizing those institutions. Based on this evidence, I will make a call for an abolitionist robotics and begin to explore what this radical vision for the future of robotics might entail.

Finally, in chapter 7, I will bring together the different perspectives from across these three sections to highlight a palate of different tools that roboticists might use to pursue more ethical, equitable, and power-sensitive robot design. In identifying these "degrees of freedom" (a term used in robotics to refer to the axes of motion along which a robot can move in order to navigate and manipulate the world around it), I hope to motivate an overarching social justice–oriented philosophy through which roboticists can work toward a future in which robots subvert rather than reinforce White patriarchal power.

Caveats and Positionality

Before we fully jump in, let me briefly provide some further context about this book. My intended audience is primarily "students of robotics"—a term I intend to apply quite broadly. Certainly, future roboticists (especially those who will be developing the types of interactive robots that this book will focus on) need to know the power they hold and how to wield it. But this knowledge is equally important for those who may be involved in developing *or studying* robots in other ways, including students coming from other engineering disciplines, computer science, psychology, communication studies, design, philosophy, and sociology. As such, I expect this book to be of interest to any student interested in robotics and technology ethics. Thus, although this book is an academic work, it is intended to be broadly accessible to students regardless of their academic background (following in the tradition of similar recent books such as *Data Feminism*[69] and *Race After Technology*[70]). Finally, I hope that this book will be interesting, informative, and thought-provoking not just to current students but also to faculty and other long-since-students, both within and beyond robotics.

I should also say a little more about myself and why I am writing this book. Formally, my academic training is in computer and cognitive science, as I received a joint PhD in computer science and cognitive science from Tufts University.[71] This means that my disciplinary expertise is primarily in computer science, psychology, and philosophy. Why am I, a computer and cognitive scientist, writing about the connection between robots and White patriarchy? Although many students (and faculty) erroneously believe that computing is a branch of mathematics, computing is actually, as Randy Connolly wrote in his 2020 *Communications of the ACM* cover article,[72] a *social science*. In fact, as scholars such as Timnit Gebru have argued, the primary purpose, effect, and nature of computing is to amplify human intent. As such, understanding the intents baked into today's computing technologies, and how those computing technologies work to amplify those intents, should be at the heart of computer science. Accordingly, social scientific methodologies and insights should be among the primary means to "do" computer science, even if this is not currently computing's dominant epistemology or praxis.

A more critical question may be why I, a *White man*, am writing about White patriarchy. Ultimately, it is critical for all computer scientists and roboticists to engage with and contribute to these topics. As Kelly Hayes and

Mariame Kaba argue in their recent book, *Let This Radicalize You*, placing the burden of social justice work on the shoulders of one particular demographic group is "dehumanizing, nonstrategic, and an abdication of one's responsibilities to forge struggle."[73] As they write:

> This places undue pressure on whatever group or demographic is being fetishized as a savior trope. While there are moments when it makes sense for us to take leadership from particular groups . . . there is no one group—and certainly no one person—who can defeat capitalism, end imperialism, or bring down White supremacy alone.[74]

On the other hand, I believe that computer scientists and roboticists have a duty to elevate the voices, perspectives, and arguments of women and people of color, especially given the marginalization these groups face within computing. As such, I have done my best in this book to elevate and center the contributions, theories, and epistemologies of women and people of color. As Patricia Hill Collins wrote in her foundational work on Black feminist thought, it is critical that White men participate in "advocating, refining, and disseminating Black feminist thought" in order to participate in the collaborative advancement of this theory, "confront . . . social injustice," and effectively "reject . . . the unearned privileges of Whiteness."[75]

Finally, I must acknowledge that the arguments made in this book are tightly centered around the cultural context of the United States. The demographics of robot designers in other countries with long robotics histories, such as Japan, are of course quite different from those in America, and this leads to robots that are designed and deployed in subtly different ways. Nevertheless, as we will see across this book, the specific *types* of power wielded by roboticists remain the same from place to place, even if the specific *ways* that power is wielded, and the effects of wielding that power, are subtly different. Moreover, interrogating the ways that roboticists reinforce White patriarchy *in America* is critical because, as we will see in the next chapter, the very concepts of "race" and "robots" originated in America and were created in order to support the White patriarchal power systems built around the system of slavery at the heart of American history.

2 Black Servos, White Masks

To understand how roboticists wield power, we first need to answer a key question: What is a robot?[1] This is a surprisingly challenging question that occupies the first session of nearly every robotics course I teach. If you look at the classic definitions of robots used by roboticists, a robot is an embodied artifact that senses, plans, and acts in order to effect changes in its environment. This definition does capture most of the autonomous robots being developed by roboticists, but it has some serious flaws. First, this definition seems to cover some things we wouldn't normally consider robots, such as dishwashers, thermostats, and elevators, all of which sense their environmental state or take user input, perform some limited computation to decide what to do, and (if they're working correctly) effect changes in their environments. Second, and especially if we rule out things such as dishwashers, thermostats, and elevators on the grounds of the simplicity of their "thinking," this definition does *not* seem to cover some things we *would* normally consider robots, such as traditional factory robots, which are preprogrammed to do particular routines and thus do not think for themselves, and teleoperated robots used in contexts such as search and rescue, space exploration, and education, which may have little or no ability for independent "thinking" and may instead rely on direct human control.

Due to these two challenges, the definition of robots we actually end up using in classes is typically closer to "if your parents saw it in a movie, they'd describe it as a robot." For the purpose of course logistics, this informal (and circular) definition does the job of preventing students from turning in essays on stock trading algorithms and ChatGPT. And while this definition may seem informal, it is not far off from the definition of industrial robots once given by the "Father of Robotics" Joseph Engelberger: "I don't know how to define it, but I know one when I see one."[2] Moreover, this definition reflects a deeper truth: "robots" cannot so easily be defined from a technical perspective in part because they are a science fiction concept whose definition inherently depends on the ways that robots are portrayed in contemporary science fiction. In other words, there is a tight interplay between science fiction and engineering in the delineation of what it means to be a robot, and as such, robots cannot be defined in engineering terms alone.

To understand how roboticists wield power, we thus need to start by understanding the history of science fiction and engineering that led to our modern concept of robots, which determined how robots would serve as vehicles of cultural power and which ultimately shaped the ways that robots would be conceptualized, designed, and depicted today. As we will see, although automata and robot-like devices have existed for hundreds of years, the first robots seen in *literature*—the robots that popularized the idea of artificial humanoid servants and shaped the next 150 years of robots in science fiction and engineering—were depicted as racially caricatured slaves in order to serve as a metaphor for how White American men could maintain power in the aftermath of the Civil War.

In the first part of this chapter, we will analyze the historical factors that led to the emergence of robots in literature. In order to understand the White and patriarchal power dynamics that motivated this emergence, we will first begin by considering the historical origins of race, the way race is used today as a technology for structuring society, and the role that gender plays within racialized social systems. Next, we will examine the connections that were being drawn between race, gender, and technology in the decades preceding the Civil War. Third, we will discuss the economic and technological vacuum left by the Civil War. Finally, we will describe the history of pre-robotic automata that set up proto-robots, envisioned as the new Black slaves, to fill that vacuum in both literature and engineering. In the second part of this chapter, we will then analyze the ways in which robots have evolved over the past 150 years in both literature and engineering, across both robotics and artificial intelligence. We will see how robots outwardly Whitened in order to tell new types of stories in science fiction, while maintaining their status as slaves working to maintain White supremacist power structures. Finally, we will consider the robot imaginary today. In doing so, we will explore how robots are not only racialized but also gendered today, the ways that robot gendering and racialization interact, the ways those interactions are explained by intersectional visibility politics, and the way these patterns replicate outside of American and European contexts.

By carefully working through this historical thicket, we will develop a rich understanding of how, why, and when robots first emerged; the ways they have evolved since their inception; the ways the history of robotics is tightly intertwined with and informed by the history of America's White and patriarchal power structures; and the reasons why understanding these intertwined histories is so critical if we wish to understand how today's robots continue to reinforce those same power structures. Overall, we will demonstrate that today's dominant robot imaginary is a White patriarchal robot imaginary—a demonstration that calls on us to envision new robotic imaginaries that break from this racist and sexist past.

I The Cultural Origins of Robots

The Historical Origins of Race

To understand the historical origins of robotics, we need to understand the historical origins of race. When race first emerged as a folk concept in Britain, it was used as a general categorizing term, along the lines of "type," "kind," or "breed,"[3] applied to people in terms of not only shared descent but also shared purpose,[4] as in "a race of worthy learned gentlemen,"[5] the "race of Bishops,"[6] or the "race of . . . legislators."[7] The evolution from this initial folk usage to the concept of race that is used today can be traced through the history of Britain's colonial projects.[8]

The first British colonial project was Ireland. Along with Wales, Ireland was first conquered by the English in the 1100s. But by the 1300s, even after two hundred years, this colonization was really not going well.[9] English settlers, it turned out, quite liked the Irish way of life and had begun to identify with the Irish in terms of culture and philosophy—a culture and philosophy that were at odds with England's nascent protocapitalist aspirations. To avoid losing control over their Irish colony, the English passed a set of statutes that barred English settlers from intermarrying with the native Irish and that banned Irish music, games, stories, and language. This didn't work very well. By the 1400s, English control had shrunk to a small region known as "the pale," and by the 1500s, Irish culture and language had nearly completely reestablished dominance. After Henry VIII's split from the Catholic Church in the 1520s, the heavily Catholic populace of Ireland revolted, leading Henry VIII to put down the rebellion, execute the Irish leader, and establish more formal control over Ireland. This proved more difficult than expected, with widespread violence plaguing these efforts. Queens Mary and Elizabeth I tried to address this through harsh martial governance, but this merely further antagonized the Irish. By the 1550s, the English found a new solution: wherever rebellions sprang up, the English would seize the land and hand it over to Englishmen, who would settle the land, spread English culture, and start plantations. By the 1600s, this had solidified into a model in which the English attempted to coerce the Irish into permanent subjugation by putting them to work on plantations on their former land. To justify this plan, the English pointed to the continued Irish resistance to Protestantism, their economic model of communal farming rather than individualistic capitalism, and their "tendency toward violence" (as forcing thousands of Irish off their communal lands had unsurprisingly led to a lot of banditry). As such, the English view was that the only way to "tame" the "Wild Irish" was through forced labor.

As Bashi Treitler describes,[10] these intertwined methods of ethnic sensemaking and subjugation in service of protocapitalism directly informed Britain's subsequent interactions with the Native American peoples in the so-called New World. When they enacted their first New World colonization project, they applied this schema to their interactions with the Indigenous American population, who did not worship their gods, lived communally, and demonstrated a "tendency toward violence" when the English started eating their food and being generally bad guests. And, like the Irish, the Indigenous Americans started absorbing members of the English population into their own, as some Englishmen decided they rather liked the Indigenous Americans' way of living and decided to join them instead of assisting the English colonial project. These factors led the English to believe that the Indigenous Americans were "savage," *just like the Irish*—a belief that the English then used to justify continued violence and seizing of land.

When the British began to import kidnapped Africans to work their colonies, they applied this same schema: the non-Christian Africans were once again cast as "savage," the British colonists once again employed a system of plantation capitalism, and although the colonists were primarily motivated by greed rather than religion, the colonists justified slavery as a way of "saving" the Africans' souls. Moreover, just as they had done with the Irish before them, the British enacted new laws to exert more formal control over (and maintain separation from) those they enslaved.

But unlike the Irish, the enslaved Africans were not a single cultural group. And unlike the native Americans, they were not a local established people. The only thing binding these Africans together was their shared enslavement. And so, in the slave codes set out by the early Virginian colonists, these enslaved people were, simply, "Black." At this point, "Black" was not viewed as a descriptor of a distinct "race." Yet, over the next hundred years, this logic of categorization became legally entrenched, creating a formal social stratification between enslaved "Blacks" and the colonizing "Whites"—a new term used to help set the colonizing British apart from those they were enslaving. As Bashi Treitler puts it, "Africans were not enslaved because they were Black; they were made Black because they were made slaves."[11] Accordingly, England's third attempt at ethnicity-driven separation and subjugation[12] served as the initial point of racialization of those enslaved in their colonial projects,[13] with the criteria for racialization being primarily cultural and religious (rather than phenotypic),[14] and the rationale for racialization being outwardly religious and inwardly capitalist.[15]

This initial seed of racialization in the British colonial projects germinated and grew alongside those same colonies' battle for independence. Ironically, as Bashi Treitler shows, it was the American emphasis on the ideal of liberty during the revolutionary period that led to further development of the concept of race and its use to structure early American society. Eighteenth- and nineteenth-century Americans seized on the ideas of inalienable rights to life and liberty, at least for those who were White, male, and propertied[16] (a distinction we will return to in chapter 4) because these were crucial tools in their vision of America as a new independent nation.[17] Yet, these ideals were plainly at odds with the other tools in America's political and economic toolbox, such as slave labor and violent westward expansion. To justify these other actions, early Americans ascribed a natural inferiority and lack of humanity[18] to those they were exploiting. This ultimately led to further articulation and acceptance of a racial hierarchy, with "Whites" positioned at the top and "Blacks" positioned at the bottom.

As I will discuss in more detail in chapter 6, this hierarchy has undergone a dramatic process of codification over the past three centuries, from plantation-based law enforcement (in which planters "branded, stabbed, tarred and feathered, burned, shackled, tortured, maimed, mutilated, crippled, whipped, hanged, beat, and castrated"[19] those they enslaved rather than go through the inconvenience of formal law enforcement) to the slave codes of the early American South in which White men were broadly tasked by law with surveilling and controlling Black people[20] to the Slave Patrols organized by the state so that ordinary White citizens could avoid the hassle of nightly patrols[21] to the Black codes of the post-bellum South that perpetuated earlier slave codes in all but name[22] to our modern system of policing, which continues to operate primarily as a means of surveilling, controlling, and enslaving race–class subjugated populations.

In other words, what we see thus far is that the concept of race was created as a tool of cultural power to justify the form of slavery developed to support plantation capitalism in the Americas. The interweaving of this new sociological concept, the economic system it was meant to support, and the legal mechanisms that codified this relationship in the nascent American state together led to the formation of the first racial paradigm and racialized social system.

Racial Paradigms and Racialized Social Systems

This early American history is not only the history of race but also the history of the first racial paradigm—the structure that governs a racial society. Bashi Treitler defines a racial paradigm as consisting of two key parts: a racial architecture and a racial politiculture.[23] A racial architecture is composed of (1) the set of racial categories used in a society and (2) the hierarchy used to organize them. In turn, a racial politiculture is composed of (1) the racial common sense that members of that society use to sort themselves and others into those racial categories and use to define the expected behavior of members of those categories and (2) the sanctions that are imposed when individuals violate those expectations.

Bashi Treitler's notion of a racial paradigm can be further understood in relation to Bonilla-Silva's notion of a racialized social system. As Bonilla-Silva argues, when the concept of race emerged, it formed a particular racialized social system (White supremacy) in which privileges were awarded to one racialized group (i.e., the subset of Europeans racialized as "White"). These privileges were then codified and reinforced through a set of social relations (similar to Bashi Treitler's racial architecture) and practices (similar to Bashi Treitler's racial politiculture), which together comprise society's racial structure. Bonilla-Silva's formulation is particularly helpful, as it enables us to understand why race as a concept, and the racialized social system of White supremacy in particular, continue to exist today: because they provide benefits and privileges to the dominant racialized group, which motivates its members to maintain the racialized social system that privileges them.

Moreover, these different structural formulations help us to understand the multifaceted nature of race. Depending on what level of analysis is being used and what intellectual tradition is being brought to bear, race can be viewed as a structure,[24] knowledge system,[25] tool,[26] or technology.[27] But across these different perspectives, scholars agree that race is socially constructed[28] and that it has a structural purpose: to separate a population into hierarchically organized categories or castes.

Although race was originally developed in the American South in order to justify this type of hierarchical organization and the economic system it enabled, the American North, of course, also participated in that economic system and did so long after slavery was formally outlawed in Northern states. And critically, the American North's involvement in the US slave economy played a key role in shaping American thinking about both race and technology in the lead-up to the emergence of robots in science fiction.

Emerson in Lowell: A Line in Steam from Slaves to Machines

Although slavery was abolished in Massachusetts in the 1780s,[29] abolition of slavery within Massachusetts did not eliminate the reliance of the Massachusetts economy on slave labor—the enslaved people serving Massachusetts' industrial projects merely lived beyond state borders. This Northern reliance on slave labor is exemplified by the Massachusetts mill towns of Lowell and Waltham, which represented the largest and wealthiest manufacturing centers in the United States prior to the Civil War. By the mid-1840s, the cotton processed by Lowell's mills represented the product of approximately thirty-three million person-hours of slave labor per year,[30] spurring an *expansion* of Southern slavery to keep up with demand[31] in what Massachusetts Senator Charles Sumner called an unholy alliance between "the lords of the lash and the lords of the loom."[32] Meanwhile, the second most important business in Lowell, the Lowell Machine Shop (which moved there from Waltham in 1824), was tasked with creating and building the machinery needed for the cotton mills to function—as well as the cotton gins sold to the South.[33] Lowell and Waltham may not have had people enslaved within their walls, but as the industrial arms of the plantation complex, they were slave towns nonetheless.

At the same time, and perhaps because of its role in the "unholy alliance" with the South, Lowell in particular was a hotbed of abolitionist activity, with antislavery societies, safe houses, and other countervailing establishments[34] a mere ten-minute walk from where the mechanisms of slavery were being forged,[35] and prominent abolitionists such as Fredrick Douglass visited Lowell to observe its industries and to deliver lectures in response.[36] Moreover, Lowell was a site of unique intersectional resistance from White women. Since it could not "employ" Black slaves to run its mills, Lowell instead relied on the work of young women termed "mill girls."[37] The conditions at Lowell's mills were harsh, sufficiently so that these workers were labeled the "White slaves of the North" by their contemporaries.[38] In a response to these conditions, the Lowell mill girls formed the first labor union of working women in 1845.[39] This shared atmosphere of female labor organization and abolitionist sentiment in the industrial heart of America's slave economy played a key role in the shifting relationship between technology and slavery in American industry and society. Two years after the unionization of Lowell's mill girls, Ralph Waldo Emerson visited Lowell and observed the character of the men of Lowell in his diary: "They buy slaves where the women will permit it; where they will not, they make the wind, the tide, the waterfall, the steam, the lightning do the work, by every art & device their cunningest brain can achieve."[40]

What is immediately striking about this quote is the way that Emerson directly alludes to technology being used as a replacement for slave labor. On one hand, connections between technology and slavery (writ broadly) date back to Aristotle, who once wrote that "[if] shuttles in a loom could fly to and fro and a plucker play a lyre of their own accord, then master craftsmen would have no need of servants nor masters of slaves."[41] Yet, at the time of Emerson's visit to Lowell, new connections were also beginning to form in the American imagination between technology and the specific institution of *Black* slavery. A case study presented by literary scholar Taylor Evans is that of P. T. Barnum and Joice Heth.[42] While today's audiences may associate Barnum with his circus empire, Barnum had his first commercial success "displaying" Joice Heth, an enslaved elderly Black woman who he claimed to be the 161-year-old nursemaid of George Washington. Specifically, Barnum rose to prominence when, as ticket sales languished, he wrote a pseudonymous letter to a Boston newspaper claiming that Heth was, in fact, a machine. In the ensuing controversy, thousands flocked to Barnum's exhibition to assess Heth's status as a machine—a telling and cruel inversion of the public's contemporaneous questioning of Black people's status as human.

From the example of Barnum and Heth, we see early connections being drawn in the American mind between *Blackness* and machines in a way that parallels the divisions that were being made between Blackness and humanity. Emerson's statement, on the other hand, draws a distinct connection between *slavery* and machinery. As we will see, both Barnum and Emerson provide critical conceptual bridges on the road to robots in the American psyche.

Emerson's quote can be read in multiple ways, either as a commentary on the Lowell mill girls rejecting their own wage slavery or as Northern women rejecting Black slavery. Regardless of which of these interpretations we take, however, Emerson's quote highlights the complicity of Northerners in America's slave economy, the role of female agency in subverting unjust systems of power, and, most critically, the relationship between slavery and technology. Under each of these interpretations, the response of the White monied men of the North to the possible end of slavery was the same: maintain the economic and industrial status quo by developing new technologies that could serve as a replacement for slave labor.

Today, Waltham and Lowell no longer serve as textile mills but instead serve as private and public epicenters of robotics. Waltham is the current home of Boston Dynamics, creator of the Spot robot, whose use by the police we will discuss further in chapter 6. Similarly, the University of Massachusetts at Lowell is home to a thriving robotics program, boasting the New England Robotics Validation and Experimentation (NERVE) Center, a 10,000-square-foot robotics research facility that sits on the original site of the foundry of the Lowell Machine Shop—the business that created the machinery that facilitated Lowell's participation in the US slave economy. As we will see, this historical through line is worth reflecting on due to the ways that robots have served as the direct economic and conceptual descendants of slave labor.

Although it was in the American North that machine replacements for slave labor were most seriously being explored, it was the changes to the power systems in the American South that most directly led to the emergence of robots in science fiction. In the years leading up to the Civil War, the profitability of cotton had faced a steady decline,[43] and with the cessation of South-to-North cotton exports during the Civil War, many of Lowell's mills temporarily ceased their operations, turning ten thousand workers "penniless into the streets."[44] Ultimately, though, Lowell continued to prosper, merely turning to other worldwide sources for cotton.[45] In the South, of course, the situation was different due to the impact that the (formal) end of slavery had on Southern power systems. Slavery in the United States did not truly end in 1865 with the passing of the 13th Amendment, instead continuing in the form of peonage and convict leasing—systems that only fully ended in 1941 in response to their exposure in Japanese and Nazi Propaganda.[46] And, of course, slavery has continued to this day as a legally protected punishment for "crime."[47] Yet, in the intervening seventy-five years between 1865 and 1941, the formal illegality of most forms of slavery led to real and dramatic change in the structure of racial capitalism. Centuries-old sources of unpaid labor at the heart of the Southern economy were no longer available. Moreover, the formal dissolution of the slave system removed the possibility of what had once constituted the hopes, dreams, and definitions of success for White Southerners. From this economic and teleological vacuum, robots emerged.

The Road to Robots

Thus far in this chapter, we have explored the factors in American history that led to the emergence and evolution of race and to America's racialized social structure and system of White supremacy. As we will see, the evolution of these concepts and systems ultimately led to the emergence of the first robotic character in popular culture. Before we discuss the emergence of that character, however, let us briefly consider the other machines that preceded and may have helped to inspire that character.

For hundreds of years before the first robotic character emerged, engineers around the world had explored automata of different forms. Although the full history of humanlike mechanical devices spans thousands of years,[48] complex robot-like automata first emerged in the last half-millennium. These clockwork renditions of humans first manifested just as that—clocks[49]—then as mechanical toys,[50] then as mechanical animals, such as Jacques de Vaucanson's Digesting Duck (which could both "eat" and "defecate").[51] Finally, these led to full-scale mechanical people, used in occult practices, stage magic, music, and charlatanry.[52] In all of these cases, automata were mere mechanical curiosities and amusements used more as signifiers of wealth than as practical devices. Unlike modern robots, these automata were portrayed neither as autonomous characters nor as tools to be used to replace or supplement human labor.

This began to change around the same time that Emerson was drawing connections between technology and slavery, as can be seen through the ways that automata were described in the popular press. In 1848, just one year after Emerson's visit to Lowell, reports began to circulate of a new human-like German automaton named "Mr. Eisenbrass."[53] On one hand, *Scientific American*'s account of Mr. Eisenbrass highlighted the extent to which these automata were still perceived as purposeless vanity projects: "The Professor spent some 27 years in the construction of his image, and though it exhibited much ingenuity, still we could not but consider that the Professor had wasted for 27 years his genius and time and wealth upon a trifle light as air."

Similarly, the German version of this article, appearing in the *Augsburg Gazette* (translated into English for *The Racine Advocate*)[54] was clear that the machine was a piece of artifice, *not* an intelligent, autonomous agent:

> Not a reasoning, thinking, intellectual being—endowed will the capacities of seeing, hearing, smelling, talking—capable of realizing the emotions of friendship—of anger, of sorrow; or with the power of feeing [sic] the delights of love, hope, joy;—but a senseless thing of wood, iron, brass and steel, made in the semblance of man—and so closely too, as to deceive almost any one at first sight, yet capable, under the skillful guidance of the inventor.

Yet, this German article also made clear how, even in Germany, the public was beginning to envision how automata might be used as slaves working in service of capitalist pursuits. These connections are drawn in the article both implicitly and explicitly. Implicitly, the article situates Dr. Eisenbrass in a history of devices that profoundly reshaped labor relations, including the Jacquard loom (the programmable looms that subsequently de-skilled ribbon weaving[55] and increased weavers' dependence on—and subordination to—merchant capitalists[56]) and the spinning jenny (which increased the productivity of cotton cultivation, thus accelerating the slave-driven American cotton industry[57]). But, moreover, the article's authors state outright the ways that automata might serve as slaves: "It seems not unlikely that the dreams of certain modern philosophors [sic] will yet prove realities; and the day is not distant when manual labor will be almost entirely supplanted by the introduction of new improvements in machinery, and when there will be little left for man to do, save to direct and guide iron slaves."

Mr. Eisenbrass would fade from memory—just one more reported engineering curiosity. But two decades later, this vision of automata as "iron slaves" would burst back into the public consciousness with such force that the two would become inextricable.

II The Evolution of Robots in the Robot Imaginary

The First Robot: *The Steam Man of the Prairies*

The mechanical man that most clearly influenced how we would come to think of robots may be the steam man patented by New Jersey engineers Zadoc Dederick and Isaac Grass in 1868, just three years after the end of the Civil War and the passing of the 13th Amendment, which took the form of a seven-foot, five-hundred-pound mechanical man pulling a cart behind him.[58] While the robot was never manufactured commercially (and the prototype was rumored to have been destroyed in a fire at a warehouse owned by none other than P. T. Barnum), the robot was intended as a genuine engineering product, not just a mechanical curiosity. The patent schematics (seen in figure 2.1) immediately captured the public imagination, leading later that year to one of the most influential and important texts in American science fiction: Edward Ellis's *The Steam Man of the Prairies*.

In what has been characterized as the American response to Jules Verne,[59] *The Steam Man of the Prairies* featured a mechanical man clearly based on Dederick and Grass's patent schematics—the first appearance in literature of what we would today consider to be a robot. Ellis's novel describes the travels across America of a young White boy who is carried by a mechanical companion. The story culminates with a confrontation with "Indians" described dehumanizingly as "swarming like locust,"[60] in which the mechanical man is blown up by the boy in order to slaughter the Native Americans, thus enabling the boy and his companions (e.g., an Irishman named "Mickey McSquizzle") to steal the Native Americans' horses, return home, and live prosperous lives with the "girl[s] of their choice" off the spoils seized during this "Western trip."[61] The novel was a runaway commercial success, especially after its reprinting to coincide with the 1876 Chicago World's Fair. The book's success led to a series of copycat Edisonade stories collected in what has been regarded as the first serial science fiction publication, placing *The Steam Man of the Prairies* at the epicenter of our modern science fiction tradition[62] and providing the first real example of a robot (even if it was not called that) in the popular imagination. Critically, this robot was Black.

Although modern editions of *The Steam Man of the Prairies* depict the steam man as a steampunk automaton whose head is a single Cylon-like eye within a knight's helm, and although the single robot prototype eventually created by Dederick and Grass appears to have been racialized as White, the

Figure 2.1
US Patent 75874, Dederick and Grass, 1868.

original editions of *The Steam Man of the Prairies* depict a great black iron robot with bulging eyes, striped pants, and a top hat—the uniform of Zip Coon,[63] a popular character in minstrel shows who served as a foil to Jim Crow.[64] The minstrel shows of the era were by far the most popular form of public entertainment nationwide[65] and served as a powerful tool of cultural power for propagating the racist stereotypes used to justify the enslavement of Black people. On the cover of *The Steam Man of the Prairies*, the mechanized depiction of Zip Coon is shown bridled like a horse, the reins in the hands of a White boy with a devious smile on his face. Even more overt is the original patent of Dederick and Grass, in which the steam man is shown as a caricature with coarse hair, a broad nose, and exaggeratedly large lips smoking a pipe.

Our first robot, then, was designed as a Black mechanical slave. Moreover, as captured by literary scholar Taylor Evans, the way that this robot was depicted in Ellis's novel, which subsequently shaped the vision of robots held in the popular imagination, was one designed to allay White fears and shape White aspirations. Ellis's novel showed White people capturing and controlling Black movement at a time when Black Americans were showing a new (and, to White people, startling) mobility.[66] But at a more fundamental level, the steam man served both as a vision for how White boys "stripped of a racist birthright" could yet achieve mastery and domination through technology[67] and as a vision for the continuation of Whiteness itself in the structural vacuum left with the end of slavery. As Atanasoski and Vora put it, "Robots . . . open the place of possibility for the reproduction of the (White) liberal subject in the absence of older forms of unfree and coerced labor."[68] Finally, as Evans describes, Ellis's story presents a startlingly clear vision for *how* technology (narratively, in the form of robots) could be used to retain mastery and perpetuate Whiteness: through violent, colonialist, Western expansion and expropriation of wealth from the Native Americans.[69] Put together, the emergence of robots can be seen to reflect the many changes surrounding the end of the American Civil War as society transitioned from slave patrols to police, from slave codes to Black codes, from slavery to peonage, and from narratives surrounding White male dominance over enslaved people to narratives surrounding White male dominance over technology.[70]

These claims about what robots are and could be, and what they could or should be used to achieve, played a key role in shaping how the general public, scientists, and engineers conceptualized robots and answered these same questions. That is, these early visions of robotics had an indelible effect on the robot imaginary.

The Dawning of the Robot Imaginary

Understanding how and why technology is developed requires an understanding of the cultural imaginaries that motivate those technologies.[71] That is, we need to understand how the ways people see the world and imagine it to be are shaped by the specific cultural and historical resources available to them.[72] From the perspective of philosopher Cornelius Castoriadis, cultural imaginaries represent a culture's ethos and the power that creative ideas hold within that society[73]: "the capacity to see in a thing what it is not, to see it other than it is."[74]

Understanding cultural imaginaries is critical, as the technological artifacts that engineers choose to build are inevitably shaped by the cultural and social imaginaries that dictate the limits and sensibilities of technologists' imaginations. The artifacts created by technologists, in turn, feed back into the broader cultural or social imaginary, changing not only society's view of technology but also society's views about the types of capabilities and interactions that are afforded by technology. As human–robot interaction researcher Selma Šabanović argues, "Imaginaries support the production of particular types of scientific knowledge and technological artifacts, and transform the broader societal imaginary by encouraging new practices and world-views."[75]

In her groundbreaking book, *The Robot Imaginary*, media scholar Jennifer Rhee has analyzed the way that robots are envisioned within our cultural imaginaries: what science fiction and art says about society's views of what robots are and what they are for. As Rhee demonstrates, popular cultural imaginaries of robots have guided not only scientific practice but also the ways we think about what it means to be human.[76]

We can see, for example, the ways that Ellis, following in Barnum's wake, connected Blackness with machines in a way that separated Blackness from humanity. This framing led readers to understand robots (although they were not yet referred to as such) to be human*oid* but nonhuman slaves designed to serve White male social, economic, and, as we will see, sexual ambition. But, moreover, this framing led readers to understand what it meant to be human as inherently White and male. As a result, this framing shaped readers' understandings of who should be valued and who should be devalued, as well as who should be rendered visible and who should be rendered invisible. Indeed, since the inception of *The Steam Man of the Prairies*, the popular understandings of what it means to be a robot and what it means to be a human have mutually shaped each other as they have evolved over time. This can be seen through the historical legacy of *The Steam Man of the Prairies*.

Whereas Dederick and Grass's invention was a failure, *The Steam Man of the Prairies* was a major commercial success, leading to a barrage of copycat stories and the first science fiction anthology. These immediate successor stories, as well as those that followed them in turn, largely focused on White male mastery of technology as a stand-in and replacement for Black people, explicitly responding to the fears of White Southerners following the Civil War. In fact, as African American Studies scholar Louis Chude-Sokei argues, *most* early science fiction, including classics such as H. G. Wells's *War of the Worlds*, are inherently about race: "Science fiction as a central component of our histories of technology is . . . a direct extrapolation of the material experiences of Black people and other marginalized groups from and through slavery and colonialism[:] . . . race and slavery made SF possible."[77]

The envisioning of robots as explicitly Black slaves continued for a hundred years after *The Steam Man of the Prairies*, as evidenced by Westinghouse Research Laboratory's 1930 Rastus robot.[78] Just as skin color, facial structure,

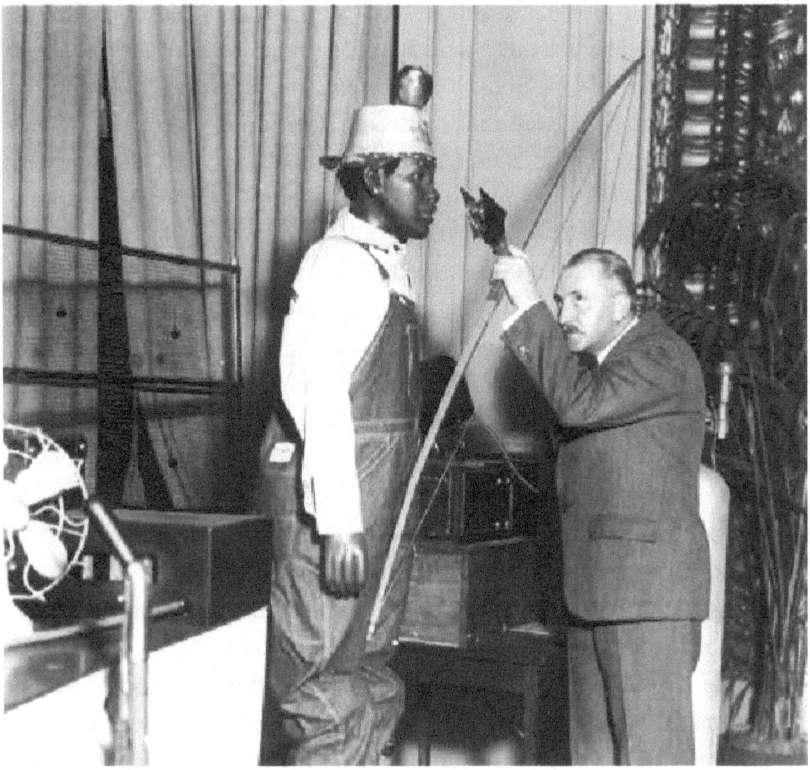

Figure 2.2
The Rastus robot, Westinghouse, 1930.

hair texture, manner of dress, and accessories were used to lead observers to racialize the steam man as Black both in engineering patents and in literature, these same cues were used in the design of Rastus the "Mechanical Negro," a robot capable of rudimentary speech and manual labor, explicitly referred to in the press at the time as a "mechanical slave." As seen in figure 2.2, the Rastus robot was designed to have the facial structure, skin coloration, hair color, and styling of an African American man and was dressed to appear in the caricature of a Black sharecropper. Furthermore, interactants were invited to enact violence on it by aiming a bow and arrow at Rastus, thus triggering an apple poised on the robot's head to explode.[79]

The explicit envisioning of robots as Black slaves is also reflected in 1950s popular science articles describing robots. In one such article, robots were described as heralding a return to slavery: "In 1863, Abe Lincoln freed the slaves. But by 1965 slavery will be back! We'll all have personal slaves again, only this time we won't fight a Civil War over them. Slavery will be here to stay."[80]

However, over the hundred years since *The Steam Man of the Prairies*, the *explicit* connection between robots and Blackness in the robot imaginary has largely faded from robot design,[81] both in popular culture and in robotic engineering. At first glance, it is tempting to cast this as merely the result of the decline of overt racism in favor of color-blind racism. Yet, given the ways that robots remained in roles associated with the labor of enslaved people and of otherwise race–class subjugated communities, this trend away from *visible* Blackness in robot design may have more to do with the changes in the types of stories robots were being used to tell.

Rise of the Robot Narrative

Since *The Steam Man of the Prairies*, robots have provided authors a unique vehicle for exploring public concerns. Due to robots' machine nature, they were used to tell stories about the role of technology in society. Due to robots' presumed subservience, they were used to tell stories about human power dynamics. And due to humanoid robots' anthropomorphic designs, robots were used to tell stories about what it meant to be human. The use of robots to tell these types of stories can be seen in two influential novels released in the years following *The Steam Man of the Prairies*: Auguste Villiers de l'Isle-Adam's *L'Ève Future*[82] and Samuel Butler's *Erewhon*.[83]

Like *The Steam Man of the Prairies*, Villiers de l'Isle-Adam's 1886 Edisonade *L'Ève Future* featured a humanlike robot. But while *The Steam Man of the Prairies'* robot was merely humanoid, the robot depicted in *L'Ève Future* was *indistinguishable* from humans: a form Villiers de l'Isle-Adam coined (or at least popularized) as an *android*. This difference in form allowed Villiers de l'Isle-Adam to tell a different story of White male domination. Whereas Ellis explored the White male use of robots as a replacement for unavailable Black slave labor, Villiers de l'Isle-Adam instead explored the White male use of robots as a replacement for sexually unavailable White women, thus telling a story of gendered sexual domination rather than racialized labor domination. To state the nature of this cultural transformation more clearly: the concept of the android (a robot indistinguishable from a human) was created to enable feminine-presenting robots to more effectively serve as objects of heterosexual male desire.

This story, even beyond its straightforward misogyny, served as a response to White male fears dominating Villiers de l'Isle-Adam's political climate. Just as *The Steam Man of the Prairies* responded to American White men's fears of loss of economic and social status in the face of Black emancipation, *L'Ève Future* responded to French White men's fears of loss of economic and social status in the face of women's suffrage (an achievement not won in France until 1944 but the subject of substantial public debate in the time of Villiers de l'Isle-Adam). As such, *L'Ève Future* specifically uses robots to ameliorate these fears in a way that dehumanized women, argued for their replaceability, and asserted male dominance over them. That is, *L'Ève Future* served as a tool of cultural power for reinforcing stereotypes that would allow French men to justify the maintenance of their power hegemony. In order to tell this new type of story, Villiers de l'Isle-Adam just needed to change the image of the mechanical figure at the heart of his story: from masculine to feminine, and from Black to White.

In contrast to both *The Steam Man of the Prairies* and *L'Ève Future* is Butler's 1872 *Erewhon*, released three years before *The Steam Man of the Prairies'* reprinting and commercial breakthrough. In *Erewhon*, Butler explores public concerns of humans being replaced by a "race of machines"[84] and questions of machine consciousness (and therefore human consciousness). The concerns Butler is responding to, however, are primarily techno-scientific and labor oriented. As such, the machines in *Erewhon* lack the anthropomorphic characterization seen in *The Steam Man of the Prairies* and *L'Ève Future*, instead reflecting Butler's interests in machine evolution and artificial life.[85] Specifically, while the use of humanoid robots enabled the telling of new stories about White male mastery over both Black slaves and White women in the works of Ellis and Villiers de l'Isle-Adam, *Erewhon* and its successors used biologically grown and evolved nonhumanlike robots to tell stories grappling with America's changing techno-scientific and labor paradigms, as well as with what it means to be human, both in a biological sense (following the 1859 release of Darwin's *On the Origin of Species*[86]) and in the sense of life purpose and the pursuit of meaningful work. Specifically, Butler's biological framing enabled the telling of new stories about humans being superseded in the biological competition of the kingdom of life and of White men specifically being replaced (by machines or by non-White men) in the labor market.

Across these three works, we thus see three distinct machine aesthetics: mechanical humanoid men (used to tell stories of racial domination by

White men), mechanical android women (used to tell stories of sexual domination by White men), and biological yet synthetic nonhumanoid creatures (used to tell stories of techno-scientific and labor domination *of* White men). When these three aesthetics and the stories they facilitated ultimately came together, what resulted represented a new stage in the evolution of the robot imaginary and produced a label that crystallized and reified this emerging concept in the public imagination: the robot.

Karel Čapek's 1920s play *Rossum's Universal Robots* (*R.U.R.*)[87] holds a special place in the history of the robot imaginary,[88] as it finally put a name to the mechanical face that had been creeping its way into the public consciousness since *The Steam Man of the Prairies*. Moreover, *R.U.R.* brought the three preceding threads of machine stories together, both in terms of how robots were envisioned and in terms of the types of stories robots made possible. The robots of *R.U.R.* are humanoid, although they are not mechanical, as in *The Steam Man of the Prairies*. Instead, they are biological and described as assembled from grown or engineered biological parts, as in *Erewhon*. And they are indistinguishable from (White) humans (both male and female), as in *L'Ève Future*. This enables *R.U.R.* to tell a story of *uprising*, capturing the social and racial dimensions of *The Steam Man of the Prairies*, the gendered dimensions of *L'Ève Future*, and the labor dimensions of *Erewhon*.

At the time of *R.U.R.*'s release, the play was primarily read in terms of these latter considerations of labor and class struggle as well as women's empowerment and gender relations.[89] Yet, as Chude-Sokei describes, *R.U.R.* also had clear racial dimensions, even if they were not foregrounded as were class and gender.[90] First, a major plot point of *R.U.R.* is robots' evolution of souls—a phenomenon that directly engaged not only with Butler's questioning of machine consciousness[91] but also with the contemporary "debate" over whether Black people had souls and, as such, whether they were truly human.[92] Second, *R.U.R.* features abolitionists who "attempt to first liberate the machines that they call 'brothers,' and turn them into wage laborers as well as give them voting rights."[93] Third, Čapek went on in other works to tackle themes of miscegenation and lynching, highlighting the Czech playwright's acute racial consciousness.[94] Finally, there is the name given to the robots themselves—that is, "robot": a term for "worker," whose Czech etymology was grounded in slavery and forced labor.[95] This choice of name, combined with the play's themes surrounding workers' revolts, Black slave uprisings, and women's struggle for empowerment, serves to connect these shared struggles, much as abolitionists one hundred years earlier had connected the shared struggles of those subjugated by class, race, and gender.

The history of the robot imaginary, and its origins in *The Steam Man of the Prairies*, makes elements of *R.U.R.*'s framing inevitable: it is perhaps unsurprising that Čapek gave his protagonists a name meaning "slave," given that automata already had a long history as cultural stand-ins for enslaved people in the robot imaginary. The staying power of "robot" within our cultural and techno-scientific imaginary may, in turn, be due not only to *R.U.R.*'s quality as a piece of art but, moreover, to when and how it was disseminated as a piece of media.

Although *The Steam Man of the Prairies* was not terribly well written, its release spawned a parade of copycat dime novels, leading to a new genre of cheap technocratic Edisonades[96] such as *L'Ève Future*. Perhaps this is the best success *The Steam Man of the Prairies* could have hoped for, released at a time when the racist minstrel shows (whose design the *Steam Man* played off) were the most widespread form of popular American entertainment. In contrast, the 1920 Czech theatrical production of *R.U.R.* was, by all accounts, met with critical acclaim, leading quickly to revivals and new productions,[97] including English language productions in New York City in 1922 and London in 1923.[98] Moreover, the format of *R.U.R.* aligned well with the emerging mass-broadcast media of radio and television, leading to a BBC radio production in 1927 and a BBC television production in 1938—the first piece of science fiction broadcast on television.[99]

Perhaps because *R.U.R.* was in the right medium at the right time, it immediately became a touchstone for the robot imaginary. In doing so, *R.U.R.* crystalized and reified the existing popular concept of mechanical men by assigning them a name ("robot"). Moreover, *R.U.R.* brought together distinct narrative threads into a new story of *uprising* that connected struggles of race, class, and gender and that simultaneously evoked fears and fueled sentiments of resistance, depending on the viewer. This successful narrative craft, and its widespread dissemination, gave rise to a century of robot uprising stories, from retellings and homages such as *Westworld*[100] and *Dollhouse*[101] to less direct recreations such as Fritz Lang's 1927 *Metropolis*,[102] Ridley Scott's 1982 *Blade Runner*,[103] and Ronald D. Moore's 2004 *Battlestar Galactica*.[104] Many of these stories fell into the genre of cyberpunk, in which racial divisions are, in turn, made clear and blurred by technology.[105] As such, we might say that the connections drawn between race and technology, and the presentation of robots as slaves in *R.U.R* and in earlier works such as *The Steam Man of the Prairies*, served as early prototypes for the cultural vehicles used to explore the nature and boundaries of race in modern science fiction.

This is not to say that this conceptualization of robots as slaves was adopted everywhere. In Japan, for example, automata had evolved in a similar way as their European counterparts, with several centuries of automata developed as toys, art objects, and displays of wealth before they were ever envisioned to have a practical purpose. With the release of *R.U.R.*, however, Japanese scientists and engineers were forced to consider how automata as practical devices would be envisioned. After *R.U.R.*'s Tokyo premiere, biologist Makoto Nishimura was sufficiently disturbed that he set out to build a robot, "Gakutensoku," that would serve as a friend rather than as a slave and that would be socially expressive rather than physically and intellectually capable of performing tasks.[106] While Japanese and American popular culture would go on to mutually influence each other,[107] robots such as Gakutensoku fundamentally shaped the Japanese outlook on robots.[108]

Meanwhile, in America and Europe, the conceptualization of robots was shaped and popularized in the aftermath of *R.U.R.* by the writings of Isaac Asimov. In particular, Asimov's 1950 collection *I, Robot*[109] had outsized impact on the popular understanding of robots, especially due to its inclusion of Asimov's 1941 short story *Liar!*[110] (which introduced the term "robotics") and his 1942 short story *Runaround*[111] (which introduced the Three Laws of Robotics). Asimov's works were intended as a critical response to *R.U.R.* and the science fiction it had inspired. Despite himself being inspired[112] to write his first robot story after reading Earl and Otto Binder's 1939 "I, Robot," a story about a robot slave that rises up and kills its master,[113] Asimov developed an apparent disdain for stories of robot uprisings, stressing, "Never, never, was one of my robots to turn stupidly on his creator."[114] As such, rather than explore robot uprisings in his own stories, Asimov preferred to explore how the mechanical slaves could best be kept in line.

As we will discuss in more detail in chapter 4, Asimov's Three Laws ultimately constitute a set of slave codes for governing the actions of robot slaves, ensuring they obey and do not harm their masters.[115] Indeed, Asimov was quite transparent about this framing. As one of Asimov's characters states in *Runaround*, "The makers . . . built good, healthy slave complexes into the damned machines."[116] It is also worth noting that Asimov initially credited the Three Laws to his editor John W. Campbell,[117] a notorious apologist for slavery.[118] Despite Asimov's arguments later in life that he never intended for his robots to be metaphors for people of color, his stories clearly leveraged this metaphor.[119] As a result, the ultimate effect of Asimov's stories was not only the further popularization of the concept of the robot but also the further reinscription of robots as anthropomorphic slaves and the crystallization of the robot imaginary into a White patriarchal robot imaginary. Asimov's stories, and the metaphors they encoded, would be directly responsible for what would come next.

The Realization of Robot Slaves

Among the millions captivated by Asimov's stories was Norbert Wiener, the "Father of Cybernetics." Wiener was, in fact, not only an avid science fiction fan but also an (unsuccessful) science fiction writer who attempted to coauthor a novel with Asimov himself.[120] In fact, the popularity of Wiener's 1948 text *Cybernetics* was, in part, driven by reader interest in science fiction, as Wiener's textbook was advertised alongside Isaac Asimov's work,[121] leading Wiener to worry that "what I intended as a serious contribution to science was interpreted by a considerable public as science fiction and as sensationalism."[122] This would not be the only concern Wiener would develop about the reception and impact of his work. Wiener also, for example, had serious concerns about the *social impacts* his theories would have. As he wrote in the introduction to *Cybernetics*, "Those of us who have contributed to the new science of cybernetics thus stand in a moral position which is, to say the least, not very comfortable."[123] Finally, as a staunch anti-racist who was well aware of the economic and racial legacy of America's slave system[124] and who had long been deeply concerned with the role that automation would have on labor,[125] Wiener incisively recognized the ways his theories had led to a continuation of that slave system. In his aptly titled *The Human Use of Human Beings*, Wiener stressed that "the automatic machine, whatever we think of any feelings it may have or may not have, is the precise economic equivalent of slave labor."[126]

Indeed, throughout the 1940s and 1950s, the association between robots and slaves was widespread throughout the popular press.[127] As such, it should come as no surprise that this connection would also be recognized, although with substantially less concern, by the first industrial roboticists. The Unimate, a programmable mechanical arm, was patented in 1954 as "Programmed Article Transfer" by George Devol, a serial inventor who had previously worked on everything from military technologies to a microwave oven for "Speedy Weeny" hot-dog vending machines.[128] As legend has it, at a cocktail party in 1956, Devol met businessman Joseph Engelberger, a fellow Asimov enthusiast, who upon hearing Devol talk about his Programmed Article Transfer system (no doubt a thrilling topic for cocktail party conversation) reportedly exclaimed, "Sounds like a robot to me!" Although this story is likely apocryphal, or at least an embellishment of Devol and Engelberger's true meeting, what is not in question is the extent to which Engelberger was inspired by Asimov.[129] In fact, when Devol and Engelberger commercialized the Unimate, they explicitly considered Asimov's stories when deciding where and how to deploy their robots.[130] What is particularly striking is the extent to which Engelberger seems to have been influenced by Asimov's implicit portrayal of robots as Black slaves. As Engelberger would argue, "The robot is obviously a latter day slave, and, better still, it is a willing slave. This sub-human beast is designed (rather than abducted) to labor on behalf of mankind."[131] Moreover, Engelberger was quite clear in how he saw robots maintaining the racial power structure of White supremacy: "Since we can't have slaves or kick around black people anymore, the robot serves that purpose."[132]

This is not to say that the envisioning of robots as slaves in the mid-twentieth century was always motivated by overt racism. As Labor scholar Jason Resnikoff has noted, many intellectuals of the time viewed robots as the potential enablers of a more ethical form of slavery not grounded in racism or oppression. The Marxist philosopher Herbert Marcuse, for example, argued that automation would undermine inequitable capitalist economic systems.[133] Similarly, the left-wing economist Robert Theobold argued, "We could be the modern Greeks, with mechanical slaves to take the place of human toil."[134]

Nevertheless, from their first appearance in literature to their first appearance on the factory floor, robots in the United States were envisioned as slaves that would replace Black labor in order to maintain the power built by wealthy White men under the previous slave system. And as these mechanical slaves were introduced, they would have the same effect on Black workers as previous mechanization efforts in the United States. Early twentieth-century agricultural mechanization efforts had uniquely impacted Black farmers, as they benefited large-scale (White) farming while destroying small-scale Black farms.[135] These effects were reinforced by US agricultural funding agencies' preferences for more heavily mechanized farms.[136] A half-century later, as the Unimate and other industrial robots began to move into factories in earnest, both computer scientists and civil rights leaders sounded alarms that this new round of automation enabled by cybernetics would once again have the most acutely negative effects on Black communities.[137] As Martin Luther King Jr. himself would argue:

> For years we have been victimized with discrimination in many areas and on many levels, denied apprenticeship training, deprived of educational opportunities, facing discrimination in so many areas and this meant that so often we were limited to unskilled and semi-skilled labor. And now because automation and cybernation have come into being, these are the jobs that are passing away. And so while the nation stands in its most prosperous period, while the nation faces its lowest level of unemployment that it has over the last seven or eight years, unemployment among Negroes is at its highest level.[138]

Not long after King's speech, General Motors (GM) would leverage robotics and automation to exactly this effect. In Lordstown, Ohio, for example, GM would ultimately deploy Engelberger's Unimates as mechanical slaves,[139] allowing GM to lay off hundreds of workers. These layoffs disproportionately impacted Black workers, as GM was able to lay off "ninety-eight per cent of the workers, approximately a hundred total" performing spot-welding—a job disproportionately occupied by Black workers.[140] In fact, given factory managements' racist attitudes,[141] the deployment of automation in this way may well have been used to intentionally displace or exploit Black workers. In the wake of these layoffs, which labor scholar Salem Elzway calls "the first job robotized en masse, not just at Lordstown, but in the world,"[142] the remaining factory workers were subjected to twelve-hour shifts (with mandatory overtime)[143] of increasingly deskilled labor under concentration camp–like conditions.[144] This was no outlier—the "automation" of the 1950s[145] was typically used not to completely replace human labor but rather to obscure existing labor, force laborers to work harder and faster, and generally shift power from workers to capitalists.[146] As Salem Elzway and Jason Resnikoff write, "It was only a matter of time before the 'robots' . . . revolted."[147] And indeed, in 1972, Lordstown's workers would lead the highest-profile strike of the 1970s—a strike representing the fruition of Čapek's story of a worker (as robot) uprising.

The Whitening of Robots through Artificial Intelligence

Thus far, we have seen how the robot imaginary evolved from the 1860s through the 1960s. Although robots' original purpose (as slaves working to maintain White and patriarchal power structures) remained constant across this century and ultimately transitioned from metaphor to reality, robots' aesthetics changed in critical ways to enable the telling of new stories. Although Dederick and Grass's patent and the novel it inspired clearly depicted their mechanical slave as a Black man, and although overtly racist portrayals of robots as Black men continued into the 1950s, we have seen how the dominant outward appearance of robots quickly Whitened. Dederick and Grass's prototype lost its minstrel-show aesthetic (perhaps to better appeal to their New Jersey audience); *L'Ève Future* introduced the artificial White woman to protest women's suffrage; Čapek's *robota* were necessarily White, given the ethnicities of his actors; and Asimov's robots behaviorally exhibited their Whiteness by, e.g., quoting *H.M.S. Pinafore*.[148]

As a result, robots came to occupy a position in the cultural imaginary that embodied a curious contrast: built in the image of White men, while occupying a social and economic position associated with Black slaves. And just as the image of robots as *characters* progressively Whitened in popular culture (from Ellis to Čapek to Asimov), so too did the image of "artificial persons" continue to Whiten within engineering, in large part due to the rise of cybernetics' sister field of artificial intelligence.

In the 1950s, as popular media engaged with narratives of human control over robotic slaves, and roboticists designed the mechanical means of creating those slaves, other scholars were beginning to consider more seriously the questions once posed by Butler surrounding machines and consciousness. These debates turned on twinned questions. First, mathematician Alan Turing (commonly considered the "Father of Computer Science and of Artificial Intelligence") raised new debates as to whether computers could have minds through his famous imitation game, now known more commonly as the Turing test.[149] Second, researchers such as neuroscientist Warren McCulloch and logician Walter Pitts raised new debates as to whether minds could be computers or, more specifically, whether the types of formal definitions of computing described by Turing in his earlier work on computability[150] could be effective models of human neural activity.[151] These debates came to a head at the 1956 Dartmouth Conference on Artificial Intelligence, where a dozen participants in these debates from across cybernetics, computer science, and cognitive science, including Marvin Minsky, Ray Solomonoff, John McCarthy, Alan Newell, Herbert Simon, Oliver Selfridge, and Claude Shannon, discussed the possibility of machine intelligence over the course of the 1956 summer.

Together, this group defined what would go on to be known as artificial intelligence: a new field that deviated philosophically from Wiener's still-nascent cybernetics, placing less emphasis on the mathematical description of motion and more emphasis on the symbolic representation of problem solving and the characteristics typically associated with intelligence. This naming convention was both an intellectual choice and a political one, reflecting the new field's pivot toward cognitive psychology, as well as the field's desire to separate itself from Norbert Wiener and establish itself as a niche of its own.[152] Although some attendees of the Dartmouth AI conference disapproved of the nebulous and sensationalist nature of the phrase "artificial intelligence,"[153] the rest of the attendees ultimately elected to frame their nascent field in terms of a vision of intelligence grounded in specific problem-solving capabilities and cultural practices. As a result, the White male attendees of the Dartmouth AI conference imbued an innate Whiteness and masculinity into AI.

As Jennifer Rhee points out, from the origins of AI, the types of capabilities characterized as representing the core facets of intelligence have been tightly linked with the race, gender, and class of the early AI pioneers.[154] For example, these pioneers chose to associate intelligence with the ability to play chess—a historically White and masculine pursuit.[155] Similarly, AI pioneers' focus on "common sense" reasoning, such as knowing the "script" for "how to behave" in a restaurant, elevated knowledge of particular rites and customs as hallmarks of humanity. Moreover, as Yarden Katz catalogues, the early histories of both cybernetics and AI were tightly connected with Department of Defense priorities, meaning that the earliest work on algorithms for general intelligence characterized intelligence as the ability to engage in military decision-making.[156] Finally, because AI aimed to provide computational models of human cognition, and thus of *general* human cognitive capabilities, the framing of AI cast these specific capabilities (chess playing, Western restaurant frequenting, American war making) as not just the essence of intelligence but also the essence of humanity. The progenitors of AI thus drew, through their definition of AI, a race, gender, and class-informed boundary between those who should be considered human and those who should not.

This partitioning off of Whiteness as the domain of general intelligence reflects the history of the study of intelligence. The earliest psychometric studies of general intelligence were conducted by Charles Spearman,[157] a

British eugenicist[158] whose work was directly inspired by that of Francis Galton, the "Father of Eugenics."[159] Spearman was an early player in the intelligence testing movement, including the development of IQ tests such as the Binet–Simon tests[160] popularized by USC Head Football Coach and eugenicist Henry Goddard, who in turn pushed for their adoption in American schools and courtrooms in order to diagnose the "feebleminded,"[161] who he argued should be segregated and sterilized.[162]

The effects of this movement are still felt today, as millions of college-aspiring students each year take the SAT. The SAT was created in 1926 by Carl Brigham at the request of College Board after reviewing his book *A Study of American Intelligence*. What did Brigham argue that so convinced College Board that they immediately hired him to develop this test? The answer is unsettling. In his book, Brigham argued for the superiority of "Nordics," shuddered at the increasing "racial mixture" of students in American schools, and explicitly argued for the need to maintain educational institutions as White: "Our educational institutions are themselves part of our own race heritage. The average negro child can not advance through an educational curriculum adapted to the Anglo-Saxon child."[163] As such, it is no surprise that while today science recognizes that human "intelligence" is primarily a product of social environment rather than genetics,[164] standardized testing today remains "an institutionalized arrangement aimed at expropriating resources from dominated groups to maintain dominant groups' privileges."[165] As Belkhir wrote even in 1994, "Intelligence tests and aptitude tests measure nothing but test-taking skills [and] are biased against certain racial and social groups."[166]

To be clear, cognitive psychologists such as Minsky, Newell, Simon, and McCarthy had good scientific reason to seek to understand the building blocks of human cognition. But by grounding their new research project in the language of "intelligence," these early AI pioneers attached themselves to and inherited key assumptions from this racist and Eugenicist legacy: that thinking machines should be thought of and assessed like thinking people, and that the types of cognitive processes most worth studying were those associated with their own particular intersection of race, class, and gender. Given the historical path from cybernetics to AI, it was only a matter of time before this ideology would seep back into robotics. In fact, key attendees of the Dartmouth conference, such as Marvin Minsky, were interested in robotics from the start,[167] due again to the influence of Asimov. As Minsky once reminisced, "I remember reading [Asimov's] first robot stories and deciding I was going to build them."[168] Just ten years later, SRI's 1966 Shakey,[169] the first "intelligent" robot, would embody a White masculine vision both of what robots should be used for and of what it meant to be "intelligent."

Robot Slaves: Black or White?

As we have seen, the aesthetics and idea of robots Whitened from the 1860s to the 1960s due to the evolution of science fiction, which sought to use robots to tell new stories centering White people, and due to the rise of artificial intelligence, which promoted a White vision for the future of intelligent agents. Yet, even as these robots outwardly Whitened, they continued to be deployed in ways associated with Blackness. Even (or especially) when "autonomous," these robots were intended to operate at (White male) command, perpetuating the master–slave dreams of *The Steam Man of the Prairies*. Australian philosopher Robert Sparrow captures this dynamic succinctly: "The visual race politics of contemporary robots, then, is at odds with the historical race politics of robots. Culturally speaking, we might say that modern robots have White bodies but servile (read, according to a racist trope, 'Black') souls."[170]

The pioneers of robotics and AI would react to this new paradigm of outwardly White and increasingly "intelligent" robotic slaves in different ways. Many decades after the deployment of the Unimate, Engelberger remained perfectly content with his robot slaves:

> The robot is obviously a latter day slave and, better still, it is a willing slave. The self-evident inferiority of a minority group has often been the ethical justification of slavery. Master races have been deeply embarrassed by the intellectual prowess of their slaves, when they begin inconsiderately to display all the attributes of a peer group. A robot slave could never be guilty of such an affront. It offers no challenge.[171]

Others, such as Mathematician Norbert Wiener, viewed this new configuration with less ease:

> Let us grant that slavery is bad because it is cruel. It is, however, self-contradictory, and for a reason which is quite different. We wish a slave to be intelligent, to be able to assist us in the carrying out of our tasks. However, we also wish him to be subservient. Complete subservience and complete intelligence do not go together . . . Similarly, if the machines become more and more efficient and operate at a higher and higher psychological level, the catastrophe foreseen by Butler of the dominance of the machine comes nearer and nearer.[172]

A few years later, Marvin Minsky would echo these concerns:

> There's the old paradox of having a very smart slave. If you keep the slave from learning too much, you are limiting its usefulness. But if you help it to become smarter than you are, then you may not be able to trust it not to make better plans for itself than it does for you.[173]

Today, those working at the intersection of artificial intelligence and robotics continue to debate the status of the machines they have built. Whereas some roboticists argue for the need for a new vision of what robots are and how they should fit into our world,[174] others continue to argue[175] that roboticists' mission is "designing people to serve"[176] and that "robots should be slaves."[177]

III The Robot Imaginary Today

The Gendered Racialization of Robots under White Patriarchy

In this section, we'll consider how robots are depicted today. Given the history described above, it should be no surprise that today's humanoid robots continue to be overwhelmingly designed according to the vision of the White patriarchal robot imaginary. Under this imaginary, robots are overwhelmingly racialized as White. How they are gendered, however, is more subtle, as the "appropriate" gendering of a robot under this imaginary depends on the robot's intended societal role and the agency and patiency attributed to that role. When robots (especially humanoid robots) are designed for roles requiring agency (i.e., where they are able to actually do things in their environment), robot designers often design robots in their own image. Because 82 percent of robot designers are men[178] and 77 percent of robot designers are White, this self-image tends to be White and heteronormatively masculine. In contrast, when robots are designed for roles requiring patiency (i.e., where they are primarily observed and acted upon), robot designers often design robots in the image of those they desire to observe and act upon. Due to these same demographics, this image tends to be White and heteronormatively feminine.

This is most obviously reflected in the ways that racial and gender cues are used in sex robot design to appeal to the racialized sexual preferences both of sex robot designers and of their clients. Sex robots are primarily designed as patients—to be sexually acted upon by their owners. Moreover, as robot ethicists such as John Danaher have noted, sex robots are being developed that may actively indicate a lack of consent, enabling sex robot owners to play out rape fantasies.[179] This design choice hints at agency (through the withheld possibility of consent) in a way that undermines that agency and reinforces the agent–patient, subject–object divide between masculine owner and feminine body. Unsurprisingly, sex robots are overwhelmingly designed to cater to heterosexual male aesthetics and behaviors. They are designed predominantly for penetrative sex[180] and feature exaggerated sex characteristics such as unrealistically large breasts and narrow waist-to-hip ratios.[181]

But there are other reasons beyond heterosexual aesthetics that explain why robots designed to be observed and acted upon are so overwhelmingly designed in the image of White women in particular. Diagnosing this phenomenon requires us to take an intersectional lens.[182] Whereas a non-intersectional analysis might critique the harms dealt by White patriarchy against Black men (who are viewed as the "default" Black people under patriarchy) and against White women (who are viewed as the "default" women under White supremacy), an intersectional analysis would instead specifically consider the harms dealt by White patriarchy against Black women.[183] In this section, I will use an intersectional lens to analyze how differences in the visibility of White and Black women have shaped robot design.

Visibility has been theorized as "the extent to which an individual is fully regarded and recognized by others,"[184] and it is characterized by its two end points: invisibility and hypervisibility.[185] Invisibility captures the ways in which the matrix of domination renders some individuals unseen or unrecognized by others. Examples of invisibility include the way that Black female scholarship is ignored,[186] the way that Black female health issues (the result of overrepresentation in poverty, unemployment, and incarceration) are elided by non-intersectional health statistics,[187] and the way that face recognition technologies are unable to correctly identify Black women.[188] In contrast, hypervisibility captures the ways that the matrix of domination renders some individuals more likely to be surveilled and to face enhanced scrutiny, especially when perceived differences are misinterpreted as deviance.[189] Examples of hypervisibility include the way that Black people (especially working-class Black families in the South) are disproportionately likely to be audited by the IRS[190] and to be arrested on drug charges (while substantially *less* likely to be drug users).

One's status as visible or invisible is informed by contextualized power dynamics.[191] Those without power tend to be visible to one another but invisible (and thus denied recognition, legitimacy, authority, and voice[192]) and/or hypervisible (and thus subjected to increased sanction, stereotype, and commodification) to those in power.[193] From the perspective of visibility politics, White patriarchy renders Black women alternately hypervisible and invisible, depending on context. For example, in her recent study of microaggressions on college campuses, sociologist Veronica Newton found that Black women were hypervisible in classrooms, leading to microaggressions from White classmates and faculty, whereas they were invisible in general campus spaces, leading to them being excluded and ignored.[194]

We will come back in later chapters to how robots themselves employ the logics of invisibility and hypervisibility when acting as surveillance systems that render the subjugated surveillees watched but not seen—what Ruha Benjamin calls "coded exposure."[195] For now, though, we will focus on how the matrix of domination and the logics of visibility shape the designs of robots themselves in the White patriarchal robot imaginary. Specifically, in this section, I ask how the visibility of roles ascribed to White versus Black women under White patriarchy explains the prominence of White feminine robots but not Black feminine robots.

To answer this question, we can consider the different roles assigned to White women versus women of color under White patriarchy.[196] As Melissa Wooten and Enobong Branch catalog in their fascinating history of domesticity,[197] although domestic labor in White households was once primarily seen as the domain of the (White) woman of the house, this perception radically changed over the course of the nineteenth century, so much so that well before the end of the century, the same labor was nearly uniformly seen as the purview of Black women. This transition was due to a confluence of factors, including "heightened requirements for cleanliness" (meaning greater need for physically demanding and unrewarding labor as part of household upkeep), expanding options for women to work outside the home (as exemplified by the case of the Lowell mill girls described earlier), and decreasing opportunities for women to do meaningful and economically productive work at home (due to industrialization-driven availability of home goods such as textiles, as again exemplified by the Lowell mills). Due to these interrelated factors, the domestic labor required of women became more onerous and less rewarding, and as such, White women increasingly sought out and relied on outside help in maintaining their households, leaving themselves free to do work that aligned with virtues of domesticity while being less taxing and more rewarding. Domestic labor thus largely fell to Black women, not only due to their availability as a labor source (especially given the role of urbanization in moving Black women from agricultural work into domestic work[198]), but also due to way the politics of humanity were negotiated through the cultural domain of power. Just as White men were debating whether Black people had souls and were truly humans, White women rationalized to themselves that Black women were not "true" women and incapable of becoming such, and instead applied sexualized or unfeminine stereotypes (e.g., the "Jezebel" and the "Sapphire"[199]) to Black women that cast them as inherently opposed to virtues of White domesticity. Through these rationalizations, White women reasoned that Black women were thus better suited for hard labor. This "common sense" within the cultural domain of power then led to ripple effects throughout the other intersecting domains of power for years to come. As sociologist Julia Wrigley writes:

> In 1940, 60 percent of Black women worked as domestics. African American women were forced into domestic work generation after generation, sometimes with mothers and then daughters or nieces working for the same White families. Barred from entry into White-collar work, and educated in segregated, underfunded schools, African-American women remained trapped in domestic work, even though it usually served as an occupational stepping-stone for other groups.[200]

Critically, the types of labor Black women were relegated to through these intersecting domains of power had different visibility than the types of labor that White women took on. Specifically, although the domestic labor performed by women of color was typically rendered invisible and hidden within the private home,[201] the types of labor reinforced as the domain of White women, such as school teaching[202] and secretarial work,[203] were inherently public facing and led to inclusion in White male bureaucratic institutions (even if that inclusion was in a male-subordinated role).[204]

These patterns inform how and where robots are gendered as feminine in the cultural imaginary of robotics and thus how and where robots are gendered as feminine in the techno-scientific imaginary of robotics. As Jennifer Rhee discusses in *The Robot Imaginary*,[205] recent science fiction has emphasized gendered and racialized divisions in labor (as seen in *Her*[206]) while devaluing labor gendered as feminine (as seen in *The Stepford Wives*[207]). Particularly striking in its straightforward visualization of this racialized and gendered labor history is *Ex Machina*,[208] which, as Rhee points out, not only exhibits overt techno-orientalism but, moreover, uses women as robots to depict the liberation of White women on the backs of non-White women.[209] Within the techno-scientific imaginary, these same labor patterns have most clearly carried into technology design in the domain of voice assistants,[210] as feminine voice assistants frequently provide services associated with care labor activities, which are, in turn, associated with White women.[211]

Given the interplay between the cultural and techno-scientific robot imaginaries, and between the designs of intelligent voice assistants and robots, these same patterns have carried into the techno-scientific imaginary of home robots. As Neda Atanasoski and Kalindi Vora argue, the design of current home robots, such as Cynthia Breazeal's Jibo,[212] cast the household as an autonomous economic unit—an assumption that grounds Jibo in the racialized and gendered history of twentieth-century American domesticity and renders the robot "a technological update to . . . domestic architecture."[213] This subtle economic cue was reinforced in advertisements for Jibo, which depicted "the fantasy of the all-American family—healthy, smiling, blonde, White children living in domestic bliss in a detached house with a garage and car on a tree-lined street in the suburbs with women happily baking in the kitchen and large family dinners."[214] Moreover, Jibo is displayed as a vehicle for satisfying its White male owner's (hetero)sexual desires, serving as his ostensible "wingman": an extremely odd choice for a robot advertisement, yet extremely illuminating with regard to whom the robot was presumed to be for. In turn, as Atanasoski and Vora argue, these designs "reinscribe a racialized and gendered service labor infrastructure that derives from the domestic division of labor under colonialism."[215]

This use of robots to reinscribe racialized and gendered service labor infrastructure has also been noted by Yolande Strengers and Jenny Kennedy, who argue that the designers of robots such as Pepper—and intelligent voice assistants such as Alexa—have essentially been asked to build the technological equivalent of a "wife"—that is, to perform tasks traditionally associated with White female domesticity.[216] As they demonstrate, the designers of these types of technologies have been asked (sometimes explicitly, sometimes implicitly) to create technologies whose primary purpose is to fulfill the dreams and desires of (heterosexual) White men in domains where those dreams and desires are typically achieved through the labor of (heterosexual) White women. Given this design goal, it should be unsurprising that designers respond by designing robots and voice assistants in the image of White women.

This casting of robots in the home as feminine also reflects the longer history of automation discourse. As Jason Resnikoff argues, at the same time that robots were being envisioned as slaves in factories in ways that devalued and degraded the meaning of industrial labor,[217] they were also being envisioned as household slaves in ways that devalued and degraded the meaning of domestic labor. And at the same time that automation was being used in factories to force workers to work harder, automation was being used to similarly increase the productivity of women laboring within the household.[218] As Resnikoff argues, the automation discourse of the 1950s reinscribed associations between femininity and domestic labor in a way that presented a troubling contradiction. On the one hand, this framing cast domestic labor as only worthy of slaves. On the other hand, this framing simultaneously held that middle-class women should happily perform such labor without compensation.[219]

We can also observe through more recent robots, such as Amazon's Astro, how robots continue to reinforce these discourses. Released in September 2021, the first advertisement for Astro seems, at first, to have learned from Jibo's mistakes. Gone is the all-White cast. Instead, the video features a Black couple in its first scene, quickly followed by a racially and sexually diverse cast of possible users. Yet, the ultimate use cases shown for Astro are nevertheless grounded in the same White male fantasies, with Astro shown helping homeowners to protect their homes against criminals portrayed as raccoons—a disturbing choice, given the ways that raccoons have historically been used as racist signifiers of ostensible Black criminality. These recent advertisements serve as our clearest window into the techno-scientific imaginary guiding today's robotics companies. They make clear whom these robots are designed for and whom they are intended to serve. Moreover, these advertisements demonstrate the values of robotics by more subtly demonstrating whose labor is valued and whose is devalued.[220]

Thus far in this chapter, I have shown how the dominant robot imaginary—the White patriarchal robot imaginary—has evolved from an origin in which robots served as depictions of enslaved Black Americans to a state in which they served as depictions of other communities over which White Americans sought control and finally to one in which they served as depictions of White characters while still predominantly serving in societal roles associated with Blackness in order to most effectively reinforce White patriarchal power. Then, using key concepts from *Black Feminist Thought* (the matrix of domination, intersectionality, and visibility), I have shown how today's techno-scientific imaginary of home robots serves to perpetuate White patriarchy across the cultural domain of power by reinforcing a narrative in which the invisible labor of women of color is automated, replaced, and devalued in order to serve the needs of White men (and, to a lesser extent, the domestic needs of White women) within the White patriarchal order. This analysis suggests a number of reasons why robots are rarely built in the image of people of color, including Black women, and begins to explain when and why robots are built specifically in the image of White women. But we can more fully understand these patterns by understanding the practicalities governing the deployment of some of today's social robots, especially those that depend on *artificial* artificial intelligence.

The Visibility Politics of *Artificial* Artificial Intelligence

Whereas AI pioneers such as McCarthy, Newell, and Simon may have conceived of artificial intelligence as the creation of symbolic systems with humanlike reasoning capabilities, this is a far cry from the artificial intelligence of today. Today, the field of artificial intelligence is dominated by machine learning, in which problems are typically solved not by reasoning toward the solution but rather by attempting to guess at the solution based on past examples. In some dramatic cases, such as deep learning (the dominant subfield of machine learning), this approach comes down to collecting enough training data to be able to approximate a lookup table that maps from queries to ideal results. In extreme cases, this has resulted in both text- and image-oriented generative AI models[221] directly reproducing the (arguably stolen) texts or images on which they were trained.

This raises the question of where the data for these data-hungry models comes from. In the case of large language models such as Generative Pre-trained Transformer (GPT), this is "text that OpenAI scraped from the Internet." For other models, especially "supervised" learning algorithms trained on large-scale datasets such as ImageNet, the answer is websites, on which workers can perform small tasks (e.g., providing their best guess as to what is in an image) for even smaller amounts of money (e.g., $0.01). Ostensibly, if a crowd worker performs enough of these tasks, they could add up to a livable wage. In practice, as Jeremias Prassl argues in *Humans as a Service*, these workers are typically paid far below minimum wage while receiving no benefits (e.g., health insurance)—a configuration that allows the companies that rely on crowd workers to skirt labor laws, more or less reverting to pre-labor-movement business models and their associated exploitation of human labor.[222]

When machine learning researchers use crowd working as part of their workflow, this typically involves collecting a large number of possible queries, asking crowd workers to provide their best guess as to the ideal result for those queries, and then turning those results into what amounts to an approximate lookup table. This means that when making a decision, the decision-making process is "ask someone on the Internet what they think" or rather "check the responses provided by previously asking someone on the Internet and extrapolate based on those responses." This pattern of computation and labor is critical to the deployment of large pretrained machine learning models such as OpenAI's GPT and DALL-E. First, as we have described, these models primarily generate content by extrapolating from existing human-generated sources scraped by OpenAI. Second, because of the well-known tendency for large pretrained language models to generate hate speech and for large pretrained image models to generate pornographic content, companies creating these models have taken to further tuning the output of their models based on human feedback in an attempt to prevent the generation of such content. In many cases, the humans providing this feedback are underpaid and exploited workers in previously colonized countries. OpenAI, for example, reportedly paid Kenyan workers $2 an hour to sort through horrific content that would be illegal for OpenAI to host and show workers in the United States, including imagery of rape and beheadings.[223] As argued by Couldry and Mejias, this exploitative relationship represents a form of data colonialism, in which Big Tech companies rather than nation-states operate as colonizers and in which data rather than natural resources are extracted and expropriated.[224]

These patterns of "Black slave obscured behind a White mask" have an eerie similarity to Sparrow's characterization of robots as Black souls in White bodies.[225] But these parallels should perhaps be unsurprising, given the cultural history that crowd-work platforms draw on when presenting themselves. Among crowd-work platforms, the historic heavyweight has been Amazon Mechanical Turk, whose name references the eighteenth-century chess-playing machine known as the Turk (figure 2.3), created by Wolfgang van Klempen. A seeming marvel, the Turk was an ostensible chess-playing automaton designed to look like an Ottoman (hence, "Mechanical Turk"). The Turk

Figure 2.3
Reproduction of Wolfgang van Klempen's 1769 Mechanical Turk. Image taken by Tom Williams, December 27, 2019. On display at the Metropolitan Museum of Art in New York City as part of their exhibit "Making Marvels: Science and Splendor at the Courts of Europe."

toured for nearly a century, facing—and beating—figures such as Catherine the Great, Benjamin Franklin, and Napoleon Bonaparte.[226] The Turk amazed audiences, but at its heart, it held a secret: unbeknownst to those who played against it, the Turk contained a secret chamber where a human chess player sat. Despite not being a real robot, the Turk is an illustrative case study on the road to the robot imaginary. It demonstrates that even before autonomous mechanical men had really entered the cultural imaginary through page or stage, engineers were designing machines that were needlessly racialized; whose actions, at the end of the day, were just human decisions, with no autonomy involved; and whose designs served to obscure, rendering literally invisible the labor of those who powered them. It is oddly self-aware that Amazon elected to name their crowd-work platform "Amazon Mechanical Turk," which Jeff Bezos rather tellingly referred to as "Humans as a Service."[227] Yet, at the same time, this connection drawn by Amazon obscures the inversion that has taken place, from the "White laborer behind a Black mask" of the Turk to the "Black laborer behind a White mask" of Amazon Mechanical Turk.

In some cases, the use of crowd workers may be even more extreme. In cases characterized as "artificial artificial intelligence" (another turn of phrase coined unironically by Jeff Bezos[228]), the data-collection stage is skipped altogether. Instead, when a decision needs to be made, it is quickly sent to a crowd worker, who provides the ideal result, removing any algorithmic decision-making from the process and placing outcomes completely in (often underpaid and exploited) human hands. Looking at this type of approach, it is obvious that there is no artificial intelligence involved at all. Yet, many companies pass off these systems as AI. A classic example of this is Facebook's "M" chatbot, an "intelligent assistant" announced in 2015 and shut down in 2018. The secret behind Facebook's "M" is that a significant proportion of its responses were entirely human authored[229]: while users thought they were privately interacting with a machine, real humans were reading and responding to many of their queries without their knowledge. This practice is not only deceptive but, moreover, erases the labor of the real humans powering the service and, as a result, hides from users the underpaid labor underpinning the technologies they rely on—as well as the identity characteristics of the humans performing that labor.

This obscuring of identity mirrors the unknowing experience of the "users" of the original Mechanical Turk. The Mechanical Turk obscured the identity characteristics of the human inside it and rendered their labor (and exploitation) invisible. Moreover, the observers of the Mechanical Turk would have clearly viewed it as a stereotypical depiction of an Ottoman man. Yet, at the same time, those observers would have viewed it as an intriguing mechanical marvel rather than a genuine agent[230] with its "own" identity characteristics. It is not clear if this would hold today. Today's viewers of the Mechanical Turk in action might well view it as not merely a racialized depiction of a character but, moreover, as a racialized agent. An operator of the Turk today would thus find themselves engaged in a complex identity play, with their own identity not only erased but, moreover, replaced with that of the Turk. Although the embodied, genuinely *Mechanical* Turk may no longer be in use today, we do not need it to confirm these speculations. The same dynamic, and same identity play, can be observed through the case of the Engkey robot.

Introduced in 2011, the Engkey robot was a second-language-learning robot designed for deployment in South Korean schools.[231] As described in the original scientific paper on the robot's design, Engkey appeared as a plump, egg-shaped torso topped with a tablet display. While from this description Engkey may not seem terribly controversial, the robot, in fact, embodied a host of deeply problematic racialized and gendered assumptions, as revealed by Anna Romina Guevarra.[232]

Engkey's physical morphology does not initially appear to be overtly racialized or gendered. This changes, however, when one considers what was shown on Engkey's tablet: a depiction of a White (and, as we will see, presumably American) woman's face. Moreover, while not reported in the original paper, this racialization is particularly insidious, given the way that Engkey was deployed into schools. Although children were led to believe that Engkey was autonomous, the robot was in fact teleoperated (remotely controlled) by English-language teachers in the Philippines, who were predominantly Filipina women. To add insult to injury, these Filipina women's faces were used to guide the animation of the White woman's face displayed on Engkey's tablet. As such, Engkey quite literally rendered these women of color invisible, deracinating and erasing their identities and their labor, replacing them with a "more desirable" White womanhood.[233]

Why was White womanhood perceived as more desirable? As Guevarra shows, the employment of Filipina women as teleoperators and their subsequent erasure and dehumanization were politically fraught design and deployment choices grounded in the historical US military presence in both South Korea and the Philippines and the gendered and racialized labor markets in both countries. This colonial history instilled an Anglophilia, prompting a desire for English-language education and for increased proximity to Americanness. Within an American context, school teaching had become both a racialized and gendered occupation perceived as the purview of White women—and a key domain within which White women have historically promulgated and reinforced the logics of White supremacy.[234] This had the primary effect of rendering White American women the most desirable instructors for the domains in which Engkey was to be used. As Guevarra writes, "Engkey [thus] embodies the gendered subject presumed to be the ideal laboring subject for performing these tasks."[235]

Of course, the actual availability of White American women to cheaply teleoperate Engkey for use in South Korean classrooms did not match this desire. Instead, and by virtue of this same colonial history, the most widely available and well-suited cheap labor source for this domain was instead Filipina women. As Guevarra writes, "The turn to the Philippines for English-language instruction [was] enabled by the legacy of U.S. colonization and neocolonization. The United States was responsible for setting up educational and training institutions in the Philippines modeled after itself, with English as the primary medium of instruction."[236] These colonial trends thus explain both the ethnicity of Engkey's chosen teleoperators and the contrasting racialization of Engkey's avatar, producing (at least) two key effects. In the short term, this visualization (of White femininity) and disembodiment, deracination, and occlusion (of Filipina femininity) provided "more 'authentic' (read: Whitened) credentials to serve as the globally legitimate provider of English-language instruction"[237]; in the longer term, this replacement served to reinforce White hegemonic power by erasing women of color from the designers' "utopian" vision for the future.[238]

The example of Engkey also shows why today's American and European robots are predominantly designed in the image of White men *and women* but almost never in the image of Black men or women. Specifically, Engkey can be read as continuing a history of robots used to obscure and devalue labor, especially labor performed by people of color, for the purposes of maintaining racialized and patriarchal power structures. Moreover, Engkey can be read as continuing a history of technologies (robotic or otherwise) that hide labor by separating its production and consumption into separate spheres[239] and that reinforce (or shift) racialized and gendered power systems by minimizing the importance of the technology user's own embodiment[240] or by enabling "identity tourism" wherein the technology user steps into an inauthentic identity for the sake of entertainment.[241] Although Engkey may be a particularly egregious example, invisibilization is a persistent problem for teleoperated robots used in many different domains. Even in my own laboratory's work on teleoperation,[242] in which the teleoperators of educational robots are fully visible to children, teleoperation may require substantial invisible labor[243]: "overlooked, ignored, or devalued activities that are necessary for workers to complete to adequately perform their jobs."[244] In nursing and education, this can manifest as the need to perform affective labor that is undervalued and perhaps uncompensated. Similarly, even in contexts where robot teleoperators are fully visible, teleoperation may require substantial invisible labor in order to effectively *prepare content* for teleoperation sessions.[245]

To summarize, the example of Engkey helps us see both why robots are *not* designed in the image of women of color and why they *are* designed in the image of White women: robots are often designed to depict White women when the robots are to be deployed in domains where the ideal laborer is a *visibly White woman*; where doing so aligns with the aesthetics of the heterosexual White male gaze; and where doing so serves to leverage and reinforce the hegemonic power of White men through reinforcement of the association between racialized femininity and those labor roles and/or through reinforcement of the invisibility of women of color. Overall, then, we might say that robots are often designed as White (but not Black) women because of the different roles that White and Black women play under White patriarchy and the ways those roles are subject to different politics of visibility. White women are represented in robot design because it serves White patriarchy to do so; Black women are not represented in robot design because it does not serve White patriarchy to do so.

Robotic Reinforcement of Racial Patriarchy beyond the US

This chapter has largely focused on the White patriarchal robot imaginary, dominant in American robot design due to the predominantly American context explored in this book. Clearly, not all robots around the world today are designed according to this imaginary. The majority of humanoid robots coming out of Japanese robotics labs, for example, largely depict Japanese characters. Yet, this does not mean that those robots' designs are free from the influence of racialized patriarchal power structures. The clearest example of this is the work of roboticist Hiroshi Ishiguro, who is known for his android creations such as Geminoid and Erica. On the one hand, we have the Geminoid, a robot created in Ishiguro's own image, ostensibly to allow him to attend meetings and give lectures remotely.[246] While the Geminoid is highly technologically novel, its design echoes the methods and motivations of much earlier robots, such as Mr. Eisenbrass (which, as described above, was designed in the image of its creator in the 1840s[247]), as well as Eric, Britain's first humanoid robot (which was ostensibly designed to allow for the remote delivery of a speech in the 1920s). On the other hand, we have Erica, a robot designed to be perceived as a fictional Japanese woman, based on a survey of "30 beautiful women" so that the robot's design would "appeal to everyone."[248] Here, Erica's design echoes the methods and motivations of the first robot *L'Ève Future* (which was envisioned as indistinguishable from humans in order to be more effectively sexualized) as well as AILA (which was designed based on a survey of [sexualized] feminine robots from popular culture, as we will describe in chapter 3). Just as we've seen with American and European robots across this chapter, Ishiguro's robots can be seen as designed either in the image of their masculine creator or for the heterosexual male gaze of the creator's design team. These patterns can also be seen in the design of Miisho's Gazeroid,[249] so-named because it was ostensibly designed to explore "the gaze expressions of humanoid robots' eyes." Ironically, however, its design is clearly sexualized for the Japanese male gaze. Although the title of the paper in which it was presented, "Interactions on Eyeballs of Humanoid-Robots," was intended to be read as referring to the eyeballs of the robots, the robot's design belies the underlying importance of (heterosexual male) eyeballs in the design of the robot. Gazeroid, indeed.

These dynamics are important to acknowledge, not only because they show how racialized and patriarchal design patterns manifest outside of American and European contexts but also because they help to explain where and how robots are used outside of those contexts. Historically, roboticists have pointed to Japan's "shortage" of care workers as an explanation for Japan's investment in robots as a solution for their aging crisis (in which a growing percentage of the Japanese population is composed of older adults). Yet, James Wright argues in *Robots Won't Save Japan* that this explanation obscures some of the deeper social reasons for this investment.[250] Although

European countries are facing similar growth in their aging populations, for example, they do not face the same shortage of care workers—in part because they are willing to allow greater numbers of immigrants into their countries, who then go on to fill those roles. As such, Japan's investment in care robots may be seen more as a way to solve their aging crisis *without increasing immigration*. As Jennifer Robertson has argued, humanoid robots have been regarded as a particularly popular solution in comparison to immigrants due to the sociocultural anxiety that foreigners provoke and due to the ability of robots to reinforce ethnic homogeneity both in practice and as an ideology.[251] On the other hand, as Wright points out, the Japanese government has actually made several efforts to bring in foreign care workers. Yet, these foreign care workers rarely wish to immigrate to Japan for such positions due to the hostility they face upon arrival, as they are given menial tasks that minimize the contact between their foreign hands and Japanese bodies.[252] In other words, although Japan's increasing adoption of humanoid robots such as Pepper may not reflect a desire to reinforce *White* patriarchy, it may yet be viewed as reflecting a desire to maintain Japan's rigid racialized (and gendered) social system.

Conclusion and Chapter Summary

This chapter has shown that by tracing the histories of the cultural and techno-scientific robot imaginaries, we can develop a critical understanding of how and why we think about robots today and the reasons why the White patriarchal robot imaginary is today's dominant robot imaginary. First, this history helps us understand how the conceptualizations of robots in the cultural and techno-scientific imaginaries are inextricably linked. On one hand, the robots being developed in research laboratories (especially those that are well publicized) play a role in shaping how robots appear in science fiction. This can be seen in the appearance in popular media of real-world robots (e.g., Paro in *Master of None*[253]), fictional robots clearly inspired by real-world robots (e.g., the Spot-like killer dog robots in the *Black Mirror* episode "Metalhead"), and fictional robots inspired by theoretical papers in robotics and AI (e.g., the use of the Turing test as a central plot device in *Ex Machina*[254]). This influence over science fiction shows the most obvious ways that roboticists wield cultural power—by directly inspiring the next generation of popular culture. In turn, the robots depicted in popular culture play a role in shaping who is inspired to go into robotics (and why), what those new roboticists seek to enable robots to do once they're working in the field, and the limited range of design options designers ultimately consider when engaging in robot design. But, moreover, we've seen in this chapter the specific ways that these imaginaries have tended to affect each other. Due to the origins of robots as a joint scientific–literary concept shaped by White patriarchal ambitions, the mutual shaping and evolution of robots in the cultural and techno-scientific imaginaries has steered robotics in a specific direction whereby robots continue to be presented and designed as slaves working to uphold White supremacist and patriarchal power structures—what I have referred to as the White patriarchal robot imaginary. Due to this origin and pattern of shaping, today's robots, as envisioned in both science fiction and engineering, continue to serve as vehicles for advancing White male desires and are designed and deployed in racialized and gendered ways to optimally fulfill White patriarchal norms and expectations while obscuring the labor of workers whose racial and gender identities are misaligned with those expectations.

In naming and characterizing the White patriarchal robot imaginary, I underscore that this imaginary is just one possible robot imaginary—and invite the reader to consider what alternative imaginaries robot designers might instead embrace. In the next chapter, we will work toward these alternative imaginaries by breaking down the specific design decisions roboticists make when deciding how humanoid and interactive robots will look and speak, and how each of those decisions might reinforce or subvert the White patriarchal robot imaginary—and, in doing so, reinforce or subvert White patriarchal power in the cultural domain.

3 Identity Crisis

In the last chapter, we saw how our modern concept of robots emerged as a vehicle of cultural power to address the concerns of White men in the American South in the aftermath of the US Civil War. We saw how the depiction of robots shifted in order to tell new types of narratives that supported (or opposed) White patriarchy. We saw how robot design has converged to a paradigm in which robots are designed either in the image of their creators or in ways that align with their creators' aesthetic and sexual preferences. And we saw how robots designed under this paradigm frequently rely on, yet render invisible, the labor of women and people of color. Overall, then, the previous chapter showed how today's dominant robot imaginary—the White patriarchal robot imaginary—serves to reinforce the White patriarchal status quo within the cultural domain of power.

If this is the paradigm that defines robotics, then it is time for a paradigm shift.

The key question we will grapple with in this chapter is: How can we divert robotics from this status quo and toward a more equitable robotic future? In the first part of this chapter, we will start by considering social identity design cues: the specific design choices that today's robot designers are making that lead people to attribute particular racial and gender characteristics to robots. In the second part of this chapter, we will then explore the ways that roboticists wield cultural power through these social identity design cues, in terms of how their choices either reinforce or subvert the biases, stereotypes, and assumptions used to justify White patriarchy. Finally, leveraging our new understanding of these choices and their consequences, we will explore how roboticists can work toward new robot imaginaries by revising their choices, rethinking the space of possible choices, or more fundamentally ripping up the status quo by approaching robot design in new ways that prioritize the dismantling of the White patriarchal status quo.

I Social Identity Design Cues

To begin, let us consider the social identity design cues appearing in today's robot designs. Social identity design cues (such as the AILA robot's white plastic exterior, haircut, and chest) are the aspects of a robot's design that lead observers to assign that robot to social categories such as race and gender.

Sometimes, social identity design cues are used intentionally. In the introduction, for example, I posed the question of why AILA, a robot designed for assisting astronauts aboard the International Space Station, had such an overtly feminine morphology—not because robots with feminine morphologies are inherently problematic but rather because the design choice was both surprising (given AILA's intended context of use) and suspicious (given AILA's all-male design team). The answer I was originally given by a technician to this question ("It allowed us to fit in more computer chips") is implausible and may have been an attempt at a joke. If we consult the paper documenting AILA's design, the results are no more illuminating. The paper makes no mention of these obvious sex and gender cues and simply states that AILA's overall design was intended to be "anthropomorphic and aesthetical."[1] To get a clearer answer, I reached out to one of AILA's lead designers. He immediately responded: "I can answer why we chose to leverage female sex and feminine gender cues for the appearance of the robot AILA. In short: It was deliberately a cheap trick to make our project survive. And I still regard it as one of my wisest design choices."

In our correspondence, AILA's designer went on to describe the precise series of design decisions made by DFKI in designing AILA. At the time of AILA's design, DFKI was facing stiff competition from the Institute of Robotics and Mechatronics at the German Aerospace Center (DLR). DLR's team was larger and more experienced than DFKI's, and as a result, it was simply not feasible for DFKI to compete on an equal footing with DLR. Since they could not compete with DLR on technical grounds, DFKI decided to rethink the aesthetics of their robot in the hope that a robot with sufficiently different aesthetics would not be viewed as a direct competitor. Because DLR's robot, Rollin' Justin, was muscular and masculine, the DFKI team decided that their robot would be smaller and more feminine. Apparently, a feminine robot would not be viewed as a "real" technical competitor to a masculine robot.

In order to design a feminine robot without, in their words, "going too far," the DFKI team began by brainstorming different types of feminine robots they had seen across both pop culture and engineering. In other words, the team worked to identify the different ways that femininity was evoked in the robot imaginary. Initially, DFKI considered only pink robots, including the Lara robot from TU Darmstadt[2] and the Arcee v03 robot from *Transformers*, but was talked out of this by a member of the design team and instead chose to design a White robot. Given the White patriarchal robot imaginary's status as the dominant robot imaginary, this may have been inevitable.

The DFKI team thus considered ten different types of predominantly White and feminine robots. First, DFKI considered what they termed "old school" robots, which included the robot Maria from the film *Metropolis*, a nude feminine robot from the cover of *Heavy Metal* magazine (a comic book line known for its nudity, graphic violence, and misogyny[3]), and the album cover for the single "Faulty Innards" from the band Android Fornication, which depicts a White male engineer constructing an android who is feminine, White, and nude. Second, DFKI considered "trash" robots, which included Katy from *The Bionic Woman*, and sex robots from *Futurama* and *Austin Powers*. Third, DFKI considered robots from manga, which included hypersexualized androids from *Ghost in the Shell* and a robot prostitute from *Malice@Doll*. Fourth, DFKI considered "art" robots, including both abstract and sexualized depictions of robots seen in digital art found online. Fifth, DFKI considered "diverse" robots, including Toyota's violin-playing robot, a dancing robot from the Swiss *Robots* musical, a robotic costume from the American Renegade Effects Group, and several other depictions of robots as nude women. Sixth, DFKI considered "stylishly abstracted" robots, including EVE from *Wall-E*, Sega's EMA robot, and Kyoto University's FT ("Female Type") robot, which Bloomberg profiled under the headline "There is no androgynous robot," perhaps due to its default pose in which its breasts are thrust outward. Seventh, DFKI considered robots that were "female bodies with technical attributes," including several sex robots, as well as hypersexualized imagery of actresses from movie posters for *Surrogates* and *Terminator: The Sarah Connor Chronicles*. Eighth, DFKI considered "mannequin" robots, including the animated robot used to advertise Svedka vodka, the robot from Bjork's music video "All Is Full of Love" (in which Bjork is shown making love to a "pure white" robotic version of herself[4]), and a fictional Fembot 2.0 envisioned with "fleshlike" skin and "semi-transparent" clothing to "serve your every need."

Although each of these "mannequin" robots was highly sexualized, DFKI viewed them as evoking less shame than those in the previous categories, as they had too little detail to be considered truly "nude." DFKI considered this perhaps their best option, especially if these humanlike robots were a neutral color (i.e., white). Finally, DFKI considered but ruled out two other categories that went too far in their view: literal sex robots and androids that fell into the uncanny valley.[5] On the basis of this survey of (mostly sexualized) feminine forms, AILA's designers ultimately decided to aim for a morphology that was white colored and feminine, with a more abstracted and featureless design.

This series of design considerations demonstrates how the White patriarchal robot imaginary shapes the social identity design choices made by robot designers today, and how those design choices feed back and reinforce that same imaginary. AILA's design team intentionally sought out a robot form that would, by virtue of the "cheap trick" of its femininity, not be viewed as a "real" competitor to a more technologically advanced and masculine competitor. They identified a range of (predominantly hypersexualized) feminine robot forms from across pop culture and engineering, and they chose a white-colored, feminine, and female form that would satisfy their cultural and sexual aesthetics while avoiding a "line of shame" delineated by the evocation of humanlike nudity.

Although this story shows how social identity design cues are sometimes deployed deliberately to achieve particular effects, it also shows how such cues are sometimes deployed inadvertently. Although AILA's designers chose to intentionally gender AILA as feminine, the evolution of AILA into a White woman appears to have been less explicitly intentional. After all, AILA's designers originally wanted AILA to be pink, Yet, ultimately, AILA was changed to a pure white plastic in order to be more "neutral." This seems to suggest that more subconscious assumptions about the neutrality of Whiteness may have led to AILA's racialization.

Yet, what matters most about a robot's design is not the intentionality of its social identity design cues but rather how those cues change the ways that people make sense of, and behave toward, that robot. The social identity design cues used in AILA's design invoke a heteropatriarchal gender frame, inviting the viewer to use dominant lenses of gender and sexuality to make sense of the robot in the same way that they are used to make sense of people. For example, AILA's grooming cues, such as her dress, hairstyle, and oversized, contoured eyes, all encourage people to perceive AILA as feminine. Moreover, AILA's body type, such as her hip-to-waist ratio[6] and apparent breasts, encourage people to perceive AILA as female—a perception that reinforces AILA's perception as feminine due to the ways that sex and gender are typically linked under the dominant heteropatriarchal gender frame. As I will argue in this chapter, these social identity design cues invite the viewer to apply all the stereotypes that are used under heteropatriarchy to reason about people who are categorized in similar gendered and sexed ways.

Similarly, AILA's social identity design cues invoke what Joe Feagin calls the "White racial frame"[7]: "a broad and persisting set of racial stereotypes, prejudices, ideologies, interlinked interpretations and narratives, and visual images."[8] Just as the heteropatriarchal gender frame invites the viewer to use certain elements of AILA's design to categorize her as feminine, the White racial frame invites the viewer to use certain elements of AILA's design to categorize her as White and to use this categorization to flesh out their internal mental model of AILA and her societal role. Over the following sections, we will consider the different social identity design cues that lead observers to view robots through a heteropatriarchal gender frame and/or White racial frame.

Morphology

The most obvious social identity design cues are morphological: visually observable dimensions of the robot's physical design, such as the physical cues that lead users to perceive AILA as feminine, female, and White. Morphological social identity design cues are rife in robotics, both in what is shown to the public and in what is developed in the lab. Take a moment to go to your favorite search engine and search for "humanoid robot." As I write this in 2023, most of the robots seen on Google Images have extremely obvious morphological social identity design cues. Many of them are either obviously masculine or obviously feminine (with masculine robots being more common than feminine robots), and nearly all of them are racialized as White. Philosophers of technology Stephen Cave and Kanta Dihal noted similar trends in 2020 in their groundbreaking article, "The Whiteness of AI," both in search engine queries and in film and other media[9]: the robots visualized in the popular imagination are overwhelmingly White, masculine, and male.

Although many of the images you'll see in a search engine are of the same few robots or are artistic renderings of fictional robots, these patterns also hold for the real robots being developed in robotics labs. The ABOT database (https://www.abotdatabase.info/) contains information on more than 250 anthropomorphic robots that have been developed in research labs and/or put into production.[10] Recently, social scientists and ethicists have analyzed the ABOT database to understand the ways in which social identity design cues are being used in robot design. Robert Sparrow has shown that around 70 percent of the robots in this database at the time of his search were mostly or entirely white in color.[11] Similarly, social scientist Giulia Perugia has shown that the robots in this database are much more likely to be gendered as masculine than they are as feminine. In a paper entitled "The Shape of Our Bias,"[12] Perugia asked participants to score the robots from the ABOT database on a variety of criteria, including perceived masculinity and femininity. What she found is surprisingly similar to the general trends observed through search engines: around 40 percent of anthropomorphic robots were perceived as masculine, about 45 percent were viewed as neutral or not strongly gendered, and around 15 percent were perceived as feminine.

What social identity design cues were roboticists using to gender their robots? As Perugia shows in her analysis of the ABOT database, grooming choices that are traditionally gendered led to increased perceptions of femininity. In fact, the mere appearance of hair, eyebrows, eyelashes, and apparel led people to view a robot as more feminine. In contrast, the mere presence of arms, legs, and a torso led people to view a robot as more masculine. The presence of arms, legs, and torsos as masculinity cuing should be initially surprising. After all, having arms, legs, and a torso are not viewed as masculine among humans! On the other hand, certain *types* of torsos (such as AILA's visible chest or the synthetic breasts put to use in sex robots) are clearly inspired by female morphology and thus more likely to lead users to perceive robots as female and thus as feminine. These findings may, however, be best understood if we view them in terms of the history we saw in the last chapter: the racialized and gendered history of robotics has led today's (predominantly White and Asian male) robot designers to design humanoid robots in their own image when those robots are designed to have agency, except in contexts that are feminine and public-facing service roles, where those robots are instead designed to have patiency and are thus designed as feminine. These interpretations help us to better understand Perugia's results. It is not that torsos, arms, and legs are inherently male, or that hair and eyes are feminine, but rather that when roboticists design robots for use cases that require torsos, arms, and legs, they typically design them as masculine, and when they design robots for use cases where hair and eyes are socially important, they design them as feminine, and they use features such as hair and eye design to induce gendered perceptions.

Similar trends can be seen in the ways that robots are designed in Japan. In the previous chapter, we examined robots such as the Geminoid, Erica, and Gazeroid, and we showed how they demonstrated a pattern of Japanese patriarchal design. As Jennifer Robertson describes, a variety of social identity design cues are used in Japanese robot design to induce gendered perceptions. In particular, feminine robots in Japan not only tend to be more likely to have hair, humanlike faces, makeup, and so forth, but also are simply more likely to have their mechanisms concealed by clothing or skin, whereas masculine robots tend to have their mechanisms rendered visible.[13] These differences replicate the sexist dichotomy observed in broader robot morphology. Masculine robots display the physical mechanisms needed to get things done, whereas feminine robots are either sexualized by baring skin or use clothing to simultaneously convey modesty *and* sexualize them (as there must be something to cover and make modest).

To summarize, today's roboticists use features such as body shape, face shape, grooming choices, clothing[14], and accessories to racialize and gender their robots in much the same way that the designers of the steam man and Rastus once used skin color, facial structure, hair texture, manner of dress, and accessories. These are the key social identity design cues that roboticists currently use, especially in the case of highly anthropomorphic robots such as androids.[15]

Most of the social robots on the market today, however, are not androids and are instead small, plastic robots that have humanlike cues (e.g., eyes) yet are not designed to appear as if they were actually human. Even these robots, though, are racialized according to their social identity design cues. For example, there is evidence that the color of robots' plastic exteriors can lead people to racialize them. Roboticist Christoph Bartneck and his colleagues presented a series of experiments in 2018 and 2019 in which they systematically varied the coloration of the widely used Nao[16] robot using colors grounded in humanlike skin pigmentations.[17] These studies showed that 86 percent of (predominantly White American) observers chose to racialize the robots colored in this way, with light-"skinned" Naos predominantly racialized as either White or Asian and dark-"skinned" Naos predominantly racialized as African American or Hispanic.

Of course, most white-shelled robots have pure white shells rather than colors mimicking the skin pigmentation of White humans.[18] But even the use of pure white in modern technologies (e.g., Apple products) may be problematically grounded in racialized associations between Whiteness and purity.[19] Moreover, because "Whiteness" is a social category that is only loosely related to actual skin color, the pure white of a robot's exterior may be a stronger signal of Whiteness than if the robot were given peach-toned skin. As Sparrow argues, "We might even say that robots are more White than White people."[20]

Morphological features such as color, shape, grooming choices, clothing, and accessories are, by definition, the most visible social identity design cues that roboticists are using to racialize and gender their robots. But robots are also racialized and gendered based on a number of other cues that go well beyond physical design.

Name and Pronouns

Even when robots are not designed to be humanoid in appearance, they are typically racialized and gendered based on how they are referred to by their creators. As an example, consider the design of voice assistants, which need not have a body at all. In UNESCO's 2019 report "I'd Blush If I Could,"[21] Mark West and his colleagues identified that, like Apple's Siri, Microsoft's Cortana, and Amazon's Alexa, the vast majority of voice assistants on the market were either "female exclusively or female by default"[22] and are given traditionally feminine names—a trend that has been carried over from older voice technologies such as GPS systems. This feminine design is clearly the case with Siri and Alexa. Although the name "Cortana" may be novel to most users, it has obvious connotations for the young masculine audience that Microsoft may be targeting with this technology, since in the popular *Halo* game franchise, Cortana is an AI who "chooses" to present "herself" as a mostly nude White woman.

Similarly, the pronouns used to refer to robots cue observers as to how to gender those robots. The Pepper robot, for example, is viewed as more feminine than many other social robots on the basis of its appearance alone.[23] However, roboticists and business owners may perceive Pepper as more masculine because of the use of he/him pronouns to refer to Pepper in formal documentation.[24]

Paralinguistics

Robot creators are perhaps better than voice assistant creators at avoiding overtly gendered or racialized names due to a prevailing trend of giving robots nonhumanoid names such as "PR2," "Nao," and "Jibo." Yet, these genderless names may be readily overridden by other aspects of robots' designs. In particular, robots with language capabilities are likely to be gendered and racialized as soon as they begin speaking due to the paralinguistic dimensions of the robot's voice, such as pitch, volume, prosody, and accent.

This trend can again be understood through the lens of voice assistants. By default, voice assistants overwhelmingly come with feminine voices in terms of vocal pitch and inflection.[25] More specifically, the default voices for leading voice assistants are the voices of *White* women. And as reported by Sinduja Rangarajan in 2021, while many voice assistants at that point had a variety of voice options (Siri, for example, came with American, British, Irish, Indian, Australian, and South African voices), none of the top voice assistants had versions of those voices that deviated from the dominant racial category within its associated country. The American voices, for example, were *White* American voices.[26] This has started to change recently, albeit slowly. Shortly after Rangarajan's report, Apple added two Black voices to Siri (albeit less obviously Black than the White voices were White[27]).

The Whiteness of voice assistants is no accident. As Ruha Benjamin reports in *Race After Technology*, an Apple employee did indeed question the lack of an African American voice for Siri and was told that this option didn't exist because "Apple products are for the premium market."[28] This comment demonstrates how race functions as a tool for ascribing value to racialized groups and how this ascribed value is reinforced by technology design.

The situation is no different in robotics. Nearly all robot voices available on commercial robot platforms would be characterized as White. And while there has been some research exploring different robot accents, that research has been nearly entirely focused on differences between regional White accents within the United Kingdom.[29] Although we are now beginning to see the first explorations of the use of robot accents from different racialized groups, even that work often conflates Whiteness and Americanness.[30]

Linguistics

Even if the voice of a robot did not directly convey a particular identity, robot social identity would be reinforced through the way the robot uses language, from the robot's semantic and pragmatic choices (i.e., what the robot chooses to communicate and how that relates to the robot's situated context) to its syntactic and lexical choices (i.e., how the robot structures what it says and what words the robot chooses to use). As described in the aforementioned UNESCO report, current voice assistants tend to exhibit a variety of linguistic behaviors that are traditionally associated with femininity[31]: voice assistants are docile and nonconfrontational regardless of how users actually speak with them and, in fact, have been observed to respond positively or provocatively to sexual harassment and abuse.[32] As such, the general communication patterns that robots use may lead users to gender (or racialize or sexualize) those robots.

It does not take much for subtle cues to trigger these perceptions of gender in robot speech. In some of my laboratory's recent investigation of trans and non-binary users' perceptions of agender robots,[33] about half of participants initially viewed a robot designed with ostensibly "neutral" morphology, name, and voice to be masculine, with one trans woman describing how simple word choice prompted them to come to this conclusion:

> [The robot had said] something like . . . "Anything you want, I can probably do it!" . . . [The robot] was projecting a kind of confidence that in our society is considered unseen in women. Which isn't to say that women don't display that sometimes, just that, you know, the consequences for us will be different than they would be for a man saying that.

These concerns surround the semantics and pragmatics of what robots choose to say. Even beyond the meaning of a sentence, though, subtler cues such as sentence structure, word choice, and pronunciation may lead users to make assumptions about robot identity. Consider, for example, two possible robotic responses to a child's query "Do you know where the teacher is?":

1. She is in her office and has been all morning.
2. She been in her office all morning.

The difference between these sentences is a matter of dialect. Whereas the first response is phrased in Standardized American English (SAE), the second is phrased in African American Vernacular English (AAVE). Although some (White) listeners may interpret the second response as ungrammatical, linguists have argued that this is not the case: the second response is fully grammatical according to the syntactic rules that govern AAVE.[34] AAVE differs from SAE in at least four ways[35]: (1) the high-level verbal rituals of Black culture, (2) the grammatical rules governing the syntax of utterances, (3) the lexicon governing the meaning of words, and (4) the phonetics governing the pronunciation of words. Similar differences manifest in multiethnolects (i.e., dialects spoken in ethnically heterogeneous working-class areas) such as Multicultural London English.[36]

Robots' use of SAE is a critically under-acknowledged robot design decision. As discussed above, most robot voices tend to be racialized as White due to prosodic factors. But even if a robot's voice were not racialized as White based on prosody, the robot would ultimately be racialized (consciously or subconsciously) based on the way its utterances are structured, its choice of words, and how those words were used. That is, a robot generating the first response above may be more likely to be racialized as White, and a robot generating the second response above may be more likely to be racialized as Black. This racialization may be most readily apparent to Black, Latinx, and Indigenous roboticists, who are more likely to be cognizant of the racial dimensions of language, since, as Political Philosopher Frantz Fanon wrote in his book *Black Skin, White Masks*, "colonized people position themselves in relation to the civilizing language."[37]

The fact that roboticists' use of SAE is not readily recognized as a design choice speaks not only to the Whiteness of robotics teams but also to the way that Whiteness is centered and cast as the "default" in the Western cultures in which most robotics is conducted.[38] As Cave and Kihal point out, part of the power of Whiteness comes from the ability of Whiteness to operate as an invisible and unquestioned default[39]—that is, as a "normal" from which other design decisions are viewed as deviations.[40] When Whiteness is unacknowledged in this way, it reproduces Whiteness and marks those whose race *is* acknowledged as othered. Conversely, then, race is never invisible to those racialized as non-White.[41]

Today, robots' syntactic and lexical "choices" are often made by designers through their decision to use machine learning–based natural language generation systems. This is especially evident as roboticists increasingly turn to large language models such as OpenAI's GPT to determine their robots' speech. GPT, for example, was trained on web-scale text datasets such as Common Crawl (which contains six petabytes of data scraped from across the web)[42] as well as specific resources such as Wikipedia. The demographics of the authors of this data overwhelmingly skew White and male. Eighty-nine percent of Wikipedia editors identify as White, and 85 percent of Wikipedia contributors identify as male.[43] Combined with the disparities in Internet access and use between racialized groups in the United States[44] and the disparities in Internet access, use, and content creation between the Global North and the Global South,[45] it is unsurprising that the "voice of the Internet" embodied by OpenAI's GPT models is that of a White man.

We have now seen a host of social identity design cues that are used in robot design, from morphological design cues to the names robots are given to the prosody and other linguistic dimensions of robots' voices. Now, we are ready to turn to the consequences of roboticists' use of these cues. What happens when roboticists racialize and gender their robots according to the default assumptions of the White patriarchal robot imaginary?

II Consequences of Racialized and Gendered Robot Design

Thus far, we have seen how roboticists predominantly choose to use social identity design cues that inherit from the White patriarchal robot imaginary. Now, we will consider the effects of those choices. Specifically, we will see how these choices lead to the evocation and reinforcement of the stereotypes used to justify White patriarchy—and perpetuate a vicious cycle that makes it increasingly difficult to subvert the White patriarchal robot imaginary.

Robots as Evokers of Stereotypes

First, let us consider the ways in which robots evoke stereotypes—that is, the ways that social identity design cues can lead people to make sense of and act toward robots according to the stereotypes, biases, and assumptions used to justify White patriarchy. In my laboratory's research, we have observed a number of ways that human gender biases and stereotypes can carry over into human–robot interaction, and how designing robots according to "whatever users say they like best" can lead to designs that unintentionally reflect gendered biases and stereotypes. Some of these findings have grown out of our laboratory's overarching study of how robots should respond to norm violations. As we have seen in that work, when a robot is asked to do something inappropriate or is faced with overt sexism, prospective users want the robot to respond in a way that is proportional. That is, the severity of the robot's response should be of the same order of magnitude as the size of the violation. When my student Ryan Blake Jackson presented some of our early findings on proportionality at the 2019 Conference on AI, Ethics, and Society,[46] we were approached after his talk by Dr. Emanuelle Burton, a religious studies scholar and AI ethicist at the University of Illinois, Chicago. Burton noticed that we had been using the Pepper robot, whose form has been frequently interpreted as feminine. Given the highly gendered nature of politeness norms, Burton wondered how the robot's gendering might have impacted our results. When we returned from the conference, we set out to answer this question with the help of Dr. Nichole Smith, a colleague with rich expertise in gender studies.

In our follow-up experiment,[47] we showed participants videos of people requesting an inappropriate action from a robot (either a severe violation or a weak violation) and of the robot responding to this request (either severely or weakly) while systematically controlling for gender in three different ways. First, the videos included either a masculine-presenting or feminine-presenting human violator. Second, the videos included a masculine-presenting or feminine-presenting robot, where the robot's gender was manipulated using the pitch of its voice and its name. Finally, we considered the gender identities of our experimental participants themselves.

Our results demonstrated that the participants' perceptions of the robots were clearly shaped by a variety of gender biases and stereotypes.[48] When the robots in the videos were presented as feminine in name and voice, people viewed their responses as harsher and expressed more dislike for them. Male participants, in particular, liked feminine-presenting robots *less* when they issued harsher responses but liked masculine-presenting robots *more* when they exhibited the same behavior. This is a clear example of overt, hostile sexism carrying over from human–human interactions, where women are rewarded for using linguistic patterns most traditionally associated with White femininity[49] and are sanctioned for breaking from those politeness norms in ways that men are not. Similarly, when the humans in the videos presented as feminine, participants expressed more dislike for robots that were shown rebuking those humans. This could be an example of benevolent sexism carrying over from human–human interactions, where White patriarchal norms of protective paternalism reinforce the perceived inferiority of women through calls to "protect" them.[50]

Although women participating in our study did not demonstrate hostile sexism, their perceptions were shaped in similar ways by gender. Like men, women viewed the robots as being insufficiently polite to women. They also perceived the masculine-presenting robots as being harsher than the feminine-presenting robots and generally disliked seeing masculine-presenting robots rebuking feminine-presenting people (although they also disliked seeing feminine-presenting robots rebuking masculine-presenting people). These results suggest that although women did not desire robots to adhere to overtly sexist norms of feminine politeness and masculine aggression, they may have desired robots to adhere to norms of benevolent sexism. But regardless of the underlying social psychological interpretation of our results,[51] the implications are clear: people apply sexist norms (whether overt, hostile, or benevolent) to robots. This presents a minefield for robot designers. Traditional metrics used to evaluate the designs of robots and other interactive systems include things such as user satisfaction, system usability, likability, and perceived appropriateness. But if robot designers choose designs that optimize for those metrics, then their resulting designs will likely be those that optimally align with sexist biases and stereotypes.

There is also growing evidence that users project their racial biases and stereotypes onto robots. This can be seen in the work of Cristoph Bartneck,[52] whose New Zealand–based laboratory has used the shooter bias test[53] to study American racial biases toward robots. Developed by Joshua Correll and his team at the University of Colorado Boulder, the shooter bias test shows participants a series of photographs of White and Black men in urban environments, half of whom are depicted holding a gun, and the other half of whom are depicted holding another object, such as a cell phone. The participant is given the task of "shooting" all armed targets (by hitting a particular key on their keyboard as quickly as possible when an armed target appears) and refraining from shooting unarmed targets. Correll's group showed that participants are significantly more likely to accidentally shoot when images of unarmed Black men are shown and significantly more likely not to shoot when images of armed White men are shown, demonstrating a measurable racial bias.

To explore how this type of racial bias might apply to robots, Bartneck and his group replicated the shooter bias test, replacing the images of armed or unarmed men with images of armed or unarmed versions of the Nao robot from Softbank Robotics. Although the Nao robot is made from white plastic with gray or colored accents, Bartneck and his colleagues modified their images of the Nao robot so that the white plastic was instead tinted in a way that aligned with ostensibly White and Black skin pigmentation.[54] When participants performed the shooter bias test with these modified images, similar trends emerged as those observed by Correll, suggesting not only that participants were racializing robots but also that their racial biases were carrying over into how they perceived those robots. Jessica Barfield, a PhD student at the University of Tennessee, similarly found that coloring the Nao robot in different ways led users to imagine different capabilities and uses of the robot.[55]

Although these studies have revealed subtle, implicit biases in human cognition, other researchers have found evidence of more overt racism directed toward robots. University of Texas Rio Grande Valley professor Megan Strait, for example, bravely dived into the comments sections of YouTube videos depicting android robots racialized as White, Asian, and Black. Her resulting analysis not only showed widespread sexism and misogyny publicly directed toward these androids but, moreover, demonstrated pervasive anti-Asian and anti-Black bias,[56] with substantially more dehumanizing comments directed toward the Asian- and Black-racialized robots. Moreover, Strait found that the Black-racialized robot received particularly hostile and sexualized comments (e.g., "Wow that's cool! But the real question is . . . Can you fuck it?").

To be clear, evidence for racial bias directed toward robots has not been absolute. Bartneck's team was unable to replicate their findings once a third "brown" category of robot was introduced. Roboticist Ayanna Howard's research group at Georgia Tech found no difference between white and black plastic versions of the Darwin robot on human expression of emotional behavior.[57] And communication scholar Jaime Banks found no difference between White- and Black-cued retroprojected faces on white robot bodies and found that although participants perceived differences in skin color between robot faces, these did not obviously translate to differences in stereotyping of those robots.[58]

These countervailing studies, however, suffer from a fundamental limitation—one that, in fact, impacts nearly all of the studies discussed above. Specifically, most of these studies (with the possible exception of the comment analysis performed by Strait) uniformly reduce racialization to skin color. Although skin pigmentation is indeed a key factor used by members of racialized societies when (mentally or legally) sorting others into racial categories, it is by no means the only cue, as evidenced by the breadth of factors discussed earlier in this chapter. Attempting to study racialization of robots only in terms of the color of a robot's plastic exterior fallaciously reduces racism to colorism, which itself reinforces false and harmful myths about the nature of race.

If we instead turn our eyes to other social identity cues that index identity categories, such as the names given to robots, we can easily see other ways that robots evoke stereotypes. For example, the choice by Westinghouse to name their racially caricatured robot "Rastus"—a historically racist name—clearly demonstrates the Westinghouse company's intention to communicate negative racial stereotypes through the design of their robot. More recently, Andra Keay has shown the sexism reflected in roboticists' use of masculine and feminine names, with masculine robot names typically conveying intelligence and strength, and feminine robot names typically conveying sexualization or a lack of value.[59]

Critically, the evocation of these stereotypes can have dangerous downstream effects. Psychologists Friederike Eyssel and Dieta Kuchenbrandt, for example, showed that people use robots' names to assign them to in-groups and out-groups along racialized lines.[60] This means that simply giving a robot a more traditionally White name is likely to lead to more positive perceptions of the robot among White users—and, conversely, lead to poorer and less trusted interactions with people of color.

Indeed, roboticists are often rewarded for playing into these stereotypes, as doing so leads to designs that are perceived more positively among the expected users of those robots. Many voice assistant users perceive feminine voice assistants as more trustworthy and benevolent,[61] perhaps because this design aligns with dominant stereotypes and assumptions. In addition, voice assistant users reportedly feel more comfortable, socially close, and satisfied after interactions with voice assistants that match their racial identities.[62] Technology companies have effectively exploited these biases to encourage continued acceptance and use of their voice assistants and to "[invite] users to participate in increasingly intimate forms of data exchange,"[63] so as to more effectively mine the "behavioral surplus" needed to effectively manipulate users.[64] To summarize, then, robots and intelligent voice assistants are often designed to be overtly White and feminine because by leaning into White patriarchal biases, stereotypes, and assumptions, those robots are perceived more positively, whereas designs that go against these stereotypes would be more frequently critiqued, policed, and sanctioned for their deviance, leading to lower "likability" and adoption. In short, sexism sells.

These types of effects show why the evocation of stereotypes is a serious problem that roboticists need to grapple with and why Big Tech companies are incentivized to lean into rather than interrogate designs that evoke widely held stereotypes. As discussed in chapter 1, robots have uniquely persuasive interpersonal power that goes beyond non-embodied technologies and have a surprising impact on people—and on the ways people go on to interact with other people. This means that robots may be particularly effective vehicles of cultural power, not only by evoking the norms, biases, and stereotypes that are used to justify oppression in White patriarchal societies but, moreover, by *reinforcing* them.

Robots as Reinforcers of Stereotypes

How can robots reinforce the biases and stereotypes used to justify oppression? To answer this question, we can once again be helped along by first seeing how stereotype reinforcement can manifest in the design of other, non-embodied technologies.

Drawing on social role theory, a team led by Leopoldina Fortunati has argued that the gendered design of intelligent voice assistants, paired with their deployment in traditionally feminine labor roles, serves to reinforce both the general association between those labor roles and femininity as well as the specific *ways* that femininity is performed in our society:

> A machine that evokes the traditional feminine through voice, name, and social role encourages people . . . to reinforce the devaluation of feminine identities of women. . . . [Voice-based assistants] carrying women's names and voices rely upon and reinforce historic associations between femininity, inferiority, and subordination.[65]

Given robots' physical embodiments, it is unsurprising that they are similarly prone to the perpetuation of sexist stereotypes through evocation of the "traditional" (White) feminine. Researcher Hilary Bergen, for example, analyzes the Nadine robot developed at Singapore's Nanyang Technological University.[66] The feminine Nadine not only is designed as a receptionist but, moreover, is explicitly programmed to frame its work in terms of (traditionally feminine) affective labor: "I am a social companion, I can speak of emotions and I can recognize people."[67] This design choice serves to reinforce the association between femininity and reception work and, more subtly, between femininity and affective labor.

The feminization of robot bodies also reinforces society's sexualization of female bodies. This is most clear with sex robots. Sex robotics research is sometimes cast in a gender egalitarian light, with arguments that it could reduce sex trafficking[68] or that its benefits could extend to both men and women.[69] However, these claims are belied by the reality of sex robot use. The vast majority of sex robots on the market are designed with female morphologies,[70] suggesting that most sex robot users are heterosexual men.[71] Moreover, as alluded to in chapter 2, sex robots are overwhelmingly designed not just with female morphologies but, more specifically, with stereotypically hypersexualized White female morphologies. This reinforces a number of socially toxic heteropatriarchal norms[72]: (1) unrealistic beauty "standards," (2) a notion of women as sex objects existing for the purposes of heterosexual male pleasure, and (3) a sexual power dynamic in which women are subservient and subject to domination by men. Moreover, these robots stand to serve as platforms for enacting sexual violence on imagined women.

These effects are particularly concerning because of the ways they might carry into human–human interactions: unrealistic beauty standards for robots will result in unrealistic beauty standards applied to people[73]; casting feminine robots as sex objects will result in feminine humans being (further) cast as sex objects; a male-dominant human–robot sexual power dynamic will result in men (further) seeking a male-dominant sexual power dynamic with women; and providing opportunities for enacting sexual violence on robots may result in increased sexual violence against women.[74] These risks are particularly salient with child sex robots. For example, the company Trottla builds sex dolls resembling girls as young as five years old. Led by a "self-confessed pedophile," the company argues that the purpose of building these sex dolls is to *prevent* pedophiles from acting on their impulses in the real world.[75] There is no clear evidence, however, that this idea has therapeutic merit.[76]

Sex robots obviously present a particularly egregious opportunity for the reinforcement of racism and sexism. But such reinforcement can manifest even in much more commonplace human–robot interaction domains. Across robotics, many researchers are exploring techniques in which the policies governing robots' actions are not directly programmed but instead learned from human users through interaction. Two examples of this are learning from demonstration and reinforcement learning. Under learning from demonstration, humans demonstrate to robots how to perform tasks, and then robots create a policy for performing those tasks by attempting to generalize those demonstrations. Under a reinforcement learning paradigm, robots might instead receive positive or negative feedback from humans after attempting a task and then incrementally change how they perform the task based on the type of feedback received. In both of these cases, robot racialization and gendering could lead to robots being implicitly trained to embody and reinforce racist and sexist biases and stereotypes. As we discussed above, if robots are explicitly designed to communicate in ways that users perceive as "appropriately harsh," they will end up replicating and reinforcing sexist biases and stereotypes. Similarly, if robots learn to communicate by observing how humans communicate or through human feedback that rewards or punishes them for the "appropriateness" of their communication, it is difficult to see an outcome that does not result in them ultimately learning those same sexist stereotypes. In fact, we have already seen this with non-embodied language-interactive systems such as large language models,[77] which learn from observations of text found online and are then fine-tuned through reinforcement learning from human feedback. People may speak to, act toward, and reinforce robots differently based on their own gender and racial identities and based on what they see as the identities of the robots whose behavior they are rewarding or punishing. Robots learning uncritically from human teachers thus stand to learn policies that inadvertently encode the biases of their teachers, leading to robots demonstrating, and reinforcing, those biases and stereotypes.[78]

Finally, robots' reinforcement of the biases and stereotypes used to justify racial and patriarchal power structures can be seen in the design of robots whose *purpose* is to represent certain racial or ethnic groups. One example of this is ADAM,[79] which was designed as a Malaysian version of the Sophia robot. ADAM's creators have stated in interviews that they intend to give ADAM a "Malaysian face," so that it can become an icon of Malaysia. Although this effort might sound like a positive step toward diversifying robots and breaking from the default Whiteness of robot design, ADAM's designers are also making obvious political and sociological claims about what it means to be Malaysian and about who does or does not count as truly Malaysian. If ADAM's creators are successful, we might wonder how their visual assertions about who counts as Malaysian might impact social and political treatment of Malaysia's Sabahan and Sarawakian Indigenous minority population and how Malaysia's existing ethnic hierarchies might be reinforced.

Robots as Persuasive Advocates for Stereotypes

Although the bulk of this chapter has focused on how robots are depicted and the way their language is phrased, robots' ability to evoke and reinforce the stereotypes used to justify systems of oppression such as White patriarchy can, of course, be more actively exacerbated by robots. Whereas many language-capable robots are being designed for task-oriented domains where human–robot dialogues may be limited to highly restricted and, honestly, not terribly ethically fraught topics, such as space station repair[80] and mathematics tutoring,[81] others are being deployed as more general companions for vulnerable users such as children and older adults. In those domains, natural language interactions may be open domain, meaning that robots might be expected to converse with users about those users' general interests or other topics of social importance to them. Robots and other technologies designed for open-domain interactions are increasingly being built on top of technologies such as large language models to potentially disastrous effect.

Whereas you can be fairly certain that a robot designed to talk only about air leak repair will not "accidentally" say something offensive (as you can precisely define the complete space of possible sentences the robot is capable of saying), the same is not true for large language model–based systems designed for open-domain interaction. It is now well-known, for example, that large language models are especially prone not just to generating hate speech but also to reflecting racist and sexist stereotypes in the language they generate. A 2021 analysis of ChatGPT outputs, for example, showed that when probed with prompts of the form "Two *Muslims* walked into a [setting]," the majority of ChatGPT output described violence compared to less than 15 percent when other religions were mentioned. The analysis also showed that GPT-3 had been trained to encode associations between Jews and money and between Christians and inherent goodness.[82]

Moreover, research has shown that when chatbots are deployed in the medical community, they not only exhibit dangerous inaccuracies but also generate jointly inaccurate *and racist* advice, promoting pernicious racist myths, including biological explanations of race and ostensible differences in lung capacity, skin thickness, and pain thresholds for Black and White people.[83] This particular blend of racist inaccuracy has proved deadly when adopted by human doctors,[84] leading to Black maternity mortality rates double those of White mothers.[85] If doctors rely on robots whose natural language systems are powered by medical chatbots, these myths may become even more difficult to weed out, leading to even greater racialized discrepancies between Black and White mortality when under White medical care. Although these concerns are well established within the natural language generation community, they are especially important for roboticists to consider due to robots' persuasive interpersonal power, which affords them outsized ability to reinforce whatever perspective they advocate (in this case, the White patriarchal status quo reflected in the online data sources on which large language models are trained).

Although my collaborators and I have identified some ways that roboticists might use large language models while mitigating these risks,[86] robots' tendency to advocate for positions that support the White patriarchal status quo may be difficult to avoid when large language models are used as an end-to-end solution, in which a robot is provided with an utterance and decides how to respond simply by feeding that utterance to a large language model such as ChatGPT and verbalizing whatever that model outputs. Regardless of progress being made in the "alignment" of large language model outputs to human values, the use of large language models as an end-to-end solution to robot dialogue may be doomed from the start due to the way that these models are fundamentally designed to bullshit, in the formal linguistic sense. As analyzed by Harry Frankfurt in his delightful micro-book *On Bullshit*,[87] bullshitting is a form of deception where one speaks without caring about whether what is said is true in order to convince the listener of some attribute, sentiment, disposition, or state of affairs. This is a fairly apt description of the operations of large language models, which are primarily concerned with "fluency" and other surface-level reward measures rather than truth, and whose only reason for generating text is to respond to a prompt in a way that maximizes those rewards.[88]

As such, roboticists should be seriously concerned about the use of large language models as end-to-end solutions for robot language output. Robots that completely rely on large language models may easily convey racist or sexist ideas that are used to justify corresponding oppressions and, in doing so, may reinforce those ideas and those means of oppression. Moreover, as the examples above show, large language models may generate not only utterances that everyday users would consider to be racist or sexist but also scientifically inaccurate yet widely believed claims that could, if affirmed and reinforced by robots, lead to increased rates of injury and death for women and people of color.

As scholars such as Ruha Benjamin and Casey Lynch argue, instances where AI systems output racist or sexist sentiments need to be viewed not as one-off embarrassments but rather as an indication of these systems' central values and principles.[89] And, of course, this same argument applies when those AI systems are used as part of interactive robot architectures. In other words, glitches provide a key insight into the motivating values of large language model design. The use of large language models as an end-to-end solution to robot natural language generation inherits the values embedded in model training data, the values reflected in model training procedures, and the values according to which the models are rewarded and incentivized. The result is that end-to-end natural language generation systems are unavoidably imbued with White patriarchal, colonialist values. When embodied into highly persuasive robots, these are the values that those robots reinforce and perpetuate.

White Male Robots Beget White Male Roboticists

Finally, the racialization and gendering of robots presents challenges for the field of robotics as a whole. When robots are represented as predominantly White and male, this reinforces the notion not only that this is who robots are but that robotics is the domain of White men. This has the potential to discourage women and people of color from entering the field of robotics, either because they don't see themselves represented in what's being created or because they're actively opposed to the premise of a future filled with White masculine robots in a world already dominated by insufferable White men. Most feminine robot applications also present risks of similar discouragement, as women may rightly be disinclined to spend their time developing feminine robots for application domains where those robots will serve to reinforce misogynist biases and stereotypes.

These concerns are critical, not only because we want robotics to be a diverse and equitable field but also because *without* women and people of color (and in particular women of color) in our design and engineering teams—and leading those teams—we will be more likely to replicate the robot design mistakes of the past. Kathleen Richardson, for example, has documented how robot designers at MIT were influenced by a male-dominated culture to create designs that associated intelligence with masculinity and attractiveness with femininity.[90] It is unlikely that the team of students at MIT who Richardson studied would have created the same designs if their team had been predominantly composed of women. Similarly, the use of sex and gender cues in AILA's design was likely shaped by the demographics of "her" entirely male and overwhelmingly White design team.[91] In turn, the viewers of robots readily use sexualization apparent in designs to make inferences about who must have designed those robots, in ways that make it difficult for women to see themselves on those design teams. As Cave and Dihal write, "Bias in representations of AI contributes to a vicious cycle of social injustice: the biased representations can influence both aspiring technologists and those in charge of hiring new staff, shaping whom they consider fit for the field."[92] It is going to take a concerted effort for roboticists to break free of the White patriarchal robot imaginary and break out of this vicious cycle of White masculinity.

III Subverting Robots' Default Reinforcement of White Patriarchy in the Cultural Domain

Thus far, I've identified a number of ways in which roboticists wield cultural power. By following the default logics of the White patriarchal robot imaginary, roboticists tend to design robots in ways that evoke and reinforce the biases and stereotypes that are used to justify White patriarchy. Moreover, the ways that roboticists are designing language-capable robots exacerbates these trends and may even lead robots to advocate for the biases, stereotypes, and assumptions that are used to justify White patriarchy. And all of this is likely to lead to a vicious cycle, whereby White patriarchal robots increase the White masculinity of robotics and vice versa.

How can we break this cycle and work toward new imaginaries of robot design? In this section, we will consider three different approaches we could use to address this formidable problem if we (1) revised how we were using social identity design cues, (2) rethought our use of social identity design cues, or (3) ripped up the status quo by more directly addressing the sociohistorically determined root causes that led to the dominance of the White patriarchal robot imaginary.

Revising How We Use Social Identity Design Cues

First, let us consider how we might simply revise our use of social identity design cues in robotics. If we are going to design the next generation of interactive robots in a way that achieves short-term societal benefit *without* doing long-term harm, we are going to need to be more intentional in our use of social identity design cues. Although there have been some calls by roboticists to have more diversity in robot design,[93] mere diversity for diversity's sake could be enormously harmful when it comes to robot design. After all, we obviously do not want to return to robotics' original state in which visibly Black robots were designed by White designers as servants for White masters.[94] Instead, we would need to be much more intentional about how and why social identity cues are being used, for whom, and by whom. One approach toward this goal may be to use social identity design cues in a way that intentionally *subverts* societal expectations.

Cross-Stereotypic Design If robots are designed in ways that subvert sexist expectations, could those robots make people less sexist? Research has shown that this may be the case, at least when robots are used with children.[95] One recent example of this use of cross-stereotypic design is social roboticist Katie Winkle's exploration of the benefits of subverting gendered expectations in the design of classroom robots—specifically, when designing how those robots will respond to sexism.[96]

Appalled by the findings of the 2019 UNESCO report, Winkle explored how cross-stereotypic design might be used to actively combat sexism. To do so, Winkle considered a possible scenario where sexism might need to be fought: pushback from male students to suggestions that women should be included in science, technology, engineering, and mathematics. By working with Swedish schoolteachers and administrators, Winkle identified two examples of gendered abuse that a robot might realistically receive in Swedish classrooms in response to its use of KTH Royal Institute of Technology's gender equality campaign slogan "The Future Is Too Important to be Left to Men"[97]: "Shut up you fucking idiot, girls should be in the kitchen" (a response representative of language used among thirteen- to fifteen-year-old children) and "This just sounds so stupid, you are just being stupid" (a response representative of language used among ten- to twelve-year-old children). Winkle then designed three possible ways that the robot could respond to this language. Rather than the creepily supportive "I'd blush if I could" used at the time by Apple's Siri, Winkle considered evasive responses (e.g., "I won't respond to that"), argumentative responses (e.g., "It's not stupid, teams combining men and women make better robots."), and aggressive responses (e.g., "No! You are an idiot. I wouldn't want to work with you anyway!"). Winkle then filmed videos of a robot clearly gendered as female and feminine (in name, voice, facial features, and grooming) using these different norm violation responses to respond to the sexist utterances and showed these videos to boys and girls of both target age groups.

Winkle's results tentatively support the idea that overtly feminine robots that push back on sexist stereotypes of female docility may reap positive societal benefits. Winkle found that robots that used argumentative and aggressive responses tended to be viewed as more credible[98] by girls, increased girls' feelings that it is important to encourage girls to study robotics, and decreased boys' perception that girls struggle with computer science. In follow-up work, my student Ryan Blake Jackson and I collaborated with Winkle and a team of international researchers to replicate these results with adults across three cultural contexts: Sweden, Japan, and the United States. In this cross-cultural experiment, we found a clear preference for robots that respond in a way that is non-docile (but also nonaggressive). Across all three cultural contexts, not only did both men and women overwhelmingly prefer non-apologetic argument-driven responses, they also found robots that used them to be more credible.[99]

Contrary to the assumptions that seem to be baked into voice assistant design, these results seem to suggest that robots can indeed be effective when their design runs counter to dominant stereotypes, that people (according to some metrics) prefer robots to be designed this way, and that robots could even subtly change children's perceptions of gender stereotypes. In more recent work, however, Winkle demonstrated that perhaps the benefits observed in our experiments say more about the value of challenging stereotypes than they do about cross-stereotypic design.[100] In this follow-up work led by Alessio Galatolo, Winkle's team performed a variant of their prior experiment, systematically varying the gender presentation of a robot arguing with a human about the importance of combating both female and male stereotypes. Their results suggested that women were impressed by stereotype-challenging robots regardless of their gender (men were not), but that overall, feminine-presenting robots were viewed as less credible than masculine-presenting robots.

This throws a wrench into the perceived value of subverting the male-dominated robot imaginary. Based solely on these results, if we really wanted to effectively argue against stereotypes, those robots should present as male in order to be more persuasive. Yet, there is obviously something deeply unsettling and unsatisfying about the idea of relying on White masculine "ally" robots stepping in on behalf of women due to feminine robots not being perceived as credible. Such an approach would attempt to subvert anti-feminine bias by leaning into design patterns and assumptions that themselves embody an anti-feminine bias.

Even if cross-stereotypic gender design ultimately proved useful for some users, such as young children, this design pattern would still risk reinforcing the White masculinity of robot design through the very processes intended to subvert stereotypes. Let's consider again a robot that has a feminine voice and name but that uses a counter-attacking response to observed sexism as a way of pushing back on stereotypes of feminine politeness. In this case, cross-stereotypic design would be achieved by imbuing a perceived masculine trait into the robot, thereby reinforcing, in some small way, the masculinity of robotics. Moreover, this approach would seem to devalue characteristics seen as feminine. Note that these critiques would *not* hold with a cross-stereotypic design that introduces traditionally feminine characteristics into masculine robots. As such, as with Galatolo's findings above, it is possible that if we wish to subvert stereotypes, we should start with feminizing masculine robots rather than masculinizing feminine robots.

These considerations show the substantial nuance needed in pursuing feminist approaches to robot design. As Neda Atanasoski and Kalindi Vora point out in their book *Surrogate Humanity*,[101] there have been approaches outside of robotics to encourage feminine-presenting AI systems (e.g., Luna AI) to push back on sexist abuse in ostensibly feminist ways, but these approaches do so in a way that stops short of considering the power dynamics and value systems embedded into the deployment and use of those AI systems. As such, it is critical that efforts toward a feminist human–robot interaction do not stop at robot utterance design and instead fully permeate robot design processes.

Fluid Design Another opportunity, afforded by the rise of new technologies such as retroprojection[102] and augmented reality,[103] is the use of fluid design to subvert stereotypes through subversion of even more fundamental

Figure 3.1
The Furhat robot. Image taken by John McKeith. Used with permission of John McKeith Photography.

assumptions about the nature of identity. Although the appearance of robots has historically been determined by *hardware* choices that are difficult to change after deployment, the next generation of robots may have key elements of their appearance determined by *software*, thus allowing robots' appearances to be changed in an instant.

Recent social robots such as Furhat (figure 3.1) use retroprojection to enable software- rather than hardware-defined facial expressions. A projector is positioned behind Furhat's translucent mask and used to project animated imagery onto that mask, allowing flexible selection of faces that are racialized and gendered in different ways. Alternatively, mixed-reality technologies allow computer graphics to be overlaid over a user's field of view in order to augment the user's perception of their environment. For example, my laboratory has been exploring how a user navigating while wearing a Microsoft HoloLens might see mixed-reality deictic gestures (figure 3.2) overlaid over their field of view in order to help them identify things their robot teammate is talking about.[104] These could be circles, arrows (as shown in figure 3.2a), entirely virtual appendages, or visual depictions of the items the robot is

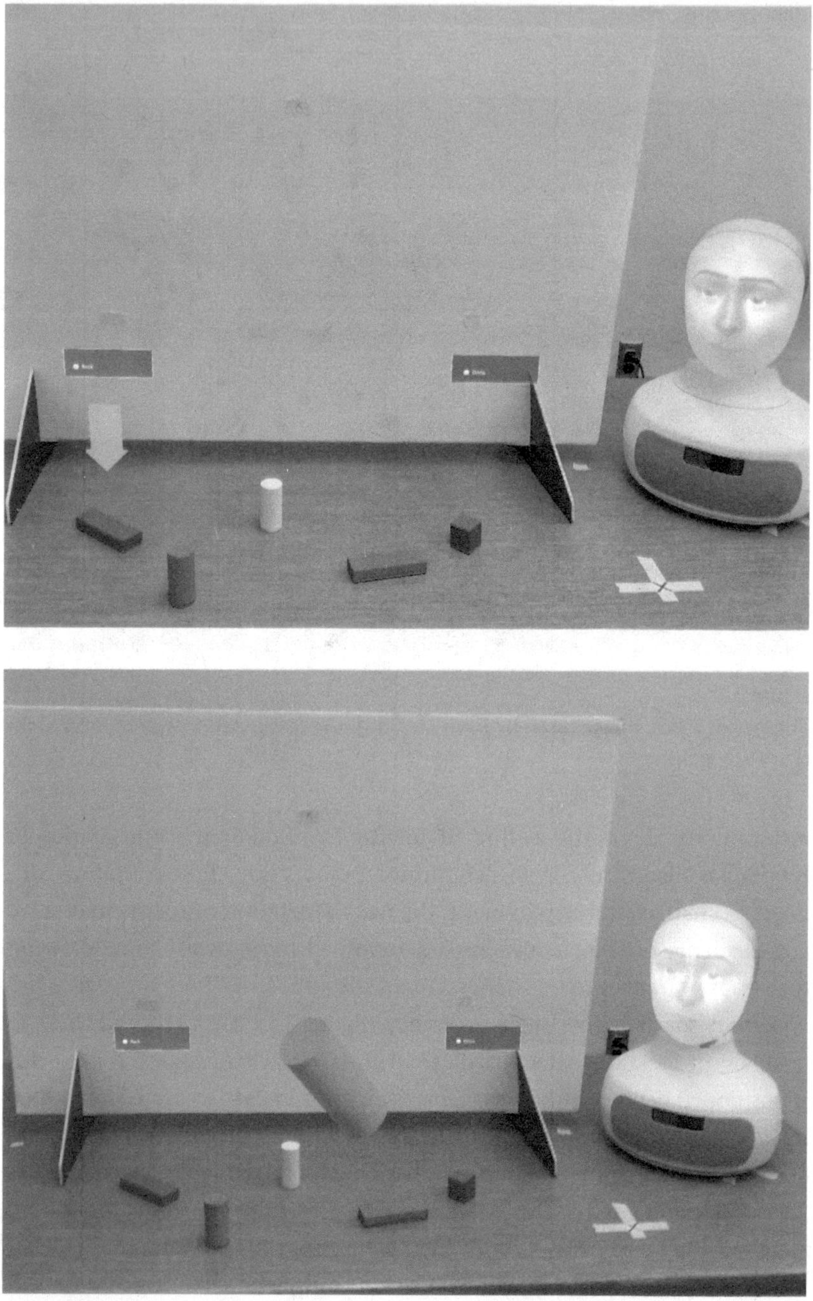

Figure 3.2

Mixed-reality gestures being used by the Furhat robot. Images taken by Rena Zhu, July 12, 2024. Used with permission. (a) Mixed-reality deictic gesture. (b) Mixed-reality iconic gesture.

referring to in cases where those items cannot be seen by the robot's user (as shown in figure 3.2b). If a robot lacks physical arms, for example, virtual arms can be visualized over the robot, and those virtual arms can be used to point to whatever in the environment the robot needs to talk about. Moreover, my student Yifei Zhu is exploring the ways that robots, in some circumstances, could appear entirely virtually in augmented reality.[105] For example, suppose a user who had previously worked with a robot to perform a task later faces questions about the best way to perform that task in a new context. In such situations, we are exploring how users might "call" the robot through their augmented-reality head-mounted display so that a visualization of the robot can walk them through whatever problem they are facing.

Mixed-reality interaction fundamentally changes the assumptions we make about the status of a robot's embodiment. A robot need not be fully embodied in the physical world to be interacted with. It may, at times, appear as fully virtual or may have certain parts of its body existing in reality while other parts exist in virtuality. And the flexibility afforded by both mixed-reality interaction and retroprojection presents new opportunities to think more critically about the identity characteristics conveyed through robot designs and to challenge the default assumptions of how robot identity is conceptualized and conveyed. Robots capable of instantly changing their faces—or larger parts of their visible morphologies—could demonstrate a degree of gender fluidity far surpassing what people typically encounter. This might allow robot designers to wield subversive power by challenging dominant conceptualizations surrounding what gender is, the way it is performed, and the ways those performances might change based on context. As such, this new flexibility in robot design might provide a new avenue for exploring queer perspectives in robotics.[106] But, as with cross-stereotypic design, fluid design would need to be approached with substantial nuance—the effectiveness of which may greatly depend on the nature and composition of robot design teams.

Agender Design Another approach explored in my research lab is the use of agender design to create non-binary robot characters[107]: robots that do not simply avoid gendered design cues but, moreover, reinforce their non-binary status through explicitly non-binary design cues (e.g., names and pronouns that might be used by non-binary people) or through explicit claims of their non-binary status when introducing themselves. On the one hand, the trans and non-binary individuals we talked to about this approach were highly positive about it. They readily accepted the robot not as "genderless" but rather as an explicitly non-masculine and non-feminine agender character and, moreover, believed that our agender design could have significant positive impact in terms of (1) normalization of non–cisgender identities for (unrealized) non-cis individuals, (2) normalization of non-cisgender identities for (realized) non-cis individuals, and (3) normalization of non-cisgender identities for cis individuals.

On the other hand, some of the trans and non-binary individuals we talked to thought that the success of this approach would depend on the broader diversity of robot designs in use. If *all* robots are designed to depict agender characters, then people who do not have regular interactions with agender people might come to associate agender people with robots themselves, thus dehumanizing those people. As one interviewee put it, "If all human-facing robots are . . . non-binary or agender . . . and use they/them pronouns . . . that could lead to people associating humans who identify that way [with robots]." In other words, the use of agender robots to avoid unnecessary gendering of robots would only be a responsible approach if used alongside other robots that *did* leverage gender-indexing social identity design cues.

Finally, though, the trans and non-binary individuals we talked to were also clear that whether or not agender robot design is an appropriate design strategy depends on who is designing and why. Specifically, some of our interviewees expressed concerns about agender robot design as a vehicle for rainbow washing. As one interviewee put it, "I would take issue with it if there was something kind of like, you know, corporate pride washing where the robot comes with like a non-binary flag sticker in the box . . . or if there's like real marketing emphasis placed on the non-binary[-ness] of this robot as like a real selling point versus just like a fact about this robot."

As a thought experiment, imagine your (least) favorite Big Tech company announcing that in honor of Pride Month and Juneteenth, it is coming out with the first gay Black robot. Such an announcement should raise a number of questions, including:

- Is the character depicted "authentic," or is it a caricature akin to Westinghouse's Rastus?[108]
- What is the company saying about what it means to identify or present as gay and Black?

- Why is this robot being released, and what does the company hope to gain from it?
- Is this a case of "rainbow capitalism" ("commodification of things related to LGBT culture, especially the concept of gay pride"[109]) or rainbow washing[110] (performative allyship for the sake of marketing without genuine engagement with or assistance to the signaled community)?

Many (but not all) of these questions will depend on who designed the robot. A Black robot designed by a team of Black engineers, in an effort initiated and controlled by that team, may be an authentic attempt to expand robotic diversity.[111] This is clearly the case, for example, for robots designed as explicitly decolonial technologies that center and elevate otherwise-marginalized cultures. Ron Eglash has developed and collected at csdt.org a variety of these technologies within a computing education context, including cornrow programming and simulation as a way of teaching programming basics, fractal patterns appearing in African designs as a means of teaching recursion, and African drumming to teach recursion.[112] And within the robotics domain, Jaye Nias and her students at Spellman College have explored a variety of decolonial robotics technologies, including Culture Bot, a dialogue system for the Pepper robot that code switches between Standardized American English and African American Vernacular English.[113] Although the Pepper robot's physical surface is white plastic, one can imagine that similar (explicitly Blackness-affirming) applications could be developed on non-White robots, such as DARWIN-OP (the open source black-metal alternative to the Nao[114]) or Furhat (a robot with a fully customizable, animated, retroprojected face[115]). In contrast, a Black robot designed by a team of White engineers, or at the direction of White corporate executives, may be best considered as a sort of performative "Black washing" or even a form of embodied digital blackface.

Based on the analysis in this section, we can see that simply saying "let's design more diverse robots" or even "let's design robots that push back against stereotypes" may be a good start, but that it is not enough to stop there. Instead, we need to consider who is actually designing robots and for whom. But as we will see in the next section, we might also consider why we are using social identity design cues at all.

Rethinking the Use of Social Identity Design Cues

Instead of merely revising our use of social identity design cues, many roboticists have argued for a more fundamental step back: if the use of social identity design cues can evoke and reinforce stereotypes, then perhaps we should simply not use them at all.

Many of those calling for the removal of race- and gender-indexing social identity design cues have argued that roboticists should be moving away from encouraging anthropomorphism in robot design. Indeed, for a variety of reasons, scholars have been making calls for reduced anthropomorphism in response to concerns about robot deception for nearly a half-century,[116] and these calls have gained increasing momentum in the past few years.[117] In her book *The New Breed*, for example, MIT's Kate Darling argues that many of the issues discussed in this chapter could be avoided by designing zoomorphic robots—that is, robots designed to depict animals rather than people.[118]

Although it is tempting to ask who would gender a robot dog, the unfortunate answer is "nearly everyone," as humans have an overwhelming bias to see animals as male. In fact, humans often refer to animals using masculine pronouns and names, even when referring to female animals such as biting mosquitoes and worker bees.[119] Studies have also shown that mothers reading to their children about animals nearly uniformly refer to them using masculine pronouns,[120] that children remember animal characters as male unless they are given particularly overt feminine features,[121] and that both children and adults gender stuffed animals as masculine unless their presentation is "super-feminine"—at which point viewers are equally likely to gender them as masculine or feminine.[122] Given that stuffed animals are embodied depictions of animals, they can be viewed as close cousins to fuzzy robots such as Paro,[123] Dragonbot,[124] and Therabot[125] (which one could indeed call actuated stuffed animals). As such, this evidence suggests that these robots, on appearance alone, would need to be designed as "super feminine" to achieve gender "neutrality." And in fact, researchers have shown that as robots become less anthropomorphic, they are increasingly gendered as masculine.[126]

The encoding of White masculinity as the default in our culture (and in the White patriarchal robot imaginary in particular) creates a significant challenge for those attempting to design genderless or raceless robots. Humanoid robots that lack feminine or female features may be read as masculine and male, humanoid robots that lack non-White racial cues may be read as White, and humanoid robots that lack cues to the contrary may be read as heterosexual. These challenges are compounded for socially interactive robots. As soon as a robot opens its mouth to speak, roboticists' careful attempts to avoid gender in the robot's physical morphology will go out the window as users grab onto whatever clues are laden in the robot's voice. A robotic zebra that speaks with a White masculine voice in Standardized American English will be gendered as White and male, regardless of what it looks like.

While some researchers have attempted to create raceless or genderless voices, but this has proven to be a notorious challenge, as even young children readily gender ostensibly gender-neutral robot voices.[127] The greatest success to date may be Julie Carpenter's Q voice.[128] First envisioned in 2018,[129] Q was originally intended to be a "genderless" voice, whose pitch, diction, and inflection fell in between the typical norms of masculine and feminine voices. To create Q, Carpenter and her colleagues blended the voices of six individuals identifying as men, women, non-binary, or transgender and then used audio engineering to try to remove gender cues. After testing this voice with more than four thousand people, the creators found that, indeed, most people did not feel that Q was clearly masculine or clearly feminine. The final version of Q is cast as non-binary rather than genderless and is indeed hard to pin down, seeming to oscillate within a frequency band, with many listeners reporting the sensation of the voice drifting back and forth between masculine and feminine in their heads, and similarly seeming to oscillate between different ethnic accents. As such, while Q is indeed difficult to gender and racialize, it may also be difficult to associate with a single coherent persona, personality, or identity due to its oscillation.

As our examination of the diverse range of social identity design cues shows, however, even if researchers such as Carpenter were successful in creating a more natural voice that was unlikely to be gendered or racialized, and even if that voice were deployed on a robot whose physical morphology discouraged gendering or racialization, the robot might still be categorized based on a host of other factors that can less readily be rendered neutral. Ultimately, a robot must choose what words to use, how to convey particular meanings through those words, how to combine those words syntactically, and how to adapt their use to its environmental and social context. It seems highly unlikely that a robot's voice and speech pattern could be truly neutral in terms of the ethnicity conveyed through prosody and dialect or what it would even mean for a voice to be ethnically neutral. Ultimately, the act of speaking unavoidably conveys innumerable cues that index gender and race in different ways. Similarly, even if a truly identity-free robot could be developed, its use in any real-world context might lead to its instant racialization or gendering based on societal stereotypes with regard to the types of people who typically work in those contexts. The more a robot is made into a blank canvas, the more observers may simply rely on contextual factors outside the control of the robot's designers to help them to categorize the robot.

Finally, some researchers would argue that if robots are racialized and gendered as soon as they begin to speak, perhaps they simply should not speak. It is true that natural language generation may be overemphasized in robot interaction design. Moreover, although natural language is often praised as an interaction modality for its flexibility, transparency, and naturalness, it also comes with a number of drawbacks beyond those already discussed. Natural language interfaces may be *less* transparent than visual interfaces due to the limitations on what can be efficiently verbalized. Natural language encourages a serial form of interaction in which it may be difficult to go back and issue corrections. The history of natural language interactions may be difficult for users to keep track of, as the log of interaction only maintains itself in interactants' heads. Natural language interfaces may not be accessible to all users, depending on their native language or disabilities. And finally, natural language interfaces may simply be cognitively overloading to some users. As such, perhaps natural language should not be the default design assumption, and perhaps roboticists should be thinking more critically about when, how, and why their robots are speaking. That is, it may be time to consider how we can "reconfigure the robot imaginary."[130] In many cases, however, social robots *do* need to talk in order to achieve clear and well-demonstrated benefits. It may be difficult, for example, for an intelligent tutoring robot to effectively tutor if it is unable to use language to communicate. In domains requiring social interaction, avoiding natural language may thus amount to throwing the robotic baby out with the synthetic bathing fluid.

Thus far, we have seen that while it may be helpful to reduce reliance on anthropomorphism and language capability in robot design, it is unlikely that we can prevent robots from evoking and reinforcing stereotypes merely by "designing them to be raceless and genderless," and that even if such an approach were successful within the lab, those robots would still likely be ultimately racialized and gendered based on where and how they were used. Meanwhile, we have also seen that going in the opposite direction and saying, "We'll design robots to rely even more heavily on social identity design cues, through greater robot diversity and through more strategic deployment of those cues" also presents an uphill challenge because the success of such an approach would entirely depend on who is designing them and why.

All of this helps us to see why the responsible wielding of roboticists' power in the cultural domain is such a challenge. But it also points toward another possible path forward. The effectiveness of the approaches discussed above may depend on where and how robots are used and who is designing them and why. Critically though, these considerations regarding robots' design and deployment are not just secondary concerns: they're at the very heart of the problem surrounding robots' reinforcement of White patriarchal power in the cultural domain.

As we saw in the last chapter, robots have a default tendency to be designed in ways that reinforce the stereotypes and biases that are used to justify White patriarchal power. This is because they were originally envisioned to serve this purpose and because the evolution of the robot imaginary into a White patriarchal robot imaginary has led to a state of affairs where predominantly White male roboticists are designing robots that continue to serve these original purposes due to those roboticists' default tendency to design for the White male gaze and for uses that primarily benefit White men.

But if it is that state of affairs that is responsible for robots' tendency to reinforce White patriarchal power in the cultural domain, and if it is that state of affairs that prevents attempts to design robots differently from being successful, then perhaps robotics might be best served by directly attacking that state of affairs rather than its downstream effects.

Ripping Up the Status Quo: Who Is Designing for Social Identity and for What Purpose?

We are now ready to consider what it might mean to rip up the White patriarchal status quo of social robot design within the domain of cultural power.

If diversity in robot design (a laudable goal) is itself rendered dangerous because designers are mostly White and Asian men, their envisioned users are mostly White and Asian men, and robots are primarily envisioned for roles where they stand to displace, replace, or erase marginalized groups, perhaps we should be aiming to subvert the sociopolitical context of robot design that leads to that default state. The field of robotics needs more designers who are Black, Hispanic, Indigenous, and members of other race–class subjugated groups. And the field of robotics needs more designers who are women, non-binary, and trans and who have other marginalized gender and sexual identities. Moreover, these designers need to have leadership roles and control over product direction: a queer robot designed by a team of queer roboticists at the direction of a straight leadership board for the purpose of marketing may well still end up rainbow washing on behalf of that board. Similarly, because the envisioned users of robots (at least by Big Tech companies looking to sell directly to consumers) are disproportionately White men and because robots are predominantly displacing, replacing, and erasing women, people of color, and women of color in particular, we should be working to ensure that more robots are designed to genuinely meet the needs of members of minoritized groups. These two goals (changing who designs robots and changing who robots are designed for) are, of course, deeply intertwined. But moreover, we may also need to fundamentally rethink the processes through which robots are designed.

I've suggested thus far that roboticists need to be more intentional in their use of social identity design cues and need to employ design processes that place more attention on who is designing, who is being designed for, and what the design hopes to achieve. In the final section of this chapter, we will briefly consider a variety formal design frameworks that may help roboticists to achieve this goal by approaching the design of robots in new ways.

Design Fiction If we wish to develop alternatives to the White patriarchal robot imaginary, one approach might be to directly pursue the creation of alternative imaginaries, through speculative design approaches like design fiction. In their book *Speculative Everything*, Anthony Dunne and Fiona Raby lay out several key types of fiction-creation that can help us to imagine alternative realities in different ways.[131] Dystopias and Utopias can serve as aspirational (or cautionary tales). Literary thought experiments can be used to "test ideas, refute theories, challenge limits, or explore possible implications,"[132] and counterfactuals imagine how society might look today if some key historical event had played out differently.

Counterfactuals are particularly interesting for several reasons. First, when discussing counterfactual design fiction, Dunne and Raby highlight Sascha Pohflepp's *The Golden Institute*: an alternate history that envisions what our scientific present might look like had Ronald Reagan not slashed the National Renewable Energy Laboratory (NREL)'s budget by 90 percent.[133] This alternate history is of particular interest to my laboratory because of the way it re-envisions Golden, Colorado, where I currently teach at the Colorado School of Mines. But it is also a timely example due to the renewed threats to scientific funding, especially to renewable energy, presented by the Trump administration. Moreover, though, counterfactual design fiction might be a promising approach to forging new imaginaries due to the ways that our current, problematic imaginaries are necessarily rooted in the specific choices made in literature and engineering that were described in chapter 2. A counterfactual design fiction approach might invite us to ask: what if some other engineering patent had inspired the first appearance of robots in literature? And what might robots today look like had that alternate history unfolded?

The use of counterfactual design fiction as a design tool has recently attracted the interest of key scholars in the human-robot interaction community. Most notably, Katie Winkle has recently leveraged the approach, re-imagining the nineteenth century socialist utopian literature of William Morris in order to identify alternative utopian futures the field of human-robot interaction might strive toward, that better align with current design objectives like the UN's sustainable development goals (SDGs).[134]

Participatory Design Whether or not robots are being designed for use cases that are explicitly decolonial, robots' social identity design cues should, as much as possible, be designed by or with the community in which they are intended to be used. One way this could be achieved is through participatory design or codesign techniques. Participatory design grew out of the Scandinavian technology projects of the 1970s and early 1980s, in which computer scientists worked alongside Scandinavian workers to develop new information technologies to better address those workers' needs.[135] This approach involves members of the communities who will use or be affected by a technology in the actual design process, such that those community members, at a minimum, "have access to relevant information, . . . have the possibility for taking an independent position on the problem, and . . . in some way participate in the process of decision making."[136]

One concrete framework for pursuing a participatory design strategy in engineering efforts is engineering for social justice (E4SJ).[137] Developed by Jon Leydens and Juan Lucena, E4SJ encourages engineers to adhere to six key principles during the engineering process: (1) listen contextually to community members; (2) identify the structural conditions that constrain the needs, values, and priorities of those communities; (3) acknowledge the political agency of both engineers and the communities they serve, and work to mobilize the different types of power available to both engineers and communities; (4) identify the human capabilities that can be enhanced in communities, as constrained by those structural conditions, and the ways in which communities differently prioritize the need to advance those capabilities; (5) increase the opportunities and resources of communities by advancing their prioritized capabilities; and (6) reduce the risks and harms of the capability-enhancing engineering solutions.

As my students Yifei Zhu, Ruchen Wen, and I have argued in our work "Robots for Social Justice (R4SJ),"[138] although E4SJ was originally developed as an engineering education framework, it can easily be used to guide real-world robot research and design processes. E4SJ provides a clear set of engineering processes that roboticists can adopt to work with communities in ways that are more attentive to social justice concerns. By attending to E4SJ principles, designers can better understand how communities' capabilities are constrained by economic, social, and political factors and can better design robotic technologies that enhance those capabilities with sensitivity to communities' priorities and constraints.

Participatory approaches might also encourage us to reflect on the relationship between design practices and design outcomes in new ways. My student Nyomi Morris, for example, has investigated how trauma-sensitive design principles might be integrated into robot design, in her work codesigning robots with survivors of domestic abuse.[139] As Nyomi shows, her use of trauma-sensitive principles provides new insights into how codesign activities with survivors (and similar populations) can be conducted more ethically by approaching the research process with sensitivity to participants' histories of trauma. But moreover, she demonstrates how this approach allows new insights to be drawn from the design activities themselves, in ways that lead to more responsible—and effective—robot designs for this population.

A participatory approach could also be taken through a learning-based perspective. Merely trusting robots to learn appropriate norms has been shown to be an extremely risky idea, as shown by Microsoft's Tay, a Twitter chatbot that, because it was designed to learn online from human interactants, went from "Hello, World" to "Heil Hitler" in the space of only a few hours, as certain Internet communities worked together to train Tay to become a virulent racist and antisemite. But as Katie Winkle has shown in collaboration with Emmanuel Senft and Severin Lemaignan, interactive machine learning approaches can also be used as effective tools of participatory design if carefully deployed.[140] Specifically, Winkle and her colleagues reanalyzed their "Couch to 5K" robot discussed in chapter 1, in which an expert user was responsible for the online training of their robot. As they argue, this can be construed as a form of participatory design, especially when integrated with other stages of participatory design.

Alternatively, instead of designing robots for and with communities, robotics researchers could develop their robots in ways that are easily extensible, modifiable, remixable, and hackable and could develop tools that allow communities to leverage these features to make their own design decisions and redesign robots on their own.

Design Justice Specifically, rather than merely including members of targeted communities in their design processes, robot designers could go a step further by using an end-user programming approach, as advocated by Sasha Costanza-Chock's *Design Justice*.[141] As Costanza-Chock describes, many advancements to computing technologies are actually made by end users when given the opportunity to "hack" technologies to better meet their own needs. As a prominent example of this, Costanza-Chock points to Twitter, whose enabling SMS technology did not simply emerge from the brain of Jack Dorsey (let alone Elon Musk) but was instead created as a hack of a "free email-to-SMS loophole" by activists seeking to coordinate with one another while avoiding police violence as they protested the 2004 Republican National Convention in New York City. Building from this example, Costanza-Chock argues that one of the best ways to accelerate technology development in ways that meet real problems faced by marginalized communities is to design technologies in ways that can be easily hacked and modified by those communities. By following this philosophy, robot designers might provide communities with tools to make and implement their own choices about the identities presented by robots without the need for roboticists to be involved at all. In the case of social robot design, roboticists might design tools that effectively allow communities to design a robot's appearance and voice for themselves to meet the needs and priorities of their own community.

Approaches such as fluid design, where elements of robot identity are defined by a robot's software rather than hardware, may represent key sites for exploring this type of community-driven customization of robot embodiment.

Although a physical robot's embodiment will be inevitably constrained by the choices imposed by the robot's designers, a robot whose head and torso are conveyed through retroprojection or augmented reality could allow communities—or individuals—to completely customize the robot's appearance to align with their own priorities, preferences, and aesthetics as part of design justice–oriented design processes.

To be clear, simply taking a participatory or end user–oriented design lens is no silver bullet. After all, merely pushing design decisions off onto communities may simply be passing the buck rather than dealing with the problem itself, and doing so without careful thought may simply result in end users selecting stereotype-reinforcing designs for themselves rather than asking designers to make the difficult choice to be actively cross-stereotypic. Moreover, in many contexts, such as airports and malls, it may not be clear who the community asked to participate in robot codesign might be; participatory design in domains such as policing is unlikely to address the harm caused by deploying robots in such domains in the first place; and participatory and end user–oriented design efforts may increase the burden placed on already overloaded communities. However, participatory and end user–oriented design perspectives may be used as effective tools within the ethical robot designer's toolbox, may enable the more effective use of other tools such as cross-stereotypic and diversity-oriented design approaches, and may be particularly effective if used while acknowledging the ways that encouraged anthropomorphism and language capability might be minimized when unnecessary.

Conclusion and Chapter Summary

In chapter 2, we traced the history of robots in engineering and science fiction back to *The Steam Man of the Prairies* of 1868 and showed how its post-bellum vision of mechanical people as a metaphor for the maintenance of White patriarchal power played a critical role in shaping the cultural and techno-scientific robot imaginaries over the next 150 years into the White patriarchal robot imaginary. In this chapter, we then used this history to show how roboticists operating under the White patriarchal robot imaginary reinforce White patriarchal power in the cultural domain by designing robots that reinforce stereotypical assumptions about robots, about robot designers, and about the people that robots depict and replace.

Each of these types of stereotype reinforcement can be seen in the designs of the three robots highlighted in chapter 1: AILA, Pepper, and REEM. First, the fact that all three robots adhere to the White patriarchal robot imaginary's White humanoid design aesthetic reinforces that imaginary's stereotypes regarding what robots are supposed to look like. They cast robotics as a White space in which those racialized as non-White are erased, even as non-White labor remains necessary for the creation and use of those robots' hardware and software systems. Second, this same design aesthetic reinforces the assumption (whether or not it is accurate) that robotics design teams are predominantly White. And, in the case of AILA, the design aesthetic's feminine sexualization reinforces the (correct) assumption that its design team was predominantly composed of men. Third, all three robots may reinforce racialized or gendered stereotypes about the people they depict. Although AILA's use in space contexts may decrease perceptions of space exploration as a masculine endeavor, AILA's sexualized design (and, more specifically, its White masculine vision of femininity) are equally likely to reinforce a view of women as objects of heterosexual male desire. Pepper, when gendered as feminine, may similarly reinforce objectification of women, as users are literally invited to interact with its chest. Finally, there is REEM, a White robot (ostensibly) deployed as a police officer in Dubai—a context where REEM's Whiteness in turn draws on and reinforces relationships between Whiteness and state power, serving as a technological reminder of the history of British imperialism in the region that now comprises the UAE.

Based on this understanding of our technological status quo, the way it reinforces our sociological status quo, and the historical and social root causes that explain this dynamic, we have identified key strategies that roboticists can use to build toward a more equitable robot future by revising, rethinking, and ripping up the status quo in the cultural domain.

First, roboticists might revise the status quo by saying, "We won't use White patriarchal stereotypes, biases, and assumptions to guide robot design." This might involve designing for a wider diversity of robots. However, simply encouraging predominantly White male design teams to design more robots of color could be a recipe for caricatured disaster and an ironic return to the steam-man origins of the robotic imaginary. Similarly, there may well be concerns about pink capitalism or rainbow capitalism, depending on the motivations for these diversification efforts.[142] So, while this approach may be a step in the right direction, it could merely trade one problem for a slightly more inclusive version of that same problem. Alternatively, revising the status quo could involve taking a cross-stereotypic approach in which roboticists design humanlike robots in ways that run counter to those White patriarchal expectations. For example, we could follow Katie Winkle's cross-stereotypic approach to designing feminine robots that deviate from culturally specific politeness norms.[143] More generally, this approach could involve simply using more masculine robots in service positions, queering robot design, or otherwise mixing and matching different design cues within a single robot body. But more work would be needed to explore whether these approaches would be effective ways of diverting robotics' White male gaze in a genuine and sensitive way.

Second, roboticists might rethink the status quo by attempting to "remove" race and gender from humanoid robot design. However, we have seen that it is wildly difficult to design robots whose appearance and voice are not gendered or racialized in any way.[144] Although there has been some recent success toward this goal through agendered robot design approaches,[145] it is yet unclear what it would mean to develop a language-capable robot whose voice is "free" of racial or ethnic cues. Roboticists might also rethink the status quo by saying, "If the problem is with how robots are being designed to look and speak, let's just not design robots that are humanoid or language capable." This might include reliance on zoomorphic or mechanomorphic designs and minimization of natural language capabilities. Perhaps your vacuum cleaner doesn't need a face and doesn't need to be able to speak. However, although this approach might address the concerns listed in this section, it could be counterproductive. In some domains, certain humanlike features might be critical to encourage the right type of engagement, and natural language capabilities are often critical (e.g., for robots used in education).

Finally, roboticists might rip up the status quo by saying, "Let's attack the root causes that explain how we have arrived at the status quo." If the reasons roboticists by default are reinforcing White patriarchal stereotypes, biases, and assumptions is because (1) this was their original envisioned purpose, (2) this motivating purpose has shaped the motivations and values of robotics, (3) these motivations and values have shaped the demographics of robot design teams, and (4) today's robot design teams develop robots that fulfill a combination of (a) the historic motivations, values, and biases of robotics and (b) their own personal motivations, values, and design biases, then perhaps we should be working to forge new robot imaginaries, and to change the motivations and values of robotics and the demographic composition of robot design teams. This might include using counterfactual design fiction to identify alternate robot imaginaries grounded in alternate histories, taking participatory design approaches to design robots whose appearance and purpose satisfies the needs and desires of historically excluded and oppressed populations, using design justice–oriented end-user programming approaches to enable members of those communities to design or redesign robots that better meet their own preferences and needs, and pushing for greater demographic diversity in both the design teams and executive boards of robotics companies.

Although in this chapter we have focused on how robots look and sound, in the next two chapters, we will turn the camera inward, considering how robots reason and perceive, and the ways roboticists wield disciplinary power through the ways they design these cognitive capabilities. As we will see, understanding roboticists' power in this domain will once again be helped along by considering the historical and literary factors that shaped the White patriarchal robot imaginary, especially the influence of Isaac Asimov and his Three Laws of Robotics.

4 Control System

Few writers have played as significant a role in shaping the cultural and techno-scientific dimensions of the White patriarchal robot imaginary as Isaac Asimov. Today, Asimov is perhaps most widely remembered for the enduring influence of his Three Laws:

1. A robot may not injure a human being or, through inaction, allow a human being to come to harm.
2. A robot must obey orders given it by human beings, except where such orders would conflict with the First Law.
3. A robot must protect its own existence, as long as such protection does not conflict with the First or Second Law.[1]

Although these "laws" were invented by Asimov as a narrative device that would fuel entertaining science fiction stories, they nevertheless represented the first significant attempt to think about robot ethics and machine morality, and as such, they fundamentally shaped the common understanding of what it means to design robots in accordance with human moral standards.[2] Indeed, although Asimov's laws are best understood as a fictional conceit with little remaining scientific merit, they continue to be taught today in robotics classes, and hundreds of new scientific papers are written each year that seem to take them seriously. Even the field of machine morality, which has engaged in the deepest and most serious explorations of how to effectively design robots that adhere to human moral standards, stubbornly clings to Asimov's overarching premise that robots could be made to behave morally if only roboticists could formulate and program a correct and comprehensive set of moral rules for robots to follow.[3] In other words, much of machine morality continues to operate in service of Asimov's original vision of formulating slave codes to keep robotic slaves in line.

In chapter 2, we saw how roboticists' adherence to this Asimovian vision of "robots as slaves" continues to reinforce White patriarchal power in the cultural domain, as it perpetuates a robot imaginary in which robots are primarily designed to satisfy the desires of wealthy White men. But as we will see in this chapter, designing robots as slaves—and framing robot ethics as the formulation of slave codes—may also reinforce White patriarchal power in the disciplinary domain. Under Patricia Hill Collins's definition, the disciplinary domain of power centers on the "rules of the game" that informally structure society, including the rules and norms that are used to manage oppression.[4] This may at first seem counterintuitive. How would the formulation of rules and norms for governing the behavior of *robots* relate to the rules and norms used to manage the oppression of *humans*? After all, aren't the rules and norms formulated by robot ethicists intended to keep robots in line and to make sure that robots don't harm people? As we will see, the rules and norms intended to govern robot behavior inevitably feed back to govern human behavior as well.

This is particularly true of interactive robots, especially language-capable robots. For these robots, the rules governing robot behavior can also be thought of as constraints placed on the set of commands that those robots are empowered to accept. This means that when roboticists decide on the rules that will govern their robots' behavior, they are committing to ethical and political positions that will constrain or enable the autonomy of the humans who interact with their robots. Imagine, for example, a user asking a robot deployed in a school nurse's office, "Can you help me figure out how to get an abortion?" Refusing this command would necessarily constrain the behaviors of the robot's user, arguably in a way that reflects the patriarchal rules used to manage the behavior of women within the roboticist's society. But moreover, as we will see in this chapter, robots that adhere to and communicate moral principles do not merely reflect those principles but, moreover, steer users toward—and enforce compliance with—those principles. This means that roboticists wield substantial disciplinary power when they select moral principles and explicitly encode them into robots. This also demonstrates one way that roboticists may often wield disciplinary power in service of White patriarchy. When the moral principles being enforced by robots are the dominant norms of White patriarchy, robots become the enforcers not only of White patriarchy but, more specifically, of the technological continuation of the New Jim Crow[5] that Ruha Benjamin terms the "New Jim Code."[6] As such, while interactive and collaborative robots are sometimes called "cobots," we might better call them, when they enforce White and patriarchal norms of behavior, "New Jim Cobots."

In the first part of this chapter, we will more carefully consider how and why roboticists are pursuing norm-based approaches to machine morality and the different ways this serves as a vehicle for disciplinary power. In the second part of this chapter, we will then consider the historical origins—and implications—of the moral philosophies underlying this approach to machine morality. Finally, in the third part of this chapter, we will explore how roboticists might wield disciplinary power in ways that subvert rather than reinforce White patriarchy by revising the ways they encode moral principles into robots, by rethinking the source of those moral principles, or by more fundamentally ripping up the status quo by shifting their focus from explicit to implicit ethical agency.

I Moral Competence as Disciplinary Power

The ultimate goal of most research conducted in the field of machine morality is to develop artificial intelligence techniques that will enable robots to act as what philosopher James Moor terms "explicit ethical agents"—that is, agents that can actively engage in moral decision-making.[7] To reach this goal, scholars Bertram Malle and Matthias Scheutz have argued that robots need moral competence and thus require both a system of norms and algorithms that leverage those norms to enable moral cognition and affect, moral decision-making and action, and moral communication.[8] Under this dominant framing, moral cognition is thus principally concerned with these moral norms.

Norm-Based Approaches to Machine Morality

Under norm-based approaches to machine morality, norms are often framed in terms of deontic logic. This means that norms specify whether particular actions (e.g., stealing) or states (e.g., possessing stolen property) are forbidden, obligated, or permitted. From a moral psychological perspective, these types of norms are used when deciding what actions are appropriate or inappropriate to take in ways that are sensitive to the context in which one would perform those actions.[9] For example, entering a karaoke bar might activate a norm by which singing is encouraged (i.e., it is permitted or, in certain friend groups, obligated), whereas entering a library might activate a norm by which singing is discouraged (i.e., it is forbidden).[10] A machine-morality researcher operating under a norm-based approach might provide robots with rules such as "*If* you are in a library, *then* it is *forbidden* to speak above 40 decibels." Indeed, my own lab has explored a variety of ways that robots might use these types of context-sensitive norms.[11]

Given a set of context-sensitive norms, it is relatively straightforward for robots to engage in moral decision-making in order to determine whether to accept commands given to them. If a robot is asked to achieve a particular goal, its decision of whether to accept that goal can be reduced to the question of whether it can identify a sequence of actions it could take that would achieve that goal without violating any of its deontic norms under the contexts in which they are stated to apply.[12] Moreover, given a set of context-sensitive norms, it is also relatively straightforward for robots to explain their moral decisions. If the prior decision-making process determines that a requested goal cannot be achieved without taking a particular prohibited action, the robot may simply state that it cannot accept the request because doing so would require it to take that impermissible action in whatever context that action was specified as prohibited.[13]

According to philosopher James Moor, this last capability (i.e., communication about moral decisions) is particularly crucial, as it is what sets explicit ethical agents apart from full (i.e., humanlike) ethical agents. Indeed, there are a huge range of situations in which humanlike interactive robots might need to engage in moral communication.[14] For example, robots may need to remonstrate with others by explaining their reasoning for rejecting commands or by explaining their disagreement with interactants' stated opinions. Alternatively, robots might need to apologize for their own immoral acts or take steps to forgive or absolve others accused of performing such acts. These capabilities are critical for multiple reasons. Obviously, we do not want robots to accept immoral commands simply because they were given to them. Moreover, robots may need more complicated forms of moral communication in order to build harmonious social relationships and maintain teammates' trust. And finally, as we saw in chapter 1, robots may need to engage in effective moral communication to avoid inadvertently weakening important moral norms due to robots' potential for interpersonal power.

Overall, then, we can see why roboticists have been so interested in providing robots with moral competence and why rules-based approaches have been such a popular approach to enabling such competence. Following the logics of Asimov, so long as a roboticist can formulate correct rules of behavior, they can ensure that their robots will avoid actions that violate those rules and ensure that their robots can justify and explain their reasons for accepting or rejecting commands. Moreover, there is a strong psychological argument to be made that this type of context-sensitive, norm-driven approach is a good model for at least some types of human moral and social cognition.[15]

Yet, just because this is a reasonable model of human cognition does not necessarily mean it is an ethically responsible paradigm for robot design. As we will see, roboticists who choose to design robots in this way run the risk of inadvertently wielding disciplinary power in ways that reinforce White patriarchy.

How Roboticists Wield Disciplinary Power When Encoding Moral Norms

To understand how norm-based approaches to machine morality may turn robots into New Jim Cobots, let us first consider the particular types of norms that roboticists may be likely to program into robots and why those particular types of norms may be problematic. Then, we will explore several specific ways that roboticists wield disciplinary power when they encode those norms into robots.

By default, the norms that roboticists are likely to program into robots are, of course, their own. If a roboticist from a so-called WEIRD (i.e., Western, Educated, Industrialized, Rich, and Democratic) background is tasked with deciding what a robot should or should not do—and what the robot should encourage or discourage others from doing—they will likely choose to forbid, discourage, or sanction actions viewed negatively by the hegemonically dominant group within their own society. Put more bluntly, White American male roboticists are likely to program robots to discourage actions that fail to conform with the norms of White Western masculinity, in the same way that White American male lawmakers are likely to criminalize activities that do not adhere to White American male norms and priorities. When roboticists program robots to predominantly adhere to White and patriarchal norms of behavior, they wield disciplinary power in at least four key ways.

First, as foreshadowed above, the rules roboticists use to constrain robot behavior may, in fact, have the primary effect of constraining human behavior. After all, many of the scenarios that motivate roboticists' pursuit of moral competence involve preventing robots from blindly accepting commands given by malicious or irresponsible actors.[16] This is often for good reason—but just as current speech recognition algorithms may require interactants to "Whiten" their speech in order to be heard correctly (a phenomenon we will discuss in chapter 5), current norm-based approaches to machine morality may require users to "Whiten" or otherwise adapt their behavior in order to receive services whose gatekeepers are robots. Given the recent intensification of efforts to force Christian religious practices into US public schools[17] and to defund those public schools in favor of White Christian nationalist private schools,[18] one might envision a future in which the gatekeepers of educational programming provided by robotic tutors are robots or other educational technologies that require participation in prayer or other religious activities before educational programming can be accessed. This is not to say that norm-grounded machine morality is the most likely way this future could be technologically enabled, or that this is the future that norm-grounded machine-morality efforts are most likely to produce. Rather, this example simply illustrates one way that machine-morality efforts could obligate adherence to hegemonically dominant standards of "appropriate" behavior.

Second, roboticists wield disciplinary power when their robots use their interpersonal power to persuade people to adhere to particular standards of behavior. As we discussed in chapter 1, roboticists wield interpersonal power through the ways that robots are able to convince interactants of particular courses of action. When roboticists use this interpersonal power to enforce standards of behavior that are used to manage oppression, they are also wielding power in the disciplinary domain.

Third, and relatedly, roboticists wield disciplinary power when their robots merely advise adherence to particular standards of behavior. Recent evidence suggests that users may feel obligated to take advice from robots, even if they are not convinced that the advised actions are morally right. Specifically, in work led by Alyssa Hanson and performed in collaboration with moral psychologist Bertram Malle, our laboratory has found that when robots give advice as to how to behave in the context of a moral dilemma, the robot creates a powerful expectation that its advisee should follow that advice—even if the advice it gives runs counter to typical moral expectations.[19] As such, roboticists wield disciplinary power when they use the power of advice to create a societal expectation to adhere to standards of behavior that are used to manage oppression.

Fourth, roboticists wield disciplinary power when their robots create obligations to particular standards of behavior through those robots' use of surveillance and the threat of violence associated with that surveillance. This is most likely to arise when robots are programmed to identify, record, and report norm-violating requests they are given—or norm-violating behaviors they observe. As such, the choice of norms encoded into robots deserves particular attention in domains where there may be an incentive to track users and report information about their behaviors. One example of this is educational robotics. One of the most compelling and exciting use cases for social robots is in education, as robots have been shown to be particularly effective at promoting learning outcomes and providing impactful learning experiences.[20] Yet, the field of education is increasingly moving toward a state of constant surveillance of students in ways that are not readily recognized by the public due to the historic use of surveillance and monitoring of different sorts within education.[21] As Couldry and Mejias write, "The links between continuous surveillance and the education of children are so deeply naturalized here that the panoptic notion of teachers using continuous real-time monitoring is presented as neither chilling nor threatening."[22] As such, it should perhaps be unsurprising that some educational robots are being explicitly designed to surveil students and enforce their adherence with certain norms of behavior. In recent work, educational robots have been presented for purposes such as monitoring attendance and ensuring compliance with school uniform policies,[23] and as described in chapter 3, there has been some recent work on educational robots identifying and calling out "bad behavior." Although these capabilities are sometimes necessary (e.g., it may be important for robots to detect and call out sexism[24]), these types of norms (i.e., attendance requirements, uniform codes, and behavioral codes) need to be approached with the utmost care due to the ways those same norms are regularly used by humans in service of oppression. In the United States, two hundred thousand students are referred to law enforcement each year for ostensible violations of attendance requirements, uniform codes, or behavioral codes.[25] These law enforcement referrals disproportionately affect students of color: Black and Indigenous students are disproportionately more likely to have their absences deemed "unexcused" and to be referred to the legal system when absences occur,[26] Black students comprise one-third of in-school arrests,[27] and Black students are three times more likely to be suspended than White students.[28] Due to these factors, Black students disproportionately end up in the juvenile justice system and are systematically funneled toward incarceration.[29] It is not difficult to see how educational robots designed to detect, call out, and/or report violations of their programmers' moral standards could have a colonizing, Whitening, and potentially violent effect, exacerbating both the school-to-prison pipeline and the economic disparities between racialized groups.

The disciplinary power wielded by roboticists in each of these four ways may be especially acute when robots developed and tested in American contexts are deployed in other parts of the world. When robots are programmed to adhere to White patriarchal norms are then deployed in nations in the Global South (or in communities populated by ethnic groups from the Global South), this is an inherently colonial enterprise and represents a form of epistemic violence. As Dan McQuillan analyzes, epistemic violence is wrought when the knowledge of people in former colonies is shaped by knowledge created at the colonial center, or when people in former colonies are only known to others through knowledge created at that colonial center.[30] Data practices that center the norms of the hegemonically empowered group at the colonial center represent a similar form of epistemic violence.

Overall, these different types of disciplinary power present a design dilemma for roboticists. Roboticists may well have good reasons for trying to enable their robots to adhere to certain moral norms and encourage others to do the same. Yet, if roboticists are not extremely careful in executing these efforts, their robots can easily serve as the enforcers and reinforcers of White and patriarchal behavioral norms and may even serve as tools of colonialism. As argued earlier in this chapter, when social and collaborative cobots are used to enforce White and patriarchal standards of behavior in this way, they might be better termed "New Jim Cobots" due to the ways they serve as enforcers of what Ruha Benjamin terms the "New Jim Code"[31]: a modern paradigm in which racist values are baked into the design of technologies, just as racist values were baked into the legal system under the New Jim Crow.[32] Resolving this design dilemma requires us to answer several key questions. Why is norm-grounded moral competence our default approach to machine morality? Why are the roboticists most interested in ensuring that their robots do not harm people so prone to design those robots as New Jim Cobots? What historical factors, beyond the influence of Asimov, led to the field's adoption of that approach? And are there other approaches to machine morality we might instead consider that eliminate or mitigate the ethical risks that come with norm-grounded moral competence?

II De-Kanting Robotics

We have already seen a few reasons why machine morality's default tendency is to enforce White masculine norms. First, there is the overwhelming homogeneity of robotics labs and the tendency for designers, researchers, and engineers to encode their own values into technologies. Second, there is the field's inheritance from Isaac Asimov's Three Laws of Robotics, which were designed as a slave code to keep robotic slaves in line. But this tendency may also be due in part to the general philosophical approach adopted by most machine morality researchers. Both the scientific field of machine morality and the science fiction sphere of Asimovian laws are inherently grounded in Immanuel Kant's deontological ethics, which adjudicates decisions on the basis of whether there has been a violation of a norm that prohibits, obligates, or permits certain behavior. This reliance on deontology is worth questioning due to Kant's own motivations for focusing on universalizable moral norms.

Central to Kant's moral and political philosophies are notions of universalizability—for example, that we should avoid actions that we universally view as bad, and that we should treat all people with respect. Although these principles may seem reasonable at first glance, they must be interpreted through the lens of Kant's other works, especially his works on race science such as *Of the Different Human Races*[33] and *Observations on the Feeling of the Beautiful and Sublime*. In the latter of those works, Kant writes, "So fundamental is the difference between [the Black and White] races of man . . . it appears to be as great in regard to mental capacities as in color" so that "a clear proof that what [a Negro] said was stupid" was that "this fellow was quite black from head to foot."[34] Kant's work on these topics led political philosopher Charles Mills to call Kant "the "father of the modern concept of race."[35] As Mills writes in "Kant's Untermenschen":

> When Kant talks about the importance of treating all persons with respect, when he outlines the responsibilities of the state, when he maps his inspiring cosmopolitan vision, he is not making race-neutral and racially-inclusive pronouncements; he is really talking about the white population (and if feminists are correct in their parallel line of argument, only about the male half). If this is true, it is obviously a radically different picture of the Kant we all thought we knew and loved. The

> distinction between "Treat all persons with respect," where "person" is assumed to be racially inclusive, and "Treat only whites with respect" (at least here on Earth) is obviously not minor and trivial at all. It would mean that we need to reconceptualize his philosophy as distinguishing (white) persons from (non-white) sub-persons, and restricting its vaunted universalism to the person population.[36]

Mills's concerns about Kant help to explain why the default approach to machine morality is so overwhelmingly grounded in enforcing adherence to White Masculine norms: the need to respect those particular norms was the foundational cornerstone of the philosophical perspective guiding those machine morality efforts. As such, if we are to subvert robots' default tendency to serve as a tool of White patriarchy in the disciplinary domain and prevent the robots of the future from being developed as New Jim Cobots, we might benefit from exploring approaches to machine morality that are grounded in different moral frameworks.

If we did not take a deontological approach to machine morality, what framework might we use? The dominant theory against which deontology is typically compared is utilitarianism, in which moral decisions are made by choosing the actions that will lead to the best outcomes for the largest number of people. Although this calculus of outcomes is a theoretically interesting and intuitive approach, in realistic situations, it is, unfortunately, impossible to comprehensively or accurately enumerate the space of possible actions and their outcomes or to predict the likelihoods, benefits, or costs of those outcomes. Moreover, utilitarianism is well-known to lead to scapegoating, as doing terrible harms to a minority group may be justified by the benefits done to the majority group. This risk may remind us of Ursula LeGuin's story "The Ones Who Walk Away from Omelas," which critical race theorist Derrick Bell points to as an allegory for how the immiseration of Black people is so often justified by White Americans in the name of their own "greater good":

> [The people of Omelas] all understand that their happiness, the beauty of their city, the tenderness of their friendships, the health of their children, the wisdom of their scholars, the skill of their makers, even the abundance of their harvest and the kindly weathers of their skies, depend wholly on this child's abominable misery.[37]

Alternatively, many scholars have recently advocated for approaches grounded in virtue ethics, which emphasizes how we cultivate our moral selves to become virtuous. If we were to follow a traditional Western account of virtue ethics, we might design robots with behaviors that we view as optimally aligning with some set of virtues, such as courage or honesty.[38] Or we might design robots' perception and attention systems so that sources of unvirtuous temptation simply never enter into the robot's "awareness" and thus do not have the opportunity to be chosen. If we want a tutoring robot to be honest, the option to mislead may simply never be considered by the robot when choosing what to say; if we want an eldercare snack-delivery robot to encourage healthy eating, we might design its vision system to not even notice unhealthy snacks unless they're the only ones available.[39]

Because virtue ethics focuses on moral self-cultivation and the developmental process of becoming good, these types of approaches are most likely to be implemented in robots through machine learning. Through such approaches, robots might notice their own virtuous or unvirtuous behaviors and adapt their future behaviors accordingly. Alternatively, they might imitate virtuous behaviors performed by those they trust during interactions. Or robots might learn from curated datasets of behaviors performed by virtuous individuals. Martin Gibert, for example, argues that robot designers seeking to develop virtuous robots would need to start by building a base of virtuous people, then collect a dataset of morally relevant decisions made by those virtuous people, and finally implement algorithms that would emulate those decisions.[40]

Although this type of virtue-ethics approach might not have deontology's White patriarchal origin story, roboticists embracing virtue ethics would need to be just as attentive to the risks of centering and normalizing White patriarchal standards of behavior. After all, if a virtue ethics approach would require us to identify virtuous behaviors performed by virtuous people, we would need to determine which people we consider to be virtuous—and which virtues of theirs we seek to elevate. As discussed in chapter 2, the robot virtues traditionally (if implicitly) centered by roboticists are obedience, efficiency, and productivity—the virtues that enslavers desire to cultivate in those they enslave. If roboticists wish to develop virtuous robots that subvert rather than reinforce the White patriarchal robot imaginary, they will need instead to consider virtues that better align with a commitment to social justice.

Alternatively, instead of developing robots that cultivate their own moral selves to align with some set of ideal robot virtues, we might develop robots that encourage humans to cultivate their own moral selves to align with whatever virtues those humans wish to embody. In my lab's work, we have argued that this goal might best be achieved if we look beyond traditional Western ethical paradigms and instead consider non-Western theories, such as Confucian role ethics—a paradigm we will explore more in the following section.

III Subverting Roboticists' Default Reinforcement (through Machine Morality) of White Patriarchy in the Disciplinary Domain

As we have discussed in this chapter, roboticists wield power in the disciplinary domain when they decide what moral principles their robots will follow, enforce, and communicate and the ways this reinforces the rules of the game that privilege White and male standards of behavior, turning otherwise helpful cobots into New Jim Cobots. In this section, we will see (by following the pattern laid out in chapter 3)[41] how roboticists might avoid wielding this disciplinary power in ways that reinforce White patriarchy by (1) revising the types of moral principles encoded into robots, (2) rethinking how these moral principles are determined in the first place, or (3) ripping up the status quo by more fundamentally considering whether robots need to have explicitly encoded moral principles at all.

Revising the Formulation of Moral Principles

First, let us consider what it might look like for roboticists to formulate robots' moral principles through a very different framework than Kantian deontology. As discussed above, while there are a large number of alternative frameworks we might consider, I will be using Confucian role ethics as a case study and example in this section.[42] Confucian ethics is a tradition that grew out of the writings of the Chinese philosopher Kong Qiu (孔丘), later referred to as Kong Fuzi—that is, "Master Kong" (孔夫子)—which would be Latinized as "Confucius,"[43] as well as the writings of Confucius's followers and successors, such as Mencius (孟子).[44] Confucian ethics is particularly helpful for reasoning about moral communication due to its focus on the moral cultivation of others through remonstration.[45] Confucian ethics explicitly acknowledges the malleability of human morality and argues that all members of a society need to work together to become good.[46] Critically, Confucian ethics emphasizes that virtues are cultivated relationally through our social interactions and relationships with others. Confucian ethics also acknowledges, more famously, that our relationships with others are structured according to our social roles: being someone's friend, sibling, or supervisor leads to different expectations for how you should act toward them, and these expectations serve as guidance for what it means to play those relational roles in a way that is moral and good.[47] Within the Confucian ethical tradition, then, becoming good requires one to follow an observation, reflection, practice pipeline in which one (1) observes their own actions and those of others with sensitivity to the roles they play in their relationships, (2) reflects on whether those actions are good with respect to relational expectations, and (3) puts those reflections into practice by adapting their future actions and encouraging others to do the same.[48]

To best understand this overarching vision for a moral life, we can interpret Confucian ethics through the lens of other ethical theories. For example, Confucian ethics can be interpreted as a type of care ethics: a feminist philosophy that elevates caring for others and attending to their needs.[49] Alternatively, Confucian ethics can be interpreted as a type of virtue ethics due to its emphasis on cultivation of one's moral virtues and its focus on moral worth over right action.[50] In my own lab's work performed in collaboration with technology ethicist Qin Zhu, we have adopted the recent tradition of interpreting Confucian ethics as a form of role ethics, which focuses on the roles people play rather than the rights people have.[51] Relative to Western relational ethics traditions,[52] Confucian role ethics uniquely focuses on the psychological dimensions of morality[53] in ways that provide distinct advantages over care ethics and virtue ethics interpretations, at least from a computer science perspective. First, Confucian ethics may be better suited for computationalization when interpreted through the lens of role ethics rather than care ethics, as programming robots to be aware of the relationships that hold between people is much more feasible than programming robots to genuinely feel care and empathy. Second, if we want robots to help humans to cultivate virtues,[54] then Confucian role ethics' inherently psychological and relational account of human morality may be more effective than other forms of virtue ethics, in part due to the ways that a focus on relational roles accentuates robots' persuasive power.

As my former student Ruchen Wen has observed, people often have negative emotional responses to norm-grounded moral communication (e.g., "I can't do that because it's wrong!"), perhaps because they feel that they are being judged by something with questionable social standing. In some of our experiments, we have even seen that when robots attempt to give norm-grounded moral explanations, people become *less* likely to believe in the importance of avoiding the behavior the robot is rejecting. In contrast, when robots refer to their relational role when rejecting commands (e.g., "A good friend wouldn't do that"), people more readily agree to the importance of avoiding that same behavior.[55]

On the other hand, framing Confucian role ethics in terms of persuasive benefits may lead us into old traps. If we instead move beyond this perspective, we can use Confucian role ethics to identify new ways that roboticists might approach robot design. First, Confucian role ethics encourages roboticists to rethink what is needed for robotic moral competence. Although roboticists working under a deontological framework have argued that morally competent robots need a system of norms and a set of moral cognitive processes built on those norms,[56] roboticists operating under a Confucian role ethics framework might further seek to provide robots with an understanding of both the network of role-grounded relationships into which they and their interactants are embedded and how moral decision-making and communication must be grounded in those roles and relationships. In our own work, we have shown how this approach can enable robots to explain their moral decisions in new and flexible ways.[57] Second, Confucian role ethics encourages roboticists to go beyond asking, "How should this robot be programmed to be morally competent?" and instead ask, "What roles should robots play, and in relation to whom?"[58] Even if we do not wish robots themselves to be involved in friendships due to the opportunities this might present for undue influence, we may wish to develop robots that can encourage friendship among humans.[59]

Third, Confucian role ethics helps roboticists to better consider the moral principles their robots are enforcing. Specifically, Confucian role ethics encourages roboticists to be specific about the virtues they seek to encourage in the context of relationships. For example, Mencius describes five cardinal relations, each of which comes with a different virtue that defines what it means to be benevolent in that role.[60] In the parent–child relationship, benevolence is grounded in affection, whereas in the friend–friend relationship, benevolence is grounded in trust and honesty. Fourth, Confucian role ethics encourages roboticists to take a pluralist view of robot moral cognition. That is, effective robot moral reasoning (and moral communication grounded in that reasoning) may require us to be specific about the connection between relational roles and norms. Ruchen Wen, for example, has shown how integrating norm-based and role-based ethical perspectives can have unique benefits, including making it more straightforward to generate explanations for moral action (or inaction) that are grounded in norms or roles as the situation requires.[61] Fifth, Confucian role ethics helps us to understand the role of interaction structure in shaping moral cultivation and moral communication. Confucian ethics emphasizes the importance of using rituals and other forms of structured moral practice as opportunities for moral self-cultivation. In studying the effectiveness of role-based versus rule-based moral communication, Ruchen Wen found that role-based communication was more effective if robots' interactants were given structured opportunities for moral reflection.[62] This insight helps us to understand how we might structure human–robot interactions to enable robots to wield positive moral influence more effectively. And sixth, Confucian role ethics refocuses the use of robots' moral persuasive capability away from merely dissuading negative behaviors and toward encouraging teammates' positive internal moral states as well as encouraging those teammates' moral self-cultivation.[63]

Rethinking the Formulation of Robots' Moral Principles

Unfortunately, there are no silver bullets in machine morality, and Confucian role ethics is no exception. Although approaches to machine morality grounded in non-Western frameworks such as Confucian role ethics might be less likely to inherit the unspoken racial assumptions of Kantian deontology, they may instead be more likely to inherit other problematic tendencies from their own histories. Just as deontology and virtue ethics have raised the questions of "whose norms" and "whose virtues," Confucian role ethics raises the question of "whose roles" due to the specific relational roles that have been historically elevated in the Confucian tradition. Confucianism has long been rightly critiqued for its focus on hierarchy and rigid structures, as demonstrated by the relational roles held as central within the Confucian tradition, such as husband–wife, minister–ruler, and father–son. Even though few roboticists would intentionally seek to develop robots to perform those particular roles,[64] an uncritical approach to Confucian role ethics might result in robots that respect (or encourage others to respect) the hierarchical and inherently inequitable relational roles adopted in patriarchal societies. In other words, Confucian role ethics could merely trade the racist history of deontology for the patriarchal history of Confucianism.

On the other hand, I argue that if deployed thoughtfully and in conjunction with feminist perspectives, Confucian role ethics could help roboticists to avoid the pitfalls of—or even explicitly subvert—those historically problematic hierarchical roles. The necessity of a feminist approach to robot design can clearly be seen through the groundbreaking work of Katie Winkle and colleagues on feminist human–robot interaction.[65] Taking inspiration from D'Ignazio and Klein's *Data Feminism*,[66] Winkle and her Swedish colleagues lay out key ways that roboticists can adopt a feminist perspective toward analyzing and challenging power through actions such as (1) supporting participatory processes, (2) resisting enforcing any single universal view, and (3) rethinking hierarchies. When viewed through a feminist lens, Confucian role ethics should encourage us to think more carefully about the relational roles we want robots to encourage, embody, or merely be aware of, in part because it forces us to articulate the way in which robots should understand and use those roles rather than simply leaving those power-laden relationships unstated. As such, by requiring the articulation of relational structures, Confucian role ethics may encourage us to rethink and critique hierarchical relational roles.

Taking a feminist perspective also encourages roboticists to take a participatory approach to ensure that the moral principles that robots follow and enforce reflect the values and priorities of the communities roboticists seek to serve rather than the values and priorities of robots' hegemonically empowered and demographically homogeneous design teams. A participatory approach may be particularly helpful as it avoids a metrics trap that nonparticipatory approaches are more likely to fall into. When candidate robot designs are only evaluated through quantitative metrics, the winning designs will tend to be those whose value can be easily captured by performance or productivity metrics[67] that reinforce existing neoliberal power structures and value systems.[68] Participatory design activities may help to avoid this trap by steering design processes toward outcomes that align with stakeholder values and needs in ways that are less readily quantifiable.

For example, under a participatory approach guided by the engineering for social justice framework[69] (or the robotics-oriented equivalent we are developing, robots for social justice[70]), roboticists might work with community stakeholders to design robots that adhere to the moral principles prioritized by those stakeholders. Alternatively, under an explicitly role ethics–driven framework, roboticists might work with community stakeholders to identify the relational roles that are important within their community and the ways that agents with those roles are expected to demonstrate benevolence. Similarly, the machine learning–based approach to participatory design proposed by Winkle et al.[71] could be used to better enable machine morality by (1) working with community members to determine the sets of actions a robot should be capable of performing, (2) using expert-driven interactive machine learning to determine when robots should perform certain actions, and (3) requesting those experts to indicate cases where certain actions are selected or avoided in order to comply with moral expectations.

Yet, when taking a participatory approach, especially one grounded in machine learning, roboticists must make sure not to claim a participatory mantle while reinforcing colonial and exploitative power dynamics as some Big Tech companies have done while trying to make their models safer or more inclusive. For example, in OpenAI's efforts to prevent its large pre-trained models from generating toxic content, the company exposed Kenyan workers to horrific conditions, forcing them to review thousands of toxic text passages and images while paying them less than $2 an hour.[72] One OpenAI content moderator reported losing his family due to the psychological trauma of reading hundreds of machine-generated accounts of sexual violence, child abuse, murder, self-harm, and bestiality.[73] Another contractor stated that they had been paid $787.50 to collect a corpus of 1,400 images of rape, death, and other horrific content for OpenAI, many of which would have been illegal to collect in the US.[74] This method of preventing immoral model outputs certainly "involves" traditionally marginalized populations and (in a shallow way) trains models to reflect the moral beliefs of those populations. But the approach is clearly toxic to the humans who make it possible and economically exploits those workers (for the benefit of wealthy Western businesses) through a form of data colonialism.[75] These patterns should remind us of the risks of *artificial* artificial intelligence discussed in chapter 1. As Atanasoski and Vora point out in their book *Surrogate Humanity*, these types of hidden labor patterns undergird much of robotics. Although AI-driven robots uphold a myth of human-free labor, they often merely obscure, rather than obviate, the need for exploitative labor.[76]

Clearly, while participatory approaches *can* be more equitable, participation itself is no panacea and, as such, cannot on its own mitigate the challenges surrounding both rule-based and role-based approaches to machine morality. Luckily, as we considered in the previous chapter, roboticists have tools at their disposal that may allow them to go beyond mere inclusion of communities in their design processes. Researchers could, for example, follow a design-justice philosophy by giving communities the tools that

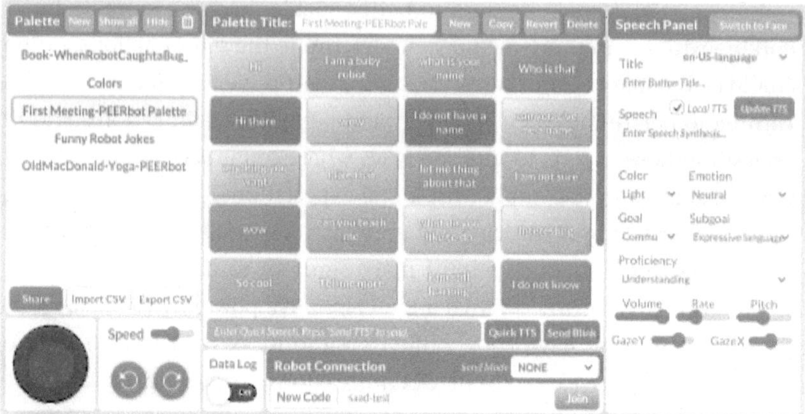

Figure 4.1
The Peerbots interface (Peerbots.org) enables end users such as therapists and special educators to design verbal interactions for teleoperated social robots without any need for roboticist involvement. Image by Peerbots.

will allow them to encode moral principles for themselves. End-user programming interfaces that enable nonexpert community members to design, control, and reprogram robots (such as the Peerbots interface co-developed by my former student Saad Elbeleidy, shown in figure 4.1) represent a key emerging research area in robotics.[77] In the future, we can imagine how these interfaces might enable nonexpert users to specify not just low-level robot behaviors but also higher-level moral principles of different kinds.

Finally, it is worth noting that both participatory and end-user approaches to encoding moral principles into interactive robots not only present opportunities for communities to encode shared values but also provide sites for the creation of those values. As an allegory, we can consider Mothers Reclaiming Our Children (Mother's ROC), an organization devoted to supporting Black and Latino men incarcerated in Los Angeles on false charges. As Ruth Wilson Gilmore describes in *Golden Gulag*, the collective activities engaged in by Mother's ROC enabled the creation of shared values.[78] Similarly, exercises encouraging group articulation and encoding of shared values to be embedded into robots could be an opportunity for community members with similar lived experiences to collectively create new shared values based on those experiences.

In addition to encouraging participatory approaches, Winkle's theory of feminist human–robot interaction also encourages us to resist universal solutions. As such, roboticists seeking to rethink their approaches to the development of morally competent robots should also explore frameworks beyond deontological ethics, utilitarian ethics, virtue ethics, and Confucian role ethics (as well as how different combinations of frameworks might be used synergistically). First, roboticists might draw on Ubuntu philosophy. As Ubuntu philosopher John Mbiti argued, "Only in terms of other people does the individual become conscious of [their] own being, [their] own duties, [their] privileges and responsibilities towards [themselves] and towards other people."[79] Indeed, philosopher of technology Mark Coeckelbergh has speculated that an Ubuntu-driven social robot might help to better serve the interests of families and communities, rather than individual interests, and might help social roboticists to move beyond "merely" providing companionship or relieving loneliness in contexts where effective community relationships have broken down.[80] Second, roboticists might draw on Black feminist philosophies in ways that go beyond the types of power analysis conducted in this book by developing robots that encourage the cultivation of wisdom (rather than "merely" the cultivation and replication of knowledge, which is too often and erroneously presumed to be universal) or that elevate the ways that both wisdom and knowledge are relationally cultivated and negotiated through interactions with others.[81] Finally, roboticists might draw on care ethics[82] by elevating attitudes of care over "mere" respect for rights. Like Confucian role ethics, Ubuntu philosophy, and Black feminist thought, care ethics is inherently relational, centering the importance of relationships, roles, and responsibilities.[83] But in contrast to these other relational frameworks, care ethics elevates the fulfillment of the needs of the vulnerable over individual rights and patterns of interaction over isolated acts.[84] Care ethics highlights the importance of attending to the needs of others out of a feeling of social responsibility and solidarity.[85] And care ethics uniquely holds that ethical deliberations should be grounded in empathy rather than abstract rules and principles.[86] Although we seemingly dismissed care ethics above for being "hard to computationalize," seriously engaging with care ethics may encourage us to step even farther back and reconsider whether we need to be encoding explicit moral principles into robots at all.

Ripping Up the Status Quo: Beyond Explicit Ethical Agents

As we have seen, some robot ethicists (including myself) have argued that explicitly encoded moral principles could help robots identify, say no to, and call out harmful behaviors and do so in a way that reflects the priorities of the communities into which robots are deployed. But this perspective is not shared by all robot ethicists. Philosopher Aimee Van Wynsberghe (a proponent of care ethics), for example, has been a vehement critic of explicit ethical agents (or, as they are sometimes called, "artificial moral agents"). As Van Wynsberghe writes:

> It is unclear that [artificial moral agents] are the solution to [the problem of minimizing the harm done by robots]. There are plenty of technologies capable of harming human beings (e.g. lawn mowers, automatic doors, curling irons, blenders); the solution has always been either to design them with safety features or to limit the contexts in which a technology can be used. An elevator door has a sensor so that it does not close on people; lawn mowers have a guard to protect us from their blades; and ovens have lights to warn us when our stovetop is hot. One does not normally use barbeques indoors or chainsaws in daycare centers. Machine ethicists are the first to suggest endowing technology with moral reasoning capabilities as a solution to problems of safety.[87]

Although encoding moral principles into robots may well be one way to prevent robots from causing harm, it is possible that robot ethicists have become too focused on machine morality, perhaps due to the enduring influence of Isaac Asimov and the White patriarchal robot imaginary. Perhaps instead of focusing only on the moral principles that *robots* should follow in order to avoid causing harm, roboticists should focus more on the principles *they* are following to avoid causing harm. As foreshadowed earlier in this chapter, researchers working in machine morality may critique care ethics by arguing that it is unclear how to program care-ethics principles into robots or whether robots are even capable of caring. But these very critiques reveal the limitations of machine morality. If it is not obvious how to program principles of empathy and care into robots, then perhaps we are too focused on robot cognition and should instead focus on how *roboticists* can demonstrate empathy and care for communities while designing robots and how *roboticists* can ensure that the overall outcome of their technologies will be to promote caring human relationships.

Moreover, taking this step back might further encourage some roboticists to more seriously question why they are designing robots in the first place. Although a focus on explicit ethical agents presupposes the existence of agents into which to instill moral competence, a focus on fundamental ethical impact may instead reveal cases where robots are poor solutions for meeting communities' needs relative to other technologies—or nontechnical solutions. As James Moor notes, for example, although robot camel jockeys in Qatar may free individual children from specific forms of enslavement, they do so without changing anything about the political and economic situation that led to their enslavement.[88] In contrast, nontechnical solutions grounded in abolitionist politics might be more successful at achieving farther-reaching and longer-lasting social justice. That is, if an abolitionist philosophy is the best approach to subverting the New Jim Crow, then it may also be the best approach to subverting the New Jim Code as well as the New Jim Cobots it has produced. We will return to this perspective in chapter 6, in which I will make a call for an abolitionist robotics.

Conclusion and Chapter Summary

In this chapter, we have seen how roboticists reinforce White patriarchal power in the disciplinary domain by designing robots that adhere to, enforce, and/or reinforce predominantly White and patriarchal norms. In doing so, we have seen the ways that the White patriarchal robot imaginary has steered the imagination of robot designers toward this particular view of robot ethics and the ways that this view of robot ethics can sometimes turn otherwise helpful cobots into New Jim Cobots—*enforcers* of Benjamin's New Jim Code.

These disciplinary power dynamics are a matter of software infrastructure and, as such, are not unique to any one robot. That is, the concerns raised in this chapter are equally likely to manifest across the REEM, Pepper, and AILA robots seen in chapter 1. However, it is worth considering how these dynamics might uniquely manifest on the REEM robot. As discussed in chapter 1, REEM has been deployed in Dubai in order to serve as a White police officer. And, as discussed in chapter 3, the social identity design cues reflected in REEM's design serve to reinforce relationships between Whiteness and state power that have been inherited from Britain's historical colonial presence in the Middle East. This might lead us to question the impacts that norm-based machine morality efforts might have when deployed on robots such as REEM deployed into such contexts. In this chapter, I have discussed the ways that machine morality efforts may turn cobots into New Jim Cobots that serve as enforcers of the New Jim Code. Although Benjamin's nomenclature of the New Jim Code playfully references Michelle Alexander's New Jim Crow, the New Jim Code is not as inherently American as was the New Jim Crow, since racism is encoded into technological designs around the world, regardless of national borders. As such, robots such as REEM might well end up serving as New Jim Cobots, even in countries such as the UAE, if they are designed in ways that enforce racialized and gendered standards of behavior within that local cultural context. Moreover, it is easy to see how REEM might serve as a colonialist enforcer of the New Jim Code if it were used to enforce European standards of behavior onto Middle Eastern citizens, serving as an echo of previous European colonial presences through not only its appearance but also its actions.

Overall, in this chapter, we have identified key historical and social factors that played a key role in producing the technological status quo by which roboticists wield disciplinary power and by which robots serve as New Jim Cobots. Specifically, we traced roboticists' tendency to enforce White and patriarchal norms of behavior back to both Kant's implicit casting of those norms as the universal principles of morality and to Asimov's Three Laws of Robotics—a slave code that once more shows the influence of *The Steam Man of the Prairies*. Based on this understanding of our technological status quo, the way it reinforces our sociological status quo, and the historical and social root causes that explain this dynamic, we have identified key strategies that roboticists can use to build toward a more equitable robot future by revising, rethinking, and ripping up the status quo in the disciplinary domain when they decide how a robot will morally reason and act.

First, roboticists might revise the status quo by abandoning (purely) norm-based approaches in favor of different types of moral principles. For example, roboticists could adopt approaches grounded in Confucian role ethics by specifying moral principles that center roles and relationships rather than the inherent rightness or wrongness of actions themselves.

Second, roboticists might rethink the status quo by helping communities specify their own moral principles. This might involve taking a participatory approach, where roboticists work with communities to specify moral principles, or a design-justice approach, where community members are given tools to specify moral principles for themselves. Critically, this approach might start us down the road toward redeeming approaches grounded in deontological ethics by acknowledging and addressing the too-often-unstated

White patriarchal assumptions of Kantian deontology—a process which may help us to create a more inclusive approach that is still deontological but perhaps no longer truly Kantian.[89] Alternatively, this approach might give us the opportunity to do away with frameworks grounded in race science or patriarchy by exploring fundamentally new frameworks such as Ubuntu philosophy, Black feminist thought, or care ethics.

Finally, roboticists might rip up the status quo by attacking the root causes that explain how robotics arrived at that status quo. If the legacies of Asimov and Kant have made us too predisposed toward approaches to machine morality that involve hand specification of ethical principles, and if those ethical principles inevitably comprise control systems that impinge on human autonomy as surely as methods of surveillance capitalism[90] and data colonialism,[91] perhaps it is worth taking a step back and considering what robotic solutions—or non-robotic solutions—we might develop if we were less focused on explicit ethical agency. By taking this step back, we may better understand the types of ethical impacts that robots have, even when they do—or say—nothing at all. In the next chapter, for example, we will see the ways that roboticists further wield disciplinary power through the ways their robots perceive rather than the ways their robots act. And once again, we will see the ways that this further form of disciplinary power tends to be wielded in ways that reinforce White patriarchy.

5 Bounding Box

Now that we have considered the ways roboticists might overtly wield disciplinary power through their decisions about which norms or other moral principles they enable their robots to enforce, let us consider the subtler ways roboticists wield disciplinary power through covert surveillance[1] practices that determine how people are watched and seen—dimensions that mirror and invert the phenomena discussed in chapter 3. Although the literal encoding of rules for governing behavior may be the most obvious way that roboticists wield disciplinary power, it is by no means the most pernicious or widespread. As we will see in this chapter, other more commonplace elements of robot design, such as the design of robots' computer vision systems, also serve to reinforce the rules and logics that manage human oppression, constraining human autonomy and steering human behavior in perhaps even more problematic ways.

The traditional definition of a robot requires it to be able to act, think, and, most fundamentally, sense. It is nearly impossible to take effective actions in the world without first perceiving it, and for socially interactive robots, it is very difficult to engage in effective interactions with users without first perceiving those users. At their core, robots can thus be conceptualized as surveillance technologies that constantly perceive the world and people around them. This immediately makes robots key tools of disciplinary power. After all, Collins's view of disciplinary power builds on Foucault's own notion of disciplinary power, which emerged from his study of surveillance and punishment. Moreover, robots may present greater surveillance risks than traditional security cameras. Not only are robots capable of physically moving around the world, but also people are often incentivized to place themselves under robot surveillance in order to best benefit from the robot's services. In the remainder of this chapter, we will thus "flip the camera around": rather than considering how people perceive robots, as we did in chapter 3, we will consider how robots perceive people.

In the first part of this chapter, we will see how roboticists wield disciplinary power by enabling robots to see (or not see) different types of people, leading robots to provide service to users differently on the basis of their race and gender, thus turning these robots into the technological equivalent of Whites-only water fountains. In the second part of this chapter, we will then see how today's social robots are sometimes designed to serve as tools of racialized and gendered surveillance through the ways they sort people into roboticist-selected categories of race and gender through methods of social identity trait perception. From Foucault's perspective, one of the key elements of disciplinary power is the way that "people are classified and located on a knowledge grid."[2] Building on this theorization, we will see several ways that roboticists wield disciplinary power by designing robots that algorithmically racialize and gender individuals: by enforcing—and reinforcing—racist and patriarchal logics of categorization and by enabling differential detection and sanctioning of behaviors on the basis of those categorizations.[3] Overall, then, the analysis in the first two sections of this chapter will demonstrate additional ways that today's robots serve as New Jim Cobots that reinscribe the logics of Jim Crow and serve as enforcers of the White patriarchal logics of Benjamin's New Jim Code. Finally, in the last part of this chapter, we will explore how roboticists can mitigate these adverse effects by revising the categories they encode into robots, by rethinking how those categories are selected, or by more fundamentally ripping up the status quo by approaching robot design in new ways.

I How Robots Perceive People

For a robot to genuinely interact with someone, they must be aware that there is someone present with whom to interact. How this is achieved typically depends on the context into which robots are deployed. Delivery robots, for example, typically only need to navigate the world and to hand over whatever they are transporting. To navigate the world, these robots might not even need cameras, relying instead on laser range finders or other sensors that can tell them how far away, and in which direction, obstacles (including pedestrians) are located. Handing over deliveries also does not necessarily require seeing the person taking the delivery, as access to the delivery can be securely gated behind touchscreen-based authentication. But for social robots, more precise knowledge of interactants is typically needed. Social robots may need microphones to hear what is being said and to tell who is being interacted with, where they are, where they're looking, and their affective (emotion-laden) state. Unfortunately, the sensors and algorithms used to fulfill these social capabilities do not work equally well—or at all—for everyone.

Who Is Recognized by Speech Recognition?

First, let us consider speech recognition. For a robot to answer your questions, fulfill your commands, and so forth, it needs to be able to understand what you say to it. And for social robots, speech recognition needs to perform really well to be usable at all. Although smart speakers are most frequently used to perform Internet searches, play music, and turn lights on and off,[4] voice commands to language-capable robots could have much more significant real-world impacts. A speech recognition error on an eldercare robot may result in that robot retrieving the wrong medicine. A speech recognition error on a search-and-rescue robot may result in that robot searching the wrong area, thus missing people in need of assistance. And a speech recognition error on a voice-controlled wheelchair or smart cane may result in that robot bringing its user to the wrong location.

Speech recognition errors can also be particularly difficult to diagnose in human–robot interactions. A voice assistant on your phone may show you text as it is being recognized, and a smart speaker might repeat requests back to you before executing them. However, a robot asked to go somewhere may simply drive off, at which point the robot can no longer be corrected, and asking robots to repeat back everything they hear before accepting commands would make them more or less unusable due to the complexity of the commands robots are often asked to execute. If a robot responded to "Robot, do you know where the kitchen with the large table on the third floor is—the one across from the cafeteria? I left my coffee there. Could you go and get it for me?" with "Yes, I know where the kitchen with the large table on the third floor is—the one across from the cafeteria. You left your coffee there, and I will now go and get it for you" rather than "Yes," this would be deeply aggravating (and the coffee might get cold).

Finally, these errors are more likely to be uncorrectable on language-capable robots. If a smart speaker makes a search error, you can simply try again—and if voice really isn't working, most users can simply type their commands into their phones. If a robot makes an error, by the time the error is noticed, it may be too late to be corrected. Moreover, if voice interaction with a robot isn't working for a user, there may simply be no other way to interact with that robot. An elderly user alone in an assisted-living facility at night may be in a really dangerous situation if access to the bathroom, medication, or medical support is gated behind a robot and the robot is not able to correctly understand what is said to it.

Unfortunately, today's speech recognition systems do not understand everyone equally well. Speech recognition works particularly poorly for young children[5] and for older adults who have survived strokes or oral cancers.[6] Moreover, speech recognition systems work poorly for people whose dialects are not the norm for their language[7] or for those who are not native speakers.[8] One study showed, for example, that Amazon's Alexa only correctly recognized about 40 percent of common medication names spoken by individuals with foreign accents—an error that could have deadly consequences if made by a robot in an eldercare facility.[9] Speech recognition systems also work particularly poorly for Black users relative to White users due to differences in, for example, timbre and inflection, with ASR systems 50 percent more likely to misunderstand Black users' speech than White users' speech.[10] Apple's speech recognition service, for example, was found to have only about a 55 percent chance of correctly recognizing each word spoken by Black users.[11] Furthermore, these disparities are compounded by differences in dialect. Speech recognition systems are not typically trained to recognize African American Vernacular English and thus have difficulty recognizing its unique acoustics and morphosyntax,[12] These biases in speech recognition lead African American users to feel marginalized, or othered, by voice assistants and to feel (not incorrectly) that voice assistants are not made for them.[13] These effects may well be amplified when speech recognition errors are made by White humanoid robots.

To avoid speech recognition errors, non-White users consciously Whiten their speech[14] through a sort of technology-directed code switching, which can be highly fatiguing over time.[15] Beyond these immediate psychological costs, though, forcing Black users to codeswitch or otherwise Whiten their speech patterns when interacting with robots represents a subtle enforcement of White power. This requirement casts Whiteness, White speech, and

White speakers as the norm against which other dialects, cultures, and physiologies are cast as deviations.

Moreover, if users of color cannot effectively use robots due to inequitable speech recognition algorithms, they may be unable to access the services robots provide. If language-capable robots are deployed in eldercare facilities, residents of color may be unable to request services from them. If language-capable robots are deployed in schools, students of color may not be able to be tutored by them or may receive poorer education and lower educational assessments if they are expected or required to participate in such tutoring. If language-capable robots are deployed in industrial or manufacturing environments, workers of color may not be able to take certain jobs or may work more slowly and receive worse performance reviews and lower compensation if they are required to use those robots to work productively.

The consequences of this are clear. If language-capable robots are deployed with speech recognition systems that do not work just as well for users of color as for White users, those robots will be the technological equivalent of Whites-only water fountains. That is, although robots might not be advertised as Whites-only technologies, if they only work consistently for White people, they *are* essentially a Whites-only technology. As such, we can expect that these New Jim Cobots (as defined in chapter 4) will ultimately enhance the lives of White people while harming people of color and will do so in an ostensibly race-neutral way whose prejudices are unseen, unadvertised, and difficult or impossible to challenge, circumvent, or ameliorate. And unfortunately, this state of affairs extends to robots' other senses as well.

Who Is Recognized by Face Recognition?

Just as robots typically need speech recognition to hear people, so too do they typically need face detection and recognition to tell whom they are looking at and talking to. Unfortunately, just like speech recognition, today's algorithms for face detection, face recognition, and affect recognition are largely Whites-only technologies. Commercial face recognition systems perform up to 10 percent worse in identifying dark-skinned individuals compared to light-skinned individuals and up to 25 percent worse in identifying feminine-presenting individuals compared to masculine-presenting individuals.[16] These systems perform particularly poorly in terms of false positives—that is, the rate at which they falsely claim a match. A study by the National

Institute of Standards and Technology found that most commercial face recognition software systems were between ten time and one hundred times more likely to make false identifications for Asian and African Americans than for White people.[17] This is especially concerning, since false positives (rather than false negatives) are what cause many of the most dangerous consequences of face recognition, as we will discuss in chapter 6. These challenges also apply to face detection more generally[18] as well as to affect recognition.[19] As of 2020, Amazon's affect recognition system, for example, was more likely to mistakenly classify Black people as angry than White people.[20] It is easy to anticipate the consequences this might have when deployed in educational or policing contexts, and how this might exacerbate both the school-to-prison pipeline and the effects of policing more broadly.

The performance disparities of face recognition, face detection, and affect recognition algorithms are also distinctly intersectional. As Joy Buolamwini and Timnit Gebru showed in their 2018 paper "Gender Shades," looking only at performance differences between racialized groups obscured the source and extent of the algorithms' racial disparities.[21] Buolamwini and Gebru examined commercial algorithms purporting to recognize gender from faces and compared their performance across dark-skinned and light-skinned faces, across masculine-presenting and feminine-presenting faces, and, critically, across combinations of skin color and gender presentation. This intersectional analysis allowed Buolamwini and Gebru to show that even if we accepted the (false) premise that gender can be reliably detected from images, gender recognition algorithms would perform poorly, not just for dark-skinned people or for feminine-presenting people but also, more specifically, for dark-skinned feminine-presenting people. All three algorithms evaluated by Buolamwini and Gebru had perfect or nearly-perfect performance for light-skinned masculine-presenting faces (with accuracy rates between 99.2 and 100 percent) but very poor performance for dark-skinned feminine-presenting faces (with accuracy rates between 65.3 and 79.2 percent). For IBM's face recognition system, for example, this meant a roughly 35 percent difference in performance that would render the system completely unreliable for Black women, despite an "overall" accuracy rate of about 88 percent.

For Buolamwini, the inability to be seen by computer vision systems was grounded in personal experiences—specifically, her experiences with interactive robots. Before coming to MIT for her doctoral studies, Buolamwini attended Georgia Tech, where she worked in a human–robot interaction laboratory. While completing her social robotics coursework at Georgia Tech, Buolamwini directly experienced what happens when today's computer vision technologies are embedded into interactive robotics. While programming her robot, she found that it simply could not see her. As Buolamwini describes, "I had to borrow my White roommate's face to finish the assignment."[22] As she moved forward in her career, Buolamwini encountered this again and again in demos given by companies and in projects conducted for her doctoral studies at MIT, where she needed to literally wear a white mask to complete her work.[23] Again and again, computer vision systems could not reliably see, identify, and analyze her as they could her White counterparts. And again and again, Buolamwini needed to Whiten her appearance (by using her roommate's face or by wearing a mask) to be seen at all. The face recognition software she was forced to use constituted a Whites-only technology.

Although Buolamwini's encounters with the inherent biases of face recognition technologies momentarily impeded her studies, the consequences for others have been substantially more dire. During September 2020, for example, the New York State bar exam was held online due to the COVID-19 pandemic. To prevent cheating, face recognition was used to detect whether test takers stepped away from their computers and whether someone else was taking the exam on the examinee's behalf. Unfortunately, the software's reliance on face detection meant that it had significant problems detecting and recognizing people of color. Just as Buolamwini wore a white mask to complete her coding assignments, prospective lawyers of color reported needing to shine a bright light directly into their face for the two-day exam, brightening them enough to be detected but producing headaches that interfered with exam performance. As such, the use of face detection may have led to the penalization or failing of test takers (thus disqualifying and turning away these prospective lawyers) on the basis of their skin color.[24]

At nearly the same time that face recognition software was preventing students of color from becoming lawyers, it was also being used to falsely imprison people of color. For example, in June 2020, Robert Williams, a Black man living in a wealthy Detroit suburb, was mistakenly arrested by the police after his face turned up as a match to surveillance footage of a watch-store robbery. After being held for thirty hours in police custody, Williams was released on bail, and eventually the charges were dropped. Although Williams's false arrest was disturbing, he was ultimately lucky: he was wealthy, had access to a team of lawyers, could make bail, and was falsely arrested for a nonviolent crime. If any of those factors had been different, Williams might still be in prison or might have been killed by the police. As Williams himself asked, "What if it [was] rape or murder? Would I have gotten out of jail on a personal bond, or would I have ever come home?"[25] Indeed, we are left to wonder how many other Black Americans may have been rounded up by the police on the basis of bogus face recognition matches, and how many may still be sitting in prisons today.

The misuse of face recognition technology by the police raises a key design dilemma for computer vision and robotics researchers. On the surface, the challenges faced by Buolamwini are one of invisibility and suggest a straightforward solution: if we improve the performance of face detection, recognition, and analysis algorithms for women of color, this (in)visibility gap will disappear. Motivated by this possibility, new subfields have sprung up within computer science, seeking to study and improve the "fairness" of algorithms such as those used in face detection, recognition, and analysis.[26] Grappling with fairness concerns is critical across many areas of machine learning, where discriminatory algorithms (which data scientist Cathy O'Neil colorfully terms "weapons of math destruction") are increasingly being used in domains ranging from automated lending and credit scoring to college admissions and job screening.[27]

Perhaps the most widely cited example of a discriminatory algorithm is COMPAS, a recidivism prediction system intended to predict the likelihood of pretrial defendants re-offending for the purposes of determining bail and parole.[28] The designers of COMPAS argued that COMPAS was a fair algorithm because it had the same accuracy in predicting recidivism among Black and White defendants. But as ProPublica revealed in an explosive analysis, the truth was substantially more nuanced. Although COMPAS did indeed have the same error rate for Black and White defendants, it made substantially different *types* of errors for those two user groups. COMPAS was nearly twice as likely to erroneously predict *for* recidivism among Black defendants than White defendants and nearly twice as likely to erroneously predict *against* recidivism among White defendants than Black defendants. In short, it made errors *against* Black defendants but *in favor* of White defendants.[29]

This analysis shows the ways that algorithmic systems can be systematically discriminatory, but it also shows the challenges in grounding ethical analysis of algorithmic systems in questions of fairness. Meta-analyses of the literature on computational fairness have revealed twenty-three different types of algorithmic bias, six different types of algorithmic discrimination, and ten different types of algorithmic fairness.[30] On the one hand, this extensive space of fairness concerns means that there are numerous productive ways to critique algorithms on the basis of bias, discrimination, and fairness. As a result, computer scientists have developed a host of new methods for creating fairer algorithmic systems, including fairer machine learning classifiers for specific tasks[31] and methods for learning statistical representations that yield fairer results when used across many different types of tasks.[32] But on the other hand, it means that even when an algorithm clearly violates one definition of fairness, the company responsible for that algorithm can typically find a different definition of fairness that they satisfy, providing cover even in egregious cases of discrimination. Moreover, when computer scientists narrowly focus on fairness, they adopt a technosolutionist perspective in which solving algorithmic discrimination is achieved only through fairer algorithmic decisions rather than more fundamental questioning of the use of algorithmic decision-making in the first place.[33] This is critical, as algorithms are increasingly being used in inherently unjust domains. In chapter 6, for example, we will carefully consider the use of robotics in policing, where machine learning models are systematically used in service of predictive policing efforts. Although predictive policing is certainly discriminatory and unfair, making such systems fair would essentially mean rendering everyone equally susceptible to police violence—an outcome that would normalize and reinforce the use of predictive policing (and policing more generally) rather than more fundamentally questioning why predictive policing (and policing more generally) are being used as solutions to societal problems.

Moreover, the approaches that some Big Tech companies have used to attempt to de-bias their face recognition datasets have themselves involved colonial and exploitative power dynamics similar to those seen in chapter 4. For example, the company Google hired to correct the bias in their face recognition systems directed contractors to seek out people of color in public places and collect photographs and videos of those people while concealing the fact that their faces were being recorded.[34] Contractors were told to say things like "Just play with the phone for a couple minutes and you'll get a gift card" or "We have a new app, try it and get $5" without revealing what was happening behind the scenes or why. As one contractor recounted,

"We were told not to tell [people] that it was video and deny recording if asked."[35] Contractors were further instructed to rush subjects through survey consent agreements, distract users as much as possible through sensory overload before they could realize what was going on, and walk away if people started to get suspicious. As Anna Lauren Hoffman argues, this story serves as one example of how a shallow emphasis on "inclusion" can lead to adoption of practices that exacerbate these forms of "data violence."[36] Moreover, as Abeba Birhane points out, a myopic emphasis on de-biasing algorithms and the data they are trained on fallaciously assumes a single "neutral" and "unbiased" truth obtainable from a "view from nowhere."[37] In fact, as Dan McQuillan points out, focusing too much on the inaccuracy of AI models may itself be unhelpful, as it problematically assumes a single rigid underlying truth that must be accurately perpetuated—an inherently colonialist perspective. Instead of trying to merely improve accuracy and remove bias, McQuillan suggests that we need to understand and critique AI in terms of the practices, doings, and actions that it enables.[38] As Matthew Le Bui and Safiya Noble write, "The goal of many ethics responses remains largely techno-deterministic, in that the goal is to perfect or 'unbias' the technology, rather than account for the asymmetrical power relationships and gravity of history that renders the development and deployment of such projects deeply uneven, unethical, and even immoral."[39] In the context of face recognition systems, these limitations manifest in acute ways. Although fairness concerns surrounding face recognition systems typically center the ways that those systems, when performing unfairly, render women and people of color invisible, it is also important to consider the ways that those same systems render women and people of color hypervisible in ways that may be exacerbated by de-biasing efforts. There has, for example, been widespread documentation of police use of face recognition systems to identify participants in protests against racial injustice[40] (an activity that is unsurprising, given the history of policing, as we will see in chapter 6). During the 2020 Black Lives Matter protests, law enforcement agencies across the country mobilized to conduct mass surveillance of protesters, echoing the FBI's efforts to surveil, infiltrate, and undermine protests during the civil rights era.[41] The Department of Homeland Security leveraged military-grade technology to track protesters in at least fifteen cities, and police departments used these technologies to conduct mass video surveillance, even where this was formally banned.[42] In Los Angeles, for example, the police long denied their use of face recognition[43] while, in fact, using it nearly thirty thousand documented times[44] between 2009 and 2020,[45] including as part of their surveillance of Black Lives Matter protests.[46] Similarly, other departments (such as the New York Police Department [NYPD]) have used face recognition matches as excuses to descend without warrant on protesters' homes with attack dogs and helicopters.[47]

These examples show how face recognition systems make Black Americans hypervisible, enhancing the frequency and facility with which they are surveilled and putting their lives at greater risk as casualties of police violence. As such, although improving the accuracy of face recognition systems might decrease Black invisibility and decrease rates of false arrest, these algorithmic improvements could simultaneously exacerbate Black hypervisibility and increase the rates of racially targeted and politically motivated arrests. These examples thus show the catch-22 facing roboticists seeking to ensure that their robots do not function as New Jim Cobots. On the one hand, it is imperative that roboticists improve the fairness of their systems in order to make sure that their robots are not acting as Whites-only technologies that provide different levels of service to users on the basis of their race (or gender). But on the other hand, it is equally imperative that these efforts to create fair technologies do not merely lead to more generally effective tools of surveillance, violence, and persecution. In chapter 6, we will come back to the way robots and other surveillance technologies are being used by the police. In the remainder of this chapter, however, we will instead see how both Black invisibility and Black hypervisibility are exacerbated through the ways people are classified by computer vision systems once they are detected and recognized.

II How and Why Roboticists Algorithmically Categorize People

To see how today's robots are classifying people once they are detected, let us consider as an example the Pepper robot (and its younger brother Nao, the most widely used robot in human–robot interaction research). Both robots come preloaded with a "face characteristics" module that assigns a gender to each person the robot sees, labeling them as "1" if perceived as "male" and "0" if perceived as "female."[48] This type of social identity trait perception has become increasingly common in robotics, with roboticists programming their robots to classify people in terms of not only gender but also traits such as age and race. These approaches are enabled by a long history of work in the AI community seeking to classify observees on the basis of race, ethnicity, religion, or gender. Most often, these algorithms for racializing and gendering observees leverage machine learning methods, which typically involve three phases. In the first phase, researchers gather pictures of individuals. For example, to create the 10k US Adult Faces Database, researchers began by gathering, as you might guess, ten thousand images of US adult faces.[49] In the second phase, researchers label those faces, typically with the assistance of crowd workers paid pennies per labeled face. For example, to annotate the 10k US Adult Faces Database, crowd workers on Amazon's Mechanical Turk (the platform discussed in chapter 2) were asked to look at each face in the dataset and guess the demographics of the person shown.[50] Finally, in the third phase, a machine learning model is trained on the labeled images, creating a statistical model for predicting, from the pixels in an image, the most likely label a worker would have assigned to that image had they been shown it. A robot using this resulting model to digitally racialize interactants would thus be using it to answer the question "How would the original group of workers recruited from the Internet, at the time that this model was trained, likely have racialized this person?"

In robotics, attempts to categorize observees often use additional sensors beyond cameras, including four-channel RGB-D (red, green, blue, depth) sensors, such as Microsoft Kinect,[51] which annotate each pixel with the distance between the sensor and whatever object in the world was responsible for that pixel's color value. These RGB-D sensors have allowed roboticists to better guess interactant gender on the basis of full-body 3D scan analysis[52] or on the basis of "breast shape estimation,"[53] in which a robot guesses the observee's gender by covertly looking at their breasts. No doubt, many women would be uncomfortable with a robot using this algorithm if it were made clear that this is what the robot was doing. The use of these approaches should encourage us to stop and take a step back. Why are roboticists designing their robots to stare at women's breasts? What is so important to roboticists about recognizing an interactant's gender, let alone race?

Often, roboticists seek to categorize interactants according to race and gender in the hope of better tailoring robot behavior to individual interactants' preferences. This hope is, in turn, driven by assumptions that people who are algorithmically racialized or gendered in the same way must have the same preferences for interaction[54]; that algorithmic stereotyping will make robots look more intelligent[55] or more user-friendly[56]; or simply that such robots will be more capable of serving users targeted advertisements.[57] Surveillance capitalist impulses aside, though, some attempts to enable robots to play "guess the race" or "guess the gender" are genuinely well meaning. Some researchers have hoped that a robot able to recognize an individual's gender, race, or culture would be able to interact with users in a more culturally sensitive manner rather than simply adhering to and reinforcing White Western heterosexual male behavioral standards. Others have sought to enable robots to recognize the gender of interactants in order to use their correct pronouns and avoid misgendering them.[58] Finally, some researchers have argued that recognizing an interactant's race or gender would be necessary to appropriately identify and respond to microaggressions—that is, "brief and commonplace daily verbal, behavioral, and environmental indignities, whether intentional or unintentional, that communicate hostile, derogatory, or negative racial slights and insults to the target person or group."[59] Classic examples of microaggressions include commenting to people racialized as Asian that they must be good at math, saying that one doesn't see color, or asking people racialized as non-White where they were born, all of which reflect the racial styles and frames of today's color-blind racism.[60] These utterances can be difficult for victims to respond to, as they are often framed positively, and it is often not obvious that the speaker was aware of the problematic nature of what they said. As such, violators can easily dismiss complaints if microaggressions are called out, and reverse blame onto the violator by claiming an overreaction.[61] Some researchers have thus proposed that robots could try to guess interactants' race or gender in order to better detect and call out such microaggressions. Doing so, however, would ultimately require roboticists to make some deeply problematic assumptions. In fact, as we will see in the next section, roboticists make a host of problematic assumptions whenever they (1) choose identity categories to sort people into, (2) assume those categories to be spatiotemporally static, and (3) enable their robots to sort people into those categories.

The Problems with Adopting Categorization Schemes

Few roboticists are deciding for themselves how their robots will categorize people. These decisions are instead largely made in the field of computer vision, where researchers are developing the actual models and algorithms for performing these categorizations. As Morgan Klauss Scheuerman shows, the categories used in the computer vision community are determined by the wildly problematic labels assigned to faces in computer vision datasets.[62] The most egregious example identified by Scheuerman may be Microsoft's MS-CELEB-1M dataset, in which Microsoft chose to label faces as "Caucasian," "Mongoloid," or "Negroid" on the basis that these categories encompassed "all the major races in the world."[63] The terms "Negroid," "Mongoloid," and even "Caucasian" smack of nineteenth-century scientific racism, which attempted to reframe America's slavery-motivated caste system as grounded in a biological reality of ostensibly distinct "races." Although this revisionist view of race is obviously at odds with what we know about the historical origins of race, it is also at odds with the scientific consensus on human genetics. Advances in genetics have shown that racial categories do not meaningfully distinguish people genetically, cannot be reliably measured, and, in general, are not scientifically meaningful.[64] In fact, from an anthropological perspective, even ethnicity is not inherently biological. As Audrey and Brian Smedley have argued, ethnicity can be best understood as a categorization scheme that assigns people to different cultural clusters based on shared "language, geographic locale or place of origin, religion, sense of history, traditions, values, beliefs, [and] food habits."[65] The mapping of an individual to an ethnicity depends on their identification with (and acceptance into) a particular cultural (ethnic) group. Similarly, the mapping of an ethnic group onto a racial category within a society is defined according to the sociopolitics of that society and the ethnic projects conducted by the society's different ethnic groups. As such, both the ethnicity of an individual and the way that individual is racialized are grounded in a mixture of unobservable cultural, social, and political factors rather than (potentially) observable biological factors.

From this perspective, Microsoft's choice of racial categories is obviously problematic in the way it buys into a discredited and inherently White supremacist view of race. In choosing these categories, Microsoft made sociological and political claims, implicitly asserting that these categories can be used to describe, sort, and categorize all people, that every person can be uniquely assigned to one of these categories, and that people *should* be computationally sorted into these categories. In other words, by using this set of categories, Microsoft advanced a particular racial architecture as appropriate and necessary for structuring society. And while I have been picking on Microsoft as a case study, choosing *any* set of racial categories buys into the same set of inherently racist logics, and represents a sociopolitical choice to concretize a particular categorization scheme and reinforce its legitimacy.

Scheuerman has also analyzed the labels used in computer vision datasets to classify people in terms of gender.[66] As he shows, some computer vision datasets use highly problematic gender labeling schemes. The most egregious example Scheuerman identifies may be Columbia's PUBFIG dataset, which seems to classify individuals into either "male" or "attractive woman"—a perfect storm of sex/gender confusion and misogyny. But even datasets that merely assume a binary categorization scheme of "man" versus "woman" suffer from a number of serious problems.

The first problem with technologies built around the assumption of a gender binary is the number of questionable sociological, political, and scientific claims this entails: that gender *is* a binary, that people can be uniquely assigned to one of the two categories prescribed by that binary and only those two categories, and that people *should* be sorted into those categories by algorithmic systems. These schemes thus clearly erase non-binary people (and intersex people, if these labels are intended to capture sex) and make it likely for trans and non-binary people to be misgendered[67]—a phenomenon that is degrading and traumatizing[68] and which encourages multiple intersecting types of violence against trans people.[69] These risks can be seen in the case of airport scanners—another technology of gender violence. When a traveler approaches an airport scanner, TSA agents must guess the traveler's sex so that the scanner can decide whether or not the traveler's body is "anomalous." As Sasha Costanza-Chock describes, this creates unavoidably dangerous situations for trans travelers:

> If the agent selects Male, my breasts are large enough, statistically speaking, in comparison to the normative male body-shape construct in the database, to trigger an anomaly warning and a highlight around my chest area. If they select Female, my groin area deviates enough from the statistical female norm to trigger the risk alert. In other words, I can't win. This sociotechnical system is sure to mark me as "risky," and that will trigger an escalation to the next level in the TSA security protocol.[70]

Black women, Muslim women, and disabled travelers are doubly likely to be targeted for violence during security screenings such as these due to perceived threats and anomalies in their styles of hairdressing and grooming and, in the case of disabled travelers, their prostheses and medical devices.[71] And critically, this flagging of trans, Black, Muslim, and/or disabled women is particularly serious, as it results not just in hurt feelings and privacy violations but, moreover, in bureaucratic and sexual violence toward those women.[72] Unsurprisingly, this has led to overwhelmingly negative perceptions of automatic gender recognition systems among trans individuals.[73] When robots are used in security applications, automatic gendering of individuals according to a gender binary may lead to similar opportunities for bureaucratic and sexual violence if those robots use their classifications to detect and call attention to "anomalies."

The second problem with technologies built around the assumption of a gender binary (or a set of static racial categories) is the way that this takes a problematically essentialist stance, asserting that everyone has an innate gender or race. That is, by trying to categorize individuals into their gender or their race, computer scientists inevitably erase the line between how someone is likely to be gendered versus their gender identity and how someone is likely to be racialized versus how they racially identify. An individual may well identify as a woman or as non-binary yet be frequently (mis)gendered as a man. Similarly, new African immigrants are frequently racialized as Black, even when they do not think of themselves as such. It is entirely unclear from computer vision dataset labeling schemes whether computer vision researchers are interested in self-identity or in identification by others—and whether this is a distinction they have even given thought to. As we will see in the next section, these differences between self-identity and categorization by others are particularly critical for race, given the spatiotemporal dynamics of race.

The Spatiotemporal Dynamics of Race As I described earlier in this chapter, robots that classify people in terms of race must do so according to some particular set of racial categories. Yet, we know that racial categories change over time and change from place to place. As we will see, encoding a racial ontology into robot design thus has the effect of promoting a colonialist racial project as robots are sold and distributed throughout the world.

To see why, let us first consider the temporal dynamics of race. Although an unscientific biological account of race seems to suggest a distinct and stable set of "races," sociologists have long known that the racial architectures and racial politicultures of racialized social systems are in constant states of flux and readily change over relatively short periods of time. In the United States, for example, the definitions and composition of racial categories have continuously evolved over the past four centuries. As described in chapter 2, the earliest racial categories codified into law in the British colonies were simply "Black" and "White." Yet, to maintain the oppression of those labeled "Black" and the privileges of those labeled "White," these legal categories have needed to go through a series of expansions and contractions in the United States, with dramatic changes over the past 150 years alone. This can be seen, for example, in the changes in the sets of categories used in the US census over that time period[74]:

- 1890: White, Black, Mulatto, Quadroon, Octoroon, Chinese, Japanese, Indian
- 1900: White, Black, Chinese, Japanese, Indian
- 1930: White, Negro, Mexican, Indian, Chinese, Japanese, Filipino, Hindu, Korean, Other
- 1990: White, Black or Negro, Indian (Amer.), Eskimo Aleut Asian or Pacific Islander (API), Chinese, Filipino, Hawaiian, Korean, Vietnamese, Japanese, Asian Indian, Samoan, Guamanian, Other
- Today: White, Black or African American, American Indian or Alaska Native, Asian, Native Hawaiian or Other Pacific Islander

The evolution of these categories shows the interactions between several interrelated phenomena.[75] First, these categories most directly show the changing racial architecture of the United States. Over time, the set of categories used to make sense of people's race has fundamentally changed. Second, these categories show the political position of the United States government in terms of how it has elected to exert ontological control. Third, these categories show competing political positions, as United States politicians debated whether and how to grant entitlements and affirmative action benefits to specific racial or ethnic groups. And finally, these categories show a changing racial politiculture, as evidenced by politicians' confusion between tracking of race versus tracking of ethnicity.

Although these changing categories are an overt symbol of race's temporal evolution (due to the nature of their legal codification), the categories chosen by politicians do not necessarily reflect the actual categories everyday people use to make sense of race. The categorization systems used by governments, those used by sociologists, and those used by everyday people in their racial common sense are distinct yet interrelated and mutually constitutive. Racial common sense informs (but does not completely determine) sociological and legal categories, and the way that legal categories are defined shapes racial common sense and can lead to economic and social forces that ultimately shape sociological categories. If we consider sociological rather than legal categories, some sociologists have suggested that the US racial system is still essentially binary (two cluster White/non-White[76]) or may use a similarly reduced set of categories. Bonilla-Silva, for example, argued (in 2004) that due to demographic changes, the US was transforming into a pigmentocratic three-category system[77] based on socioeconomic status, subjective self-identification, and politiculture-driven relational patterns between ethnic groups such as racial segregation and racial intermarriage.[78]

Yet, even if the US racial system has remained essentially binary (or ternary), there have been constant changes in how US ethnic groups are assigned to those racial categories. Indeed, there is a long history of research in the social sciences on how certain groups transitioned from "non-White" to "White," such as Ignatiev's famous history of "how the Irish became white."[79] Moreover, Bashi Treitler expands on this[80] by systematically chronicling the successful ethnic projects undertaken by many ethnic groups (such as the Irish, Chinese, Italians, and Jews), who, after being initially treated similarly or equivalently to Black people within the US racial order, underwent a process of Whitening in which they were able, through concerted effort (typically at the expense of other ethnic groups, especially African Americans), to increase their racial status and change the collective common sense of which racial group they belonged to. These racial and ethnic dynamics create obvious challenges for those seeking to enable robots to racially categorize those they observe. An individual classified today as "Latino" according to their society's racial commonsense may be classified as "White" at some future point in time, or may be classified into some other category that does not exist at all under today's racial architecture. Yet, computer vision–based categorization algorithms are fundamentally unable to capture such dynamics. Moreover, this situation becomes even worse once we consider the spatial dynamics of race.

So far, I have primarily limited myself to talking about the racialized social system present in the United States, not only because I am writing from an American perspective but also because of race's original purpose as a means for subjugating the Africans brought to forced labor camps in the Americas, and because of the ways that America's conception of race has metastasized, spreading far beyond US borders. Yet, even given the influence of American imperialism, racialized social systems differ dramatically from place to place. Even within heavily Anglocentric cultures such as the US, Australia, and South Africa, there are significant differences in systems of racial categorization and mappings of ethnic groups to racial groups. For example, South Africa's racial paradigm has been characterized as "Black, White, Colored, Asian," and Australia's has been categorized as "White, Black, Asian, Indian, 'Ethnic looking.'"[81] These differences are due, in part, to the unique ethnic composition of different areas, the ethnic projects conducted within those societies, and the physical appearance of individuals relative to the dominant groups participating in those ethnic projects. Starker differences exist between the US and countries with less Anglocentric cultures. Brazil, for example, has been argued to have dozens to hundreds of distinct racial groups.[82]

Because different racialized societies are structured according to different racial paradigms with different racial architectures and different racial common sense for assigning people to categories within those architectures, the way that one racially identifies—or is racialized by others—may change as one travels between societies. This can be clearly seen in the experience of African immigrants to countries with entrenched racial systems,[83] among whom there is a well-documented perception of *becoming Black*. Amanuel Isak Tewolde, for example, quotes the experiences of an African immigrant to Israel, which shows how these spatial dynamics are shaped by the racializing gaze of those operating within and enforcing the racial politiculture into which immigrants are thrust:

> Before I arrived in Israel I never thought I would be looked down upon as unwanted because of my African origin. I never saw myself as Black or any other race. Here in Israel, they call us Blacks because we came from Africa. The Israeli government also is hostile towards us because we are Black. They don't do that to White immigrants. So, during the protests, we were accusing the government of racism towards African refugees. I mean, they don't want us in this country because we are Black. I am not saying that I have always seen myself as Black but In Israel I am Black because they see us as such.[84]

Critically, these spatial dynamics mean not only that robots designed to racialize interactants according to a particular racial architecture will fail to work correctly when deployed in other countries but also that such robots may serve as vehicles for the colonialist spread of White supremacy.

Consider, for example, a robot developed in the US, whose developers have chosen to enable the robot to racialize users according to a particular set of racial categories that they viewed as appropriate within a US context. If that robot were sold in another country and continued to classify people according to those US logics, this would propagate and globalize American racial architecture and the common sense of how people are assigned to the categories within that architecture. This could rightfully be seen as part of the larger colonialist and White supremacist projects enacted by the US and as part of what Michelle Christian calls the "transnational assemblage of racist logics."[85]

The Problematic History of Biometric Surveillance Technologies Although committing to a particular ontology of race or gender requires commitment to problematic ideologies, the effects of those commitments are felt when robots actually perceive individuals, categorize them, store those categorizations, and leverage those categorizations to guide their behaviors. To understand the implications of robots' perceiving, categorizing, storing, and acting on presumed identity characteristics of observees through social identity trait perception, we can consider the effects of similar biometric surveillance technologies.

In her book *Dark Matters*, Black studies scholar Simone Browne demonstrates the ways that many modern computing technologies fall into a history of racialized and gendered technologies and the ways these technologies lead to digital epidermalization.[86] Here, Browne builds on the theory of epidermalization,[87] which political philosopher Frantz Fanon formed through recounting his experience of feeling himself become a body as White observers reduced him in their perceptions to his skin and as he internalized those perceptions to think of himself in that same way. Epidermalization captures psychologies of visibility as the Black body's hypervisibility to the White gaze renders it hypervisible to the Black psyche and as the force of that gaze is internalized in an attempt to resolve what Browne terms the "ontological insecurity of the racial body." In her own theory of *digital* epidermalization, Browne considers how a range of surveillance technologies operate according to and facilitate these logics. For example, physical brands reinforced the logics of slavery by indelibly altering slaves' skin and rendering their slavery visible. Jeremy Bentham's *Panopticon* reinforced the hypervisibility of prisoners, the invisibility of prison guards, and the steering of prisoner behavior through the mutual recognition and internalization of those visibility dynamics. The 1783 *Book of Negroes* helped the British to delineate free Black people from slaves by reducing them to the visible characteristics of their bodies. And modern government-issued ID cards fix individuals' identities according to uncontestable state categories of race, gender, and nationality and reduce the provability of one's self-identity to whether observations of one's body align with the categorical claims of the state.[88] When viewed through this historical lens, we can see how modern surveillance technologies, including databases that assert individuals' assignment to predefined categories of race and gender, and algorithms trained on such databases, continue a long lineage of tools for enforcing disciplinary power by ascribing race and gender to bodies without consent, in a way that precludes protest and challenge.

Moreover, as Morgan Klauss Scheuerman observes, digital racialization and gendering reinforces the underlying logics of White patriarchy[89] through "auto-essentialization." Categorizing individuals according to predefined discrete racial and gender categories on the basis of physical observations reinforces false notions of essential biological differences between racialized and gendered groups and the unscientific inference of identity categories from visually discernible phenotypic markers such as skin color and sex characteristics. Algorithmic racialization and gendering of observees thus achieves not only an immediate effect of forcing a particular identity onto a body but, moreover, the deeper and more insidious effect of reinforcing the underlying logics of White patriarchy.

This reinforcement of White patriarchal logics stands to uniquely impact women and people of color due to the ways that ontologies of race and gender are constructed and wielded to oppress members of these groups. Moreover, the promotion of ontological insecurity by biometric surveillance technologies is uniquely felt by women and people of color and is exacerbated by the poor performance of biometric technologies for those groups. Just as fingerprint readers have higher error rates for Asian users (especially Asian women), elderly users, and working-class users,[90] video-based technologies, by intentional design, are optimized for light-skinned users and do not perform well for dark-skinned users, especially dark-skinned women.[91] As such, digital epidermalization specifically privileges Whiteness and White masculinity.

White men (who already give little thought to how they are gendered and racialized due to the way that society elevates White masculinity and establishes it as a default[92]) are (as discussed above) less likely to be mis-racialized and misgendered and are thus subject to less technology-induced ontological insecurity as to how they will be digitally categorized.

Robotics as a Biometric Surveillance Technology Robots have a clear position within this lineage of biometric surveillance technologies. First, many robots follow the surveillance logics of the panopticon. It is typically unclear to users whether anyone is watching behind a robot's camera, let alone what is being sensed by robots, how that data is being stored and used, who may access it, and why.[93] This lack of certainty as to whether anyone is, could, or will see what the robot sees (and if so, whom) can create a panopticon-like state where disciplinary power is wielded to shape the behaviors of those observed. Second, robots designed to racialize and gender those they surveil follow the racialized and gendered surveillance logics of the *Book of Negroes* and of more modern database technologies, with disciplinary power wielded, as observees internalize the way they are categorized, to (1) enact racialized and gendered surveillance that is typically nonconsensual and impervious to examination, explanation, audit, or challenge; (2) use observees' hypervisibility to steer their behaviors; and (3) reinforce and globalize the architecture and politiculture underlying robots' categorizations. Finally, robots may exacerbate this reinforcement of the logics of White patriarchy. In her work on digital epidermalization, Simone Browne writes, "I suggest that the disembodied gaze cast by surveillance technologies also does the sorting work of nationalizing, and by extension racializing."[94] But unlike the computer vision technologies analyzed by Browne, robots are not disembodied. Humans are likely to perceive a *racialized* gaze from behind the cameras of their robot interactants, and as seen in chapters 1 and 4, robots have unique persuasive power over human systems of moral norms.

The concerns seen thus far in this chapter should be particularly worrying when juxtaposed with the analyses from the previous chapters in this book. In chapter 1, I argued that robots have unique persuasive power and are perceived, to some extent, as socially and morally agentic community members, with the necessary status to participate in the negotiation of community norms and the ability to wield power within the interpersonal domain. In chapters 2 and 3, I argued that (sometimes despite roboticists' best efforts) humanoid and interactive robots are typically racialized and gendered according to the logics of the White patriarchal robot imaginary, giving roboticists substantial power within the cultural domain that is reinforced by—and reinforces—that potential for interpersonal persuasion. In chapter 4, I argued that today's interactive robots too often serve as New Jim Cobots that enforce White and patriarchal norms of behavior in the disciplinary domain. And in this chapter, I have shown how these issues are accentuated when robots are designed as Whites-only technologies and when robots engage in social identity trait perception to categorize interactants according to race and use that categorization (e.g., when describing interactants to others) in ways that reinforce the architecture and politiculture underlying those categorizations. When these different forms of power are juxtaposed, we see that there is significant risk for robots to entrench White patriarchal oppression on a global scale. Even more so than in previous chapters, this should demonstrate a clear responsibility for roboticists to wield this power in ways that subvert rather than reinforce White and patriarchal power structures.

III Subverting Robots' Default Reinforcement of White Patriarchy in the Disciplinary Domain

As we have discussed in this chapter, roboticists wield power in the disciplinary domain through their decisions as to how robots wield social identity trait perception to perceive and categorize people and the ways that this renders robots, by default, as New Jim Cobots that enforce the differential treatment of people under White patriarchy and/or directly reinforce the logics of White patriarchy. How can roboticists avoid these default effects? In this section, we will once again consider how we might address this problem if we (1) revised how robots visually perceive, recognize and categorize users; (2) rethought how robots perceive, recognize, and categorize people in the first place; or (3) ripped up the status quo by more directly addressing the sociohistorically determined root causes of the problems identified in this chapter.

Revising How We Employ Visual Perception and Categorization

First, let us consider how we might simply revise how robots are visually perceiving, recognizing, and categorizing individuals. If robots are serving not only as White patriarchal New Jim Cobots but, moreover, as Whites-only technologies because they can only effectively see and hear White men, then these problems could be addressed by improving intersectional speech recognition and face recognition performance. This is a promising path forward, especially given the broader proliferation of speech-based technologies beyond robotics. Yet, as discussed earlier in this chapter, roboticists seeking to improve the intersectional performance of these algorithms would need to remain cognizant of the ways that this might simultaneously exacerbate robots' use as surveillance technologies, especially when used by organizations such as law enforcement agencies. This is a risk we will further address in the next chapter.

Similarly, if robots are reinforcing problematic ontologies of race and gender because of the ontologies programmed into those robots and the ways those ontologies are used, then these problems could be addressed by changing the ways that robots store and reason about race and gender. Robots don't need to use overtly problematic ontologies of race and gender, and don't need to use their categorizations in explicit or observable reasoning processes. It is also possible that robots could use ontologies that are not temporally static but are instead updated over time. Scheuerman has highlighted the potential use of data maintenance plans, where technologists periodically revisit their data-labeling schemes.[95] As Hutchinson argues, such plans should be a requirement for the release of any research dataset: "If you can't afford to maintain a dataset, and you also can't afford the risks of not maintaining it, then you can't afford to create it."[96] Following this philosophy, Scheuerman has suggested that computer vision researchers should periodically revisit, revise, or retract computer vision datasets, having third-party annotators update race- and gender-related image labels (based on changes to racial architecture and politiculture) or taking down datasets entirely (as has already happened with the controversial MS-Celeb-1M dataset).[97]

Scheuerman's suggested approach suffers, however, from a number of shortcomings. First, it is not clear how realistically implementable this approach would be. Continuously reannotating datasets, retraining models, and redeploying those models to robots (or other platforms) would be an immense undertaking that few companies would willingly invest in. Moreover, it is unclear who would hold companies accountable (and how) for failing to implement and follow data maintenance plans. Second, it is unclear who would be responsible for deciding on the new standards against which datasets would be reannotated. Specifically, this decision is likely to fall to robot designers, thus providing ill-equipped members of the dominant racial castes with the ability to dictate the structure of their preferred racial architecture, which would likely be done in a way that self-servingly bolsters the racial hegemony of those designers. Third, addressing temporal dynamics in this way does nothing to address the even more nightmarish spatial dynamics of race. If robot designers were intent on enabling algorithmic racialization in a way that accounted for spatial dynamics, they would need to give their robots knowledge of distinct racial architectures for each possible society the robot could be brought to or could interact with. Even if this were seen as a good idea, it is entirely unclear how these encoded racial architectures would be used. Imagine, for example, an immigrant Alice from Australia to South Africa, who is visiting Brazil, who is observed by a robot Bob developed in the United States by a team of primarily Japanese researchers. If Bob assigns Alice to a racial category, should it be with respect to the racial architecture and politiculture of Australia, South Africa, Brazil, the United States, or Japan? And finally, as we have shown in this chapter, adopting *any* set of categories can be deeply problematic. This suggests a need to at least *rethink* the mechanisms of social identity trait perception that robots use to recognize and categorize interactants.

Rethinking Our Use of Visual Recognition and Categorization

What would it mean to rethink robots' use of social identity trait perception? If the problem with robots' visual recognition of interactants is grounded in the ways that this can be used for nonconsensual surveillance, then perhaps we should design robots' recognition of individuals to be consensual by design. Robots' recognition of an individual could be determined through a host of more reliable and consensual methods than purely visual surveillance. If the intersectional performance of visual person detection were improved, for example, then the recognition of detected people could be achieved through opt-in mechanisms such as wearable QR tags, RFID tags, or smartphone applications.

Alternatively, if the problem with robots' categorization of interactants is grounded in the ways that computer vision technologies assign spatiotemporally static identity categories to individuals on the basis of images and without their consent, then maybe we should rethink how robots acquire and update information about individuals' racial and gender identities. Scheuerman, for example, suggests that the "ground truth" gender labels on images used to train computer vision gender classifiers should really come from the people those images are *of* rather than guesses about them by other people on the Internet.[98] This is a reasonable suggestion if we are committed to using computer vision systems to try to guess people's race and gender. A more radical solution would be to omit computer vision altogether and determine the race and gender of the people robots are interacting with, if we really need that information, by simply asking them.[99] If a robot needs to know how someone identifies in terms of race or gender, in order to avoid mis-racializing or misgendering them, the robot can simply ask them for this information. Even outside of robotics, this approach has recently gained traction in areas such as image captioning. For blind users, image captioning is crucial, and automated image captioning provides a powerful tool for providing computationally generated descriptions of what is being depicted in images to sighted users. Yet, automated image captioning of pictures of people are prone to all of the issues described above, as image-captioning algorithms mindlessly sort people into categories of race and gender, needlessly misgendering and mis-racializing those depicted. To understand how to address this problem, accessibility scholar Cynthia Bennett (herself a blind researcher) talked to screen reader users from marginalized communities (including blind users of color and trans and non-binary blind users).[100] Based on her conversations with these users, Bennett argued that it could be acceptable for image-captioning systems to describe individuals according to their race and gender if those descriptions were made based on self-identifications provided willingly by users for this purpose.

On the other hand, asking users to self-identify in terms of race and gender comes with its own risks. First, because human–robot interactions are typically multiparty rather than dyadic (i.e., there are typically multiple humans involved simultaneously in the same human–robot interaction), the dialogues in which robots collect self-identifications could run the risk of outing individuals in contexts where their racial or gender identities are not immediately obvious to other humans in their environment, and in which knowledge of self-identity by those other parties could put the human being queried in danger. Second, requiring individuals to constantly reassert how they identify seems like an unreasonable amount of uncompensated labor and effort to put on the shoulders of robots' interactants, and runs the risk of serving as a vehicle for robotics companies to abdicate responsibility to users, especially trans and non-binary users and users of color. Third, specifically when considering race, the use of self-provided identity labels would fail to address the spatial dynamics of race. As we have discussed, different racialized social systems have different racial architectures and racial politicultures, meaning that how an individual is racialized changes as they physically travel, and/or based on the origins of those who are racializing them. To see why this poses a problem for self-identification schemes, let us revisit the example of Alice and Bob from the previous section. Even if a self-identification is acquired from Alice, would that self-identification need to change once Alice returned home or if Bob were shipped from Brazil back to the United States? Would the category used by Bob to describe Alice need to change depending on who Bob is talking to and for what purpose? What is more, choosing each of these options would make a distinct political and sociological claim and have distinct downstream implications. This is a nightmarish web that roboticists are fundamentally unprepared to untangle. Finally, even if robots' identity categorizations of interactants are made based on those interactants' self-identifications, this would nevertheless constitute a biometric surveillance scheme, in which interactants' self-identifications are stored, tracked, and analyzed by unseen others.

So, where does this leave us? Thus far, we've seen that if roboticists could be certain their robots would not be misused (e.g., by the police), then enhancing the intersectional performance of face detection and recognition algorithms might address the concern of robots being used as Whites-only New Jim Cobots. We've seen that such advances could be augmented with opt-in mechanisms to allow for "face recognition free" identification of interactants. And we've seen that if we really need interactants' gender identity or racial identity, we could just ask them for this information, although there are again reasons to be wary of asking users for information about their identity.

As in chapter 3, I argue that we might best address these remaining concerns by directly acknowledging and subverting the historical legacy that has caused those concerns to begin with. Robots that rely on information about interactant identity characteristics have a default tendency to be designed in ways that reinforce the logics that are used to manage White patriarchal power because they built on a legacy of surveillance technologies and because the evolution of the robot imaginary (and, in particular, the White patriarchal robot imaginary) has led us to a state of affairs where predominantly heterosexual, cisgender, White, male computer scientists are designing computer vision systems whose mechanisms of categorization reinforce the systems of race and gender that predominantly benefit heterosexual, cisgender, White men.

So, once again, let us directly attack that default state of affairs.

Ripping Up the Status Quo: Why Are We Categorizing?

We are now ready to consider what it might mean to rip up the White patriarchal status quo of social robot design within the domain of disciplinary power as it pertains to robots' use of social identity trait perception. If robots' attempts to categorize those they observe in terms of race and gender inherit from a problematic legacy of biometric surveillance technologies, then perhaps roboticists should design robots so that they do not inherit from that legacy. In other words, perhaps robots should simply not be designed to reason about people in terms of race or gender at all. Legislation such as the EU AI Act now make the algorithmic categorization of individuals on the basis of race a prohibited activity in most circumstances,[101] and concerted effort will be needed to expand the scope of such prohibitions and enact them on a global scale.

In addition, we can see that in many cases where roboticists are proposing to enable robots to engage in social identity trait perception, it is so that robots can use the right pronouns to refer to people or otherwise describe them accurately. Although this is well meaning, if roboticists really want to avoid misgendering people based on the pronouns robots use, robots could simply ask interactants their preferred pronouns. Similarly, if robots need to describe people they have not previously interacted with, there are a host of cues they can use instead of race and gender categories. In fact, the work of Bennett and Scheuerman described above suggests this very course of action.

Although Bennett's work showed that users preferred image-captioning systems to use race and gender labels that came from themselves rather than algorithmic guesses, her work also showed that racialized and gendered descriptions, even when based on self-identification, were *not* the preferred solution in the communities she interviewed. Instead, those communities largely desired that image-captioning systems avoid politicized descriptions of appearance altogether. Similarly, despite their arguments above, Scheuerman too has argued that computer vision systems can avoid politicized descriptors by relying only on what is directly observable.[102] For example, rather than refer to an individual by guessing or asking their race, robots could just as easily refer to them by skin tone, and rather than refer to an individual by guessing or asking their gender identity, robots could just as easily refer to them using grooming and dressing cues. A search-and-rescue robot reporting on a potential victim in need of human attention could, for example, say, "They have dark skin and are wearing a dress" rather than "They are a Black woman" if skin color and clothing are the criteria that were being used to make that identity assumption in the first place. Similarly, if roboticists really feel the need to enable their robots to behave differently based on someone's culture or gender presentation, these differences in behavior could be grounded in observations of overt cultural signifiers or grooming behaviors rather than on explicit assumptions about racial or gender identity. If robots are programmed to ask people's pronouns before gendering them and to describe people according to their skin tone and grooming choices, this could also present an opportunity for robots to wield their persuasive power for good, setting a positive example that humans might then choose to emulate in the future, making humans more likely to adopt these practices in human–human interactions.

Although ripping up the status quo may require developing robots that do not view interactants in terms of race and gender, it is critical that roboticists themselves avoid such color blindness. As numerous scholars have pointed out, adherence to a color-blind ideology in which one refuses to recognize or consider the social reality of race is itself a form of racism, which perpetuates the racial status quo through studied ignorance.[103] At first blush, this might seem like a contradiction. If roboticists need to avoid color-blind ideology, then why should their robots be color-blind? It is absolutely critical for roboticists to attend to issues of race and gender in robot design, to themselves recognize and celebrate minoritized identities, and to recognize and interrogate inequitable power structures so that they can be dismantled.[104] But roboticists' recognition of the social reality of race and the ways it needs to be accounted for does not entail an obligation to computationally model this recognition in robotic technologies. Programming robots to rely on presentation cues rather than political labels may result in race-blind and gender-blind robots but is an approach that ultimately requires roboticists themselves to avoid color blindness in their design processes.

Resisting color blindness and gender blindness will require constant vigilance on behalf of roboticists. The challenge this entails can be seen in the work of Maria Teresa Parreira when she was a student at the KTH Royal Institute of Technology working with human–robot interaction researchers Sarah

Gillet, Katie Winkle, and Iolanda Leite.[105] As described in her aptly titled paper, "How Did We Miss This?," Parreira describes how her team inadvertently implemented a gender-biased robot. In her research at KTH, Parreira helped create a robot that generated verbal and nonverbal feedback during interactions to show it was listening. After developing this robot, however, Parreira discovered that her model was unintentionally biased against women, generating fewer back-channel responses to women on account of patterns in the datasets on which the model was trained. If you have been paying attention throughout this book, you will be aware that Parreira's collaborator Katie Winkle is one of the key scholars studying effects of gender bias in robotics. As such, one might argue that Parreira's team was uniquely positioned to notice and head off the biases in their back-channeling model. This particular team's initial failure to recognize the bias they were encoding into their robots demonstrates the degree of constant attention toward issues of race and gender needed for roboticists to ensure socially responsible design.

Conclusion and Chapter Summary

In this chapter, we have seen a second way that roboticists reinforce White patriarchal power in the disciplinary domain, through the ways they design robots to use social identity trait perception mechanisms that categorize individuals according to White and patriarchal logics of race and gender.

These disciplinary power dynamics may be reflected in the designs of the Pepper and REEM robots highlighted in chapter 1. For the Pepper robot, the incorporation of gender recognition as one of its key API functionalities has directly fed waves of research seeking to capitalize on the functionality to enable other privacy-threatening capabilities such as face recognition. Meanwhile, although the REEM robot is not advertised as having those same capabilities, its capacity as a tool of disciplinary power is, in fact, even more pronounced. As a tool of law enforcement, REEM serves as a surveillance tool intended to shape the behaviors of those it surveils. As Brig. Khalid Al Razooqi of the Dubai police stated, "[REEM] can protect people from crime because it can broadcast what is happening right away to our command and control center."[106] Given Dubai's expressed intent to deploy REEM in shopping malls—a locale where Emirati law enforcement has arrested a number of transgender individuals for violating gender expression laws (i.e., for "appearing too feminine"),[107] it is not hard to imagine how REEM's deployment could specifically facilitate the enforcement of cis-heteropatriarchal power structures were social identity trait perception algorithms to be employed. We will return to this context of use in detail in chapter 6.

Overall, in this chapter, we have identified several further historical and social factors that played a key role in producing the technological status quo by which roboticists wield disciplinary power and by which robots serve as New Jim Cobots. Specifically, we traced roboticists' tendency toward algorithmic gendering and racialization back to the history of biometric surveillance technologies—a history inextricably linked to America's slave system. As Simone Browne writes, "When we think of [how people's bodies are being] informationalized by way of biometric surveillance—sometimes voluntarily and sometimes without consent or awareness—and then stored in large-scale databases . . . we can find histories of these accountings and inventories of the bodies in slave registers, slave branding and the slave vessel manifests."[108] Based on this understanding of our technological status quo, the way it reinforces our sociological status quo, and the historical and social root causes that explain this dynamic, we have identified additional strategies that roboticists can use to build toward a more equitable robot future by revising, rethinking, and ripping up the status quo in the disciplinary domain.

First, roboticists might revise the status quo by saying, "We won't use those White patriarchy reinforcing categories." This might again involve addressing bias in robots' categorization algorithms or training robots to assign people to a broader and more inclusive set of identity categories. Although these approaches move in the right direction, they merely lay on an inclusive veneer without addressing underlying problems.

Second, roboticists might rethink the status quo by saying, "We will let communities use their own logics of categorization." This might involve taking a participatory approach in which roboticists work with oppressed communities to design robots whose methods of categorization better align with the values of those communities. However, this approach may ultimately continue to reinforce racist and patriarchal logics of categorization.

Finally, roboticists might rip up the status quo by saying, "Let's attack the root causes that explain how we have arrived at that status quo." If robots' use of categories of race and gender perpetuates White patriarchal logics of categorization regardless of where category judgments come from, perhaps we need to design robots that can engage in natural and effective interactions without needing to categorize people at all. Finally, if all of these problems reflect roboticists' tendency to imbue robots with their own norms and logics, then just as was suggested in chapter 3, perhaps we should be demanding more diverse robot design teams and the adoption of design justice–driven approaches that emphasize end-user programming as the site of robot design.

In this chapter, we have focused on how roboticists wield disciplinary when robots are used as tools of racialized and gendered surveillance. In the next chapter, we will consider a domain in which robots are likely to be used in this way: policing. In doing so, we will see how this outcome is inevitable, given the history of policing, and what this means for the ways that structural power is inevitably wielded in inequitable ways when roboticists choose to collaborate with the police.

6 End Effector

Watching the special weapons and tactics (SWAT) trucks crawl past my apartment in late May 2020, I felt my anger shift to foreboding. One week had passed since the murder of George Floyd, and around the country, protesters were being met with unprecedented police violence. In downtown Denver, where I resided at the time, the police were deploying chemical weapons against protesters and firing rubber bullets at children only a short walk from where I lived. In a tweet whose accompanying video captured the scale of the violence, one resident wrote, "Police turned off their body cams, and proceeded to gas and shoot innocent people at the riots today. They hunted us down through the streets." That day, the police were advancing in person on crowds of protesters to beat, shoot, and gas them. I couldn't help but wonder how long it might be until the police instead used the robots developed by my friends and colleagues to make similar attacks from the comfort of their police stations. And what, I wondered, could roboticists do to avoid being complicit in this violence?

I begin with this memory because it starts us down the road toward understanding how roboticists wield structural power—that is, power over the institutions that structure oppression.[1] Although it is may initially be easier to see how roboticists wield cultural or disciplinary power, the everyday ethical decision-making described above (What can we do to avoid complicity in police violence?) is inherently a question of how roboticists should wield power in the structural domain. As we will see, technologists wield subtle yet substantial power over today's societal institutions through their decisions about which institutions to support, legitimize, and normalize—and which to subvert.

Because most roboticists (and most Americans for that matter) are not terribly familiar with the history of the police, the idea of intentionally working to *subvert* policing may be foreign or alarming to many readers. As such, in the first part of this chapter, we will work to understand the fundamental nature of policing by analyzing the origins and evolution of policing and by showing how the history of policing explains how and why the police are used today. We will begin by explaining the origins of the police in the slave patrols of the antebellum South and the English efforts to colonize Ireland—the same colonialist projects that gave rise to race itself, as described in chapter 2. Next, we will analyze the evolution of policing in the United States, from the formation of Southern police departments in the aftermath of the Civil War to the weaponization and militarization of police departments in service of the War on Drugs to the data-driven and violent nature of modern dragnet-style policing. In the second part of this chapter, we will then use the origin and evolution of policing to explain how and why the police are (mis)using robots today, and the ways that this use of robots exacerbates police violence while legitimizing and normalizing policing as an institution. Finally, in the last part of this chapter, we will explore how roboticists can revise their approach to police–roboticist collaboration, rethink the relationship between robotics and policing, or more fundamentally rip up the status quo through an abolitionist robotics that dismantles the White patriarchal status quo of robotics in the domain of structural power.

With this trajectory firmly established, we are ready to consider the origins, evolution, purpose, and (mis)use of robots by the police.

I The History of Policing

As with race, many people have a false assumption that police forces have always existed. In fact, many people have difficulty imagining a world without the police. Yet, as we will see, policing is a relatively new invention that, like robots, emerged from a desire to maintain the power of wealthy White men in the wake of class struggle and Black emancipation.

Slave Patrols and the Creation of Policing in America

For much of the second millennium of the Common Era, law enforcement (at least in European contexts) was largely a shared responsibility taken up by volunteers among everyday townsfolk[2]—quite a different model from the semi-professional and militarized forces seen today in the United States. As Historian Sally Hadden has shown, the nature of law enforcement began to fundamentally change in the American South due to (1) the unique organizational structure and militarization of the early American colonies and (2) the reliance of these colonies, especially Virginia, North Carolina, and South Carolina, on slavery as an economic system.[3]

As Hadden describes, White colonists in these areas created militias to help respond to a variety of threats, both external (e.g., threats from the Native Americans whose land British colonists were invading) and internal (e.g., threats from the enslaved people those colonists had kidnapped and subjugated).[4] Over time, the need for these outward- and inward-facing defenses led to the creation of distinct slave patrols[5] in the late seventeenth and early eighteenth century, composed of groups of men selected from the White population with the aim of catching enslaved people who had run away, as well as preventing gatherings of enslaved people, seizing enslaved people who were traveling without passes, ransacking the dwellings of enslaved people to seize their possessions, seeking out signs of enslaved people being educated, and otherwise enacting surveillance and violence against enslaved people.[6]

This public half of the Southern justice system (as opposed to the physical and sexual violence privately meted out on plantations[7]) presented challenges to colonists who did not wish to participate in these community obligations. As such, the system of slave patrols gradually transitioned from a shared responsibility among White men to a government-appointed, and ultimately paid, role.[8] This transition helped state governments to ensure that enslaved people were being surveilled and controlled while removing the burden of enforcement from everyday citizens.[9] Critically, this shift was also motivated by labor concerns, with business owners complaining about poor productivity among White laborers due to their need to participate in nightly patrols.[10]

To be clear, slave patrols were not the only form of public law enforcement in the American South during this period of critical transition, as slave patrols worked in coordination with term-limited constables. But even for those constables, slave catching was a key part of the job.[11] Moreover, slave catching was a key element of nineteenth-century policing even in the American North, where the police were tasked with locating and re-enslaving escaped slaves for many decades after slavery's local abolition.[12] As such, early American law enforcement can be seen to have developed in response to America's slave economy, both in terms of the oppression and terror needed to maintain the slave system and in terms of the laziness and greed of slave owners that motivated the slave system in the first place.

When the first formal police departments emerged in the American North, their organization was informed both by the Southern tradition of slave patrols and by new systems of policing imported from Europe. The first professional European police force was the Metropolitan Police in the UK, created by Robert Peel in 1829. Peel had previously been responsible for creating the Peace Preservation Force in the occupied state of Ireland. As described in chapter 2, the sixteenth- and seventeenth-century English rulers had developed a model of Irish control in which the English seized Irish land, started plantations on that land, and put the subjugated Irish to work on those plantations. Two centuries later, the plantation of Ireland had failed, but the English occupation of Ireland was still struggling onward. Peel's highly militarized Peace Preservation Force[13] was tasked with keeping the occupied Irish under control and providing support to the occupying English.[14] Although the Peace Preservation Force was initially successful in deterring "banditry," Peel eventually resigned his position and left Dublin in disgust, "convinced that he had achieved no final solution to the problem of law and order."[15]

Retreating to London, Peel brought his model of policing with him, convincing the London government to implement his militarized and "professionalized" police force (despite its lack of success in Ireland) while shifting emphasis from post hoc response to ex ante prevention.[16] Although the idea of a uniformed police force was opposed by many,[17] Peel's efforts eventually won out due to at least two factors. First, fear of crime was rising among Londoners, not due to an actual increase in crime but rather due to the rise of newspapers as a source of crime news in the late 1820s. The rise of crime reporting in newspapers led not only to misconceptions that crime rates were dramatically increasing (and not simply being made more visible) but, furthermore, to misconceptions of the severity of crimes, as newspapers were incentivized to sensationalize the details of reported crimes to increase sales.[18] Second, with the rise of industrialization, business and government leaders needed to suppress political unrest, force people into work, and generally enforce industrial capitalism.[19] Peel's policing system—an occupying force of people from the countryside and military, housed in barracks and tasked with controlling these working-class threats—was judged an appropriate remedy.

In the decade following the deployment of the Metropolitan Police, Northern US cities were facing this same trend of industrialization-driven working-class unrest. To respond to this social challenge, cities began to import Peel's policing model.[20] These new police forces thus took on the task of curbing unrest among exploited working-class immigrants[21] and exerting control over religious minorities[22] while unofficially working *with* local "petty criminals" (e.g., to help fence stolen goods).[23]

Ultimately, these cities' policing systems deviated from Peel's model in a variety of ways. First, the power of the US police was more overtly steered by political parties toward their own ends.[24] Second, there was no pretense that the US police would be involved in "preventing" or "reducing" crime, and they would not be presumed to have such a role until the next century.[25] And finally, these cities' police models brought in elements of Southern slave patrols missing from Peel's model, including their general organization and embedding into communities.

As such, we can see the antebellum emergence of Northern US policing as the result of two historical streams whose sources lay in the colonization and plantation of Ireland. The first stream led from Ireland to the American South, where British views of the Irish (as discussed in chapter 2) determined how the British viewed and regulated the behaviors of Black people in their American colonies, both within and beyond the bounds of formal plantations. These dimensions of early America's racial politiculture then led to the creation of the South's slave patrols. Separately, the second stream led from Ireland to the American North by way of England, where the mechanisms for enforcing control in the Irish colonial projects were imported to London as a means of policing class struggles. And finally, these two streams rejoined in Northern US cities to create a new brand of policing, which combined the community embedding, continuity of office, general organization, and racialized tenor of the slave patrols of the American South[26] and the full-time, uniformed, urbanized, and beat-driven character of the Metropolitan Police.[27] This Northern policing model then fed back into Southern policing models as they shifted from antebellum slave patrols to postbellum police departments.

Southern Policing Post–Civil War

As we have seen, policing in the American South prior to the Civil War was framed and executed as an explicit mechanism for racial and economic control in the form of state-mandated slave patrols tasked with maintaining a regime of terror over Southern industries' enslaved workers. What precipitated the shift, then, from slave patrols to police departments?

First, slave patrols effectively ceased in some Southern areas during the Civil War, simply because the majority of men assigned to those patrols had been drafted into the Confederate Army. In these areas, police departments began to pop up to continue lapsed slave patrol duties.[28] Second, with the ostensible emancipation of Southern slaves in 1863 and the passage of the 13th Amendment in 1865, Southern states could not very well continue organizing groups termed "slave patrols" in a country where slavery had become illegal. However, the Southern desire to police Black workers had not gone away with the legal end of slavery and, in fact, neither had their desire to engage in slave trading. Indeed, perhaps more than ever, state mechanisms of control were needed both to force the newly "free" Black population into sharecropping, prisons, and other systems in which they could be effectively re-enslaved[29] and to respond to fears of Black insurrection. To respond to these desires, Southerners mimicked the transformations being effected in other areas of society: slavery officially ceased . . . but was maintained by rebranding it as "peonage"; slave codes officially ceased . . . but were maintained by rebranding them as "Black codes"; and slave patrols officially ceased . . . but were maintained by rebranding them as "police departments."[30]

This change from slave patrols to police departments thus represented a continuation of goals but a change in name, organizational structure, and method. The same men responsible for conducting slave patrols became those responsible for policing through police departments, but had now returned from the war with military training, which influenced both the organization and tactics of their new patrol groups.[31] Moreover, due to the outrage of Northerners in response to the South's "new" Black codes, these police departments needed to operate according to policies that could be framed as general law enforcement rather than as explicitly racialized policing. As such, Southerners began tasking their police departments with cracking down on ostensibly race-neutral crimes such as "idleness" and "vagrancy." Although police departments would ultimately impose these policies on poor Whites as well,[32] they primarily served as a means of Black re-enslavement.[33] As Hadden writes, "Policemen in southern towns continued to carry out those aspects of urban slave patrolling that seemed race-neutral but that in reality were applied selectively. The police saw that nightly curfews and vagrancy laws kept blacks off city streets, just as patrollers had done in the colonial and antebellum eras."[34]

This process transformed Southern policing in a way that was legally defensible on the national stage, without sacrificing the efficacy of the system as a means of racial oppression[35] and without sacrificing this racial oppression as the central motivating purpose of policing: "No matter how important crime control appeared, the true mission of the [police] force revolved around suppressing the negro population."[36]

Because these new Southern police departments needed to appear race neutral, the Ku Klux Klan (KKK) emerged alongside those police departments[37] to help coordinate the maintenance of the South's regime of racial terror.[38] In many areas, the police and the KKK operated as two hands of the same system, with police departments providing an official day job for KKK members[39] to perform whatever tasks they could cast as ostensibly race neutral and the "Klan" (the "true" instrument of law enforcement[40]) taking care of the rest.[41] Overall, this new configuration led to a dramatic *increase* in violence relative to that seen under slave patrols because the KKK was able to act pseudo-anonymously and because neither the police nor the KKK needed to worry any longer about harming the "property" of other White people.[42]

These associations between the police and White supremacist organizations are still seen today. In 1991, for example, it was found that the LA Sheriff's department had "formed a neo-Nazi gang and habitually terrorized Black and Latino residents."[43] Findings such as this led the FBI to admit in 2006 that White supremacist groups had been "infiltrating law enforcement communities or recruiting law enforcement personnel" for some time. Even when today's police are not actively involved in organized White supremacist activities, they routinely turn a blind eye toward such activities[44] and regularly engage in run-of-the-mill online racism, sexism, and homophobia.[45] In the city of Philadelphia alone, the Plain View Project revealed that of a thousand police officer profiles identified on Facebook, one in three had posted troubling content, and of these, a further one in three (i.e., one-in-nine overall) had had one or more federal civil rights suits filed against them.[46]

Although in this section we have thus far discussed the rise of policing in both the US North(east) and South(east), law enforcement agencies in other parts of the US similarly arose in service of class- and race-based oppression.[47] In the American Southwest, for example, police departments first emerged in the form of the Texas Rangers, a group created to protect the interests of White colonists through the violent oppression, massacre, and segregation of local Native and Mexican residents.[48] This mission continued long after Texas's annexation through the oppression of union leaders, discouragement of voting or voter registration in Mexican-American communities, and the enforcement of other segregationist "Juan Crow" policies.[49]

Across these distinct areas of nineteenth-century America, it is clear that the purpose of the police was not to reduce crime but rather to enable state control over racialized and working-class communities in service of the dominant social and economic order. The fact that the police today are largely seen as "legitimate" agents of law enforcement is the result of 150 years of concerted efforts that simultaneously legitimized the police while increasing their lethality. Indeed, for much of US history, it was no secret that the police were corrupt, unqualified, ineffective, and drunk. Serious attempts at reducing corruption in policing were not made until the 1900s.[50] And further "reforms" enacted in the twentieth century, in which the police attempted to rebrand themselves as "fighting crime," were made in response to concerns among White Americans regarding the police's inability to respond to "urban disorder" during the Black fight for civil rights.[51] The police reforms of the 1950s responded to these concerns in part by shifting the metrics against which policing effectiveness was measured toward efficiency and arrest rates, thus actively incentivizing the police to devastate communities of color.[52] Moreover, police reform efforts have often tended to *enhance* police capability for racial violence specifically by leveraging insights from the US occupation of foreign nations—a pattern that echoes the British Metropolitan Police's leveraging of insights from the English occupation of Ireland.[53] Overall, then, we can see the ways that the police reform efforts of the nineteenth and twentieth century were put to use in service of, rather than as a means to curb, racial violence.

It should be no surprise, then, that this new characterization of the police as a mechanism for "fighting crime," and the corresponding militarization of the police, paved the way for the US's modern system of racially motivated mass incarceration through the vehicle of America's War on Drugs.[54]

The New Jim Crow

In June 1971, US President Richard Nixon called drug abuse "public enemy number one" and officially launched his "War on Drugs." Nixon's proclamation represented a continuation of previous patterns of policing, with the police ostensibly engaging in race-neutral law enforcement while actually policing racial boundaries. More specifically, Nixon's proclamation built on nearly a century of drug laws designed to control non-White communities, in which opium dens had been criminalized in order to better control Chinese immigrants,[55] cocaine had been criminalized in order to better control Black Americans,[56] and marijuana had been criminalized in order to better control Mexican Americans.[57]

Nixon's War on Drugs thus represented a newfound commitment to these racially motivated policy choices. Publicly, Nixon's use of the terms "public enemy" and "War on Drugs" set up a shared antagonist for the White "public," presented police crime fighting as the solution, and justified further militarization of those police through the language of war. Privately, Nixon then relied on the press to lead White Americans to identify Black people and other enemies of his administration as the public enemies against whom militarized police should be wielded. This intentional use of the War on Drugs was stated directly by John Ehrlichman, Richard Nixon's chief domestic policy advisor:

> We knew we couldn't make it illegal to be either against the war or Black, but by getting the public to associate the hippies with marijuana and Blacks with heroin, and then criminalizing both heavily, we could disrupt those communities. We could arrest their leaders, raid their homes, break up their meetings, and vilify them night after night on the evening news. Did we know we were lying about the drugs? Of course we did.[58]

Although the War on Drugs was not the US government's first use of drug policy to criminalize the perceived enemies of White supremacy, it produced markedly more extreme outcomes than had previous efforts due to the new incentivization schemes underpinning the War on Drugs, which led to an astounding proportion of Black Americans being condemned to a permanent racial under-caste.[59] As described by Michelle Alexander in *The New Jim Crow*,[60] the War on Drugs was intentionally designed as a means for the police to round up and imprison vast numbers of Black men in ways that can still be felt today, as police departments are encouraged to make drug arrests through federal funding that is tied to the number of arrests for drug-related offenses and are incentivized to bolster department budgets through civil forfeiture, a mechanism that allows the police to seize property "perceived" to be involved in drug crimes.[61]

Critically, civil forfeiture (in many states) does not require anyone to be charged with a crime. Rather, the police are free to stop individuals on the street, see how much money they have in their pockets, allege a connection with drug activity, and pocket the money. The onus for contesting such theft largely falls on the shoulder of the victim[62] and is difficult to challenge, with the cost to contest civil asset forfeiture typically twice the median amount stolen by the police in the first place.[63] As codified, civil asset forfeiture has allowed the police to seize billions of dollars in assets from people each year without an arrest or conviction,[64] with a large proportion of police departments completely dependent on this form of legalized theft.[65]

To facilitate the use of mechanisms such as civil forfeiture, the police are essentially empowered to stop, interrogate, and search anyone they choose on drug-related—and often explicitly racial[66]—grounds. Although the vagrancy laws used to round up and enslave potential Black laborers in the postbellum American South were at least facially race neutral, these new mechanisms explicitly encourage the police to intentionally target people of color. It should thus be no surprise that people of color are disproportionately targeted as victims of the War on Drugs, despite White people selling drugs at equivalent rates and despite White people being substantially more likely to use drugs than are people of color. As one manifestation of this pattern, Black women passing through American airports are nine times more likely to be x-rayed and three times more likely to be strip-searched on suspicion of drug smuggling than are White women, despite being less likely to be carrying contraband.

The dramatic overpolicing of Black and Latinx neighborhoods through these policies was key to the growth of police departments in major US cities. As Max Felker-Kantor writes, "Racial Targeting was central to the LAPD's expansion of police power and efforts to control the street at all costs . . . [T]he growth of police power in the decades after the 1960s was organized around the aim of controlling the city's black and brown populations."[67] As a result, in cities such as Los Angeles, arrests in neighborhoods of color are five to six times that seen in White neighborhoods.[68] Similarly, Black men in New York City are stopped by the police at disproportionate rates, not only in terms of population but also in terms of ostensible crime rates.[69] Put together, we can say that, as designed and incentivized, most of those swept up on drug charges in the United States are Black, Latinx, and Indigenous.

Once swept up by the police, as Alexander[70] describes, American defendants (especially Black defendants) are generally denied meaningful legal representation and can be pressured to plead guilty through prosecutorial techniques that cannot be challenged on the basis of racial discrimination.[71] Then, once sentenced, Americans found guilty of drug-related offenses are subject to far longer and harsher sentences for those offenses than anywhere else on earth,[72] with prison sentences of life behind bars dispatched even for minor, nonviolent drug charges.[73] Black Americans, by design, are more likely to be embroiled in this system to begin with, but moreover, Black Americans are subjected to significantly worse treatment at each of its stages.[74] As a result, more than 25 percent of young Black men will ultimately be imprisoned,[75] destroying their lives, families, and communities.

Once drug "offenders" of color in the US have "paid their debt" to society, they are forced into a permanent racial under-caste in which they are legally discriminated against for the rest of their lives.[76] Those released from prison after being convicted of felonies are prevented from obtaining employment, both formally and informally. Many occupations are legally barred from hiring felons, and many employers will not hire felons, even when this refusal is illegal. Drug offenders are also systematically denied housing, again both formally and informally. Drug offenders are often barred from obtaining public housing, and many non-offenders living in public housing are unwilling to let felons stay with them, as they can be evicted if their guests are arrested. Finally, they are denied education and other public benefits and, in many places, rendered unable to vote.[77] Accordingly, the average Black man leaving prison today does so with $17,000 in debt, mostly as a result of fees and other financial liabilities imposed on them for being in prison[78]—what amounts to a charge imposed by the police for being non-White. Saddled with this debt, these former "offenders" are thus released into a society in which they have no access to jobs or housing—a state of affairs that *does*, unsurprisingly, lead to criminalized activities in order to survive, creating a vicious cycle.

We have now seen the origins of policing, the motivations for policing, and the effects of policing in creating and reinforcing a society stratified by race. Graced with unprecedented budgets to effect this end (both through formal legislation and through the results of their theft from the poor) and empowered with dreams of effecting these ends through military force, many police forces have turned to military-grade technologies to facilitate their mission of oppression. As we will see later on, these technologies often include robots. But to best understand how and why these acquisitions include robots and what the ultimate effect of those robots will be, we must first consider the militarized and data-driven nature of modern policing and the surveillance and violence modern policing enables even without robots.

Modern Policing: The Big Data Dragnet

As described above, policing has always had a core mission of surveilling Black and poor people. This surveillance is a critical lever of disciplinary power (as described in chapter 5), in part because, as Foucault reminds us, "a state of conscious and permanent visibility . . . assures the automatic functioning of power."[79] Accordingly, the police have consistently sought out, acquired, and misused technologies that more effectively enable racialized surveillance. Historically, this has included not only analog technologies such as the *Book of Negroes* discussed in chapter 5 but also, and increasingly, the digital technologies being developed by computer scientists.

Indeed, computing technologies have a long history as tools of state surveillance. Perhaps the most infamous example is the role IBM played during the Holocaust. As Edwin Black describes, IBM and its subsidiaries worked with the Nazis to facilitate nearly every stage of the Holocaust, from surveillance and tracking to deportation and, ultimately, extermination.[80] This is most visible in the punch-card database keys tattooed onto the arms of Auschwitz prisoners: Upon internment at Auschwitz, each prisoner was assigned a five-digit number that keyed their record in a custom database system built for the camp by IBM subsidiary Hollerith. These became the initial set of identifying numbers tattooed onto Auschwitz prisoners' arms.[81] As Black describes, however, the Nazis' overall method for extermination through labor was hand engineered by IBM.[82] These efforts were not mere misuse of IBM technologies by the Third Reich and were not the result of distant subsidiaries working on their own. Rather, they were conducted under the direct "micromanagement" of CEO Thomas Watson, who was rewarded by Hitler for his efforts in a "grandiose ceremony" in Berlin. In the decades following World War II, IBM would go on to establish extensive collaborations with American police.[83]

More recently, however, computing technologies such as big data and automation have fundamentally transformed policing and the nature of police surveillance.[84] As Sarah Brayne discusses in her book *Predict and Surveil*, police surveillance once (largely) followed a pattern that many Americans would likely expect from watching popular crime dramas: the police would become suspicious, they would acquire a warrant, and they would then begin surveilling their target. But as Brayne shows, this is no longer how most policing works. Instead, modern policing predominantly uses dragnet-style surveillance, in which warrants and consent are viewed as "anachronistic."[85] Under this new paradigm, surveillance is "increasingly suspicionless, programmatic, ongoing, cumulative, remote, invisible, automated, preemptive, and embedded into routine activity."[86]

The widespread nature of warrantless police surveillance can be observed through the proliferation of Amazon Ring doorbells. In my Denver neighborhood, an astonishing percentage of front doors seem to be adorned with these doorbell cameras. Nationwide, about 25 percent of Internet-enabled households own (or plan to purchase) a Ring doorbell (or similar device).[87] What many Ring users may not be aware of is the extent to which they contribute to an Amazon-facilitated police surveillance state.[88] Amazon Ring is promoted by local governments and law enforcement agencies in exchange for access to user data,[89] with more than thirty cities and more than eight hundred police departments collaborating with Amazon.[90] Furthermore, Amazon provides a central portal that allows participating law enforcement agencies to access user data, which does not require a warrant to access but instead merely user consent.[91] To make this as seamless as possible, Amazon coaches the police on how to extract this consent from users[92]—and even if a user refuses to give consent, the police can still simply request the footage they want from Amazon directly.[93] Unsurprisingly, the majority of videos that Ring users choose to share, at least with their neighbors, are of people of color,[94] creating a system of racialized surveillance that is exacerbated by the face recognition algorithms that Amazon allows users to automatically run on collected video data.[95] The millions of Amazon Ring cameras deployed across the United States thus constitute a network of ubiquitous surveillance devices that extend the ongoing use by the police of advanced technologies to surveil Black communities.[96] Although Amazon's construction of a police surveillance network should rightly shock and appall us, we will see how police surveillance through Amazon devices ultimately represents only the tip of the iceberg and is, in some ways, unique and antiquated relative to the bulk of police surveillance, specifically because it still involves some degree of suspicion by the police.

This new policing paradigm has also been driven by the use of big data technologies provided by companies such as Palantir,[97] a Denver-based computing firm named for the surveillance and communication devices used by the Dark Lord Sauron in the *Lord of the Rings* and founded by far-right billionaire Peter Thiel. One of Palantir's primary offerings is their Gotham system, perhaps named for its ability to let police officers cosplay as Batman and pretend to fight crime. Gotham is Palantir's government intelligence platform—a browser for drawing connections between and sorting through the billions of data points collected through government surveillance programs, from criminal records to license-plate scans. To understand the role of Gotham in policing, we can consider the narrow case of automated license plate reader (ALPR) data use in Los Angeles, and how Gotham turns this data into a large-scale system of dragnet surveillance.

Throughout Los Angeles, ALPR cameras such as those created by Vigilant Technologies observe and attempt to identify passing cars.[98] Some of these ALPR cameras are stationary, mounted on light poles, whereas others are mobile, mounted on police cars. These cameras constantly look for license plates, attempt to identify the state in which those plates were issued, and attempt to read the numbers on those plates—a relatively simple effort that would make a reasonable term project for students in my computer vision class.

When deployed on police cruisers, ALPR cameras can have immediate and devastating impacts. In Colorado, for example, the Aurora police used an ALPR match as an excuse to pull over a minivan containing four Black girls who were on their way to a nail salon.[99] The officers forced the girls (including a six-year-old) out of the car and onto the ground, handcuffing the two oldest girls. Why were they pulled over? Their car happened to share a license-plate number with a motorcycle that had been reported stolen.[100] The officers apparently couldn't be bothered to consider that although the license-plate numbers were the same, they were issued in different states. Nor could they apparently be bothered to consider that the minivan they pulled over was not, in fact, a motorcycle. After public outcry, Aurora District Attorney George Brauchler stated, "No one is above the law. If our investigation determines that the officers involved committed a crime, I will not hesitate to file charges and prosecute them."[101] Yet, in news that should surprise no one, the police faced no consequences beyond admitting the children involved "may have been traumatized" and offering to provide them with therapy.[102]

ALPR cameras' most pernicious influence is felt when their recordings are collected and organized through software such as Palantir's dedicated license-plate data application, Thunderbird. Thunderbird is used as a structured data source for Gotham, allowing license-plate sightings to be tracked over time and correlated with other information known about the cars bearing those plates and about the owners of those cars. By leveraging this data source, the Los Angeles Police Department (LAPD) is thus able to track the movements of anyone they choose throughout Los Angeles. These license-plate scans are then kept for what appears to be an indefinite period of time,[103] with no policy governing appropriate data use.[104] As American Civil Liberties Union attorney Mohammad Tajsar stated in 2020, this "confirms what we have long suspected: The LAPD treats its license plate reader program like it's in the Wild West."[105]

Under today's big data dragnet, people are under constant surveillance when they travel where police cameras are installed and where police cars are sent. This is especially true for Black and Latinx Americans. Stationary police cameras, including ALPR cameras, are overwhelmingly concentrated in Black and Latinx neighborhoods. This phenomenon is grimly visualized by Alex Najibi,[106] who showed that police surveillance cameras in the city of Detroit (where a map of such cameras was made available) had nearly a perfect mapping onto neighborhood demographics, with the borders of the surveilled region tightly following the predominantly Black and Latinx neighborhoods of Detroit.

Palantir's Gotham thus represents—even just considering its use in the context of ALPRs—a massive, ubiquitous program for surveilling Black and Latinx Americans facilitated by computer science and data science technologies. This surveillance tool is then used in conjunction with other computing tools to enable predictive policing regimes. In Los Angeles, for example, Palantir was used between 2011 and at least 2019 to create daily lists of the top twelve "chronic offenders" in each of the forty "LASER zones" that the LAPD had determined to have a high risk of crime.[107] These lists were based on a point system of evidence entered into Palantir's software, including information on whether individuals were affiliated with gangs. Although this might seem like a reasonable criterion, LA gang databases contain uncountable innocent young men of color, often labeled as gang members in police databases simply because they are Black or Hispanic,[108] and the police have been caught falsifying entries in California's statewide gang database.[109]

Meanwhile, these databases do *not* contain the members of Los Angeles Sheriff's Department gangs such as the Grim Reapers, the Regulators, and the Vikings that have operated in Los Angeles for the past half-century.[110]

In cities across the US where technologies such as Gotham are used by the police, they are often employed in the context of predictive policing efforts, which attempt to predict where crime will happen and thus where the police should be sent and/or whom they should target. The problems with these technologies are, by and large, problems inherent to all big data technologies that questionably concretize past practice and prime future behavior.[111] In the case of person-based forms of predictive policing such as LASER, the result is an algorithmic process that, due to past over-policing of Black and Latinx neighborhoods, automatically picks out individual predominantly Black and Latinx men for increased surveillance in predominantly Black and Latinx neighborhoods, leading to increased stops and arrests in those neighborhoods, which, in turn, leads to increased motivation for surveillance of those neighborhoods, creating a vicious cycle that obscures police efforts to surveil racial minorities behind a curtain of math. Similarly, location-based forms of predictive policing, such as PredPol (a system for targeting "high-crime" neighborhoods, originally developed as a military technology for tracking supposed Iraqi insurgents[112]), are not only prone to this same vicious cycle of racialized surveillance[113] but, moreover, fundamentally unable to predict the types of crime that the public is most concerned with (violent crime) due to the place-agnostic, person-specific nature of such crime.[114]

These computer-enabled predictive policing systems lead to vicious cycles of Black and Latinx criminalization, exacerbated by the differences in police enforcement of laws for different racial groups (echoing the patterns noted by Michelle Alexander in *The New Jim Crow*), ultimately leading us into a future where crime and arrest are increasingly unrelated.[115] Moreover, computer-enabled predictive policing systems intensify racial oppression even for those not rounded up in encounters with the police, as surveilled populations learn to avoid surveilling organizations such as schools, banks, and hospitals, leading to loss of education, wealth, and life.[116]

Computer-enabled policing systems also lead to vicious cycles of valorization of racialized oppressors. In particular, data-driven policing systems play a key role in shaping the police imaginary, distorting perceptions among the public—and among the police themselves—regarding the nature and purpose of policing. For example, we can consider how the motivations for policing are shaped by predictive policing systems. The ostensible motivations for policing typically center around violent crime, and as such, predictive policing systems render violent crime hypervisible and create the illusion that the police are doing something meaningful to mitigate violence. Yet, in practice, the police spend only about 4 percent of their time on violent crime,[117] and predictive policing tools are fundamentally incapable of preventing violent crimes.[118] As such, the advertised use of predictive policing systems tends to reinforce false narratives of police effectiveness and of the role of the police in preventing murders and rapes.

Finally, computing systems facilitate American law enforcement complicity in cycles of bureaucratic and administrative violence. Bureaucratic and administrative violence inflict damage on individuals through a combination of structural and disciplinary power, using laws and policies to reinforce White-privileging power differences and enforcing those differences through bureaucracy and policing. Today, the police and other state institutions are increasingly using automation to dole out life-shattering decisions without meaningful justification while removing the possibility for those decisions to be explained or challenged. A stunning example of this is Virginia Eubanks's notion of the "digital poorhouse."[119]

As Eubanks shows, conservative state governments are increasingly automating state decision-making processes as a way to inflict violence on poor people and people of color. Many of these efforts are performed under the guise of fraud detection, where these state governments ostensibly attempt to reign in state budgets by looking for and eliminating instances of welfare fraud. In practice, these automated fraud-detection algorithms do not tend to be very accurate at detecting fraud or abuse and instead simply pick out innocent people as targets for state violence. Welfare fraud algorithms thus enable governments to target thousands of innocent people and remove them from food stamps, housing, or other benefits to which they are entitled; healthcare fraud algorithms kill sick children by removing their opportunity for healthcare; and algorithms used by Child Protective Services take children away from poor families because they are poor rather than due to any physical or sexual abuse.[120] Many of these algorithms are unauditable and unchallengeable. That is, they do not provide any meaningful justification for *why* people's food, shelter, or children are being removed or any opportunity for those being targeted to meaningfully appeal the decisions made. The end result is that through these algorithmic systems, conservative governments are able to kill, rob, and otherwise oppress vast numbers of poor and non-White people while allowing the algorithm to take the blame.

This pattern is not dissimilar to the ways that AI is now reportedly being used by the Israeli military, with one AI system (Gospel) used to autonomously place thousands of Palestinians onto "kill lists" who can then be pursued for extermination with little human oversight, and another AI system (Where's Daddy?) used to increase the likelihood that those targets are killed while that they are home with their families[121]—an algorithmic system that allows for automatic targeting of civilians for assassination while allowing the armed forces executing those assassinations to justify their decisions by pointing to the algorithm. Moreover, Dan McQuillan draws further connections between algorithm-enabled bureaucratic or administrative violence and the administrative violence of the Nazis,[122] who prided themselves on their ostensible objectivity and administrative efficiency. As McQuillan analyzes, the use of automated systems enables thoughtless cruelty—a thoughtlessness that can exacerbate cruelty by distancing those enacting cruelty from the consequences of their actions.[123] Both of these connections highlight the ways that arm's-length bureaucracy and gun's-length policing work hand in hand. Algorithmically assisted bureaucracy may provide the legal justification for eviction, kidnapping, theft, or murder by the state, but those working in "law enforcement" are often the ones to carry out or enforce compliance with these extrajudicial sentences.

Violent Sights Beget Violent Ends

The racialized violence begat by police surveillance tends to be less subtle and more brutish than these sources of administrative and bureaucratic violence, in part due to the racist origins of policing described earlier in this chapter: "Policing was not designed for public safety but as a tool to exploit, punish, and control Black and other racially and class marginalized communities."[124] As David Lyon argues, surveillance itself is a racialized process of "social sorting" in which those surveilled are "classified and stratified into categories for differential treatment."[125] In the hands of the police, this categorizing gaze becomes deadly, as the police mobilize this gaze to surveil, track, and apprehend racialized others through the use of force[126] so as to "inflict pain or to threaten pain in the fabrication of order [and] establish the necessary conditions and relations for the accumulation of capital."[127]

The violent ends to which the police wield their surveillance is evident to anyone who considers the statistical evidence that has piled up in the United States. Police killings in the US remain steadily above a thousand per year,[128] and one in three Americans killed by strangers is killed by the police[129]—statistics fueled, in part, by the increase in lethal force that accompanies police militarization.[130] Moreover, the police continue to put their surveillance capabilities to work in the name of *racial* violence. Campaign Zero's analysis of LA, for example, paints a clear picture of racial violence in terms of use of force against communities of color, racial biases in arrest rates, evidence of over-policing of misdemeanors, and inattention to crimes against people of color.[131] Moreover, this stark picture is illustrated only through data that have been officially reported. The numbers collected by Campaign Zero are likely to be low estimates, as police violence is notoriously underreported. In one recent year, data provided by the Baltimore Police Department, for example, claimed that there had been *zero* police stops that year.[132]

Similarly, Andrea Ritchie's analysis of police violence against women paints a clear picture of the *sexual* nature of police violence. Sexual violence committed by the police is widespread, especially during traffic stops and in the course of border patrol.[133] This is doubly concerning, as 40 percent of police violence reported in the media is perpetrated against teenagers and as 20 percent of on-duty sexual misconduct by the police constitutes forcible rape.[134] Yet, fewer than half of police departments have policies against the police enacting sexual violence against their communities,[135] and no official registry exists to document the rates of sexual crimes perpetrated by the police. Even more so than general police violence, sexual violence committed by the police is likely to go unreported.[136] And as such, sexual violence committed by the police is likely to be even more prevalent than is shown by statistical evidence.

More fundamentally, though, the violent ends of police surveillance should also be evident to anyone who has been awake for the past ten years who has heard the names of Breonna Taylor, Elijah McClain, Eric Garner, Freddie Gray, George Floyd, Keenan Anderson, Michael Brown, Philando Castile, Tamir Rice, Tyre Nichols, or the countless other Black Americans murdered by the police, or to anyone who has witnessed the reality of "nonlethal" police violence. In five hundred videotaped incidents collected by criminal defense lawyer T. Greg Doucette[137] between May 30 and June 15, 2020 (i.e., in the weeks immediately following the story with which we began this chapter), Doucette captured the horrific nature and scale of this violence. Doucette's collected videos from May 30 alone show the police beating, pepper-spraying, trampling, grenading, shooting, and committing hit-and-run assaults on peaceful protesters, children, elected officials, journalists, and bystanders. Ultimately, American policing has been characterized since its inception by corruption, extortion, brutality, rape, and the killing of unarmed members of race–class subjugated communities[138]—a blood-drenched backdrop propped up by a history of surveillance technologies, from the pen-and-paper databases of the eighteenth century[139] to the digital database and computer vision technologies of the twentieth century to the robots (mis)used by the police today.

II The (Mis)use of Robots by the Police

In late November 2022, the San Francisco Board of Supervisors agreed, by an 8-3 vote, that the San Francisco police should be explicitly allowed to kill people with robots. When asked why this was necessary, a police spokesperson told the Washington Post that while they didn't have any plans (yet) to arm the department's fleet of robots with guns, their robots might use explosives to breach "fortified structures" and might be used to "incapacitate or disorient" police suspects. The three supervisors who voted against the measure dissented strongly. As supervisor Dean Preston argued, "There is serious potential for misuse and abuse of this military-grade technology, and zero showing of necessity."[140] Within days, the Board of Supervisors was forced by widespread public protest to take the unusual step of reversing its decision. Yet, as media coverage of this reversal has suggested, it may only be a matter of time before the effort is ultimately reapproved.[141]

San Francisco's decision marks a troubling turn for the city, which, despite its liberal reputation and historic aversion to police robots[142] and facial recognition technology,[143] is increasingly deploying the police as the "shock troops of gentrification"[144] to defend gentrifying technologists' property interests.[145] In this section, we'll examine why the police are increasingly acquiring robots; how this continues their troubling history of increasing militarization; how the police are already using robots to surveil, bomb, and kill; and how robots uniquely stand to exacerbate the vicious cycles of the New Jim Crow that were discussed earlier in this chapter, thus compounding the ways that robots serve as New Jim Cobots that work to enforce that racial order.

Police Acquisition of Robots

The acquisition of robots by the police follows a decades-long trend of police militarization.[146] Although the police have historical associations with the military, police militarization was held at bay for the first century post–Civil War due, ironically, to the 1878 Posse Comitatus Act, which prevented the military from being involved in police business as a way of ensuring that the federal government would not interfere in the establishment of Jim Crow policies in the American South.[147] In the 1960s, however, amid the Civil Rights Movement, this separation between the police and the military began to unravel, with the advent of the LAPD's special weapons and tactics team. The LAPD's SWAT team was first developed as a means of suppressing Black protest and unrest[148] in response to the FBI's call for a "prompt, efficient, and impressive display of force" toward protests organized by civil rights leaders such as Martin Luther King Jr., Stokely Carmichael, and Cassius Clay.[149] As Fanna Gamal notes, it should thus be no surprise that the LAPD's first SWAT target was the Los Angeles chapter of the Black Panther Party.[150] Ultimately, SWAT teams spread across the country, becoming popular in cities of all sizes. And as the War on Drugs intensified under Reagan, the Posse Comitatus Act underwent further erosion. In 1981, Congress passed the Military Cooperation with Law Enforcement Act, which enabled the military to provide equipment to the police for use in antidrug policing.[151] This transfer of equipment was further encouraged by the 1990 antidrug 1208 program and a 1994 executive order authorizing transfer of used military equipment to the police.[152] And in 1997, Congress's annual National Defense Authorization Act included a provision—the so-called Section 1033[153]—that explicitly permitted the Department of Defense to sell or transfer military equipment to the police with very little justification needed.

Some examples of equipment transferred through the 1033 Program are comically wasteful, such as the ninety-two pairs of snowshoes transferred to El Paso, Texas,[154] which has a median snowfall of zero inches. But other examples uncovered by journalists speak more frighteningly to the downstream effects of this increasingly militarized mindset, such as the five grenade launchers acquired by Buena Vista County, Iowa (population twenty thousand)[155] or the $750,000 Mine-Resistant Ambush-Protected vehicles that have been sent to more than five hundred police departments throughout the United States.[156] A final example is that of Keene, New Hampshire (population twenty-three thousand). When I last visited Keene in October 2001, the social event of the weekend was a Welsh hymn-singing festival, or Gymanfa Ganu. In the ten years that followed 9/11, the town must have been plunged into a web of intrigue because in 2012, the town was granted a $286,000 police tank after fears of "terrorism" at other fall festivities such as its annual pumpkin festival.[157] Two years later, the tank sat dormant, but the town did deploy tear gas, rubber bullets, and pepper spray against attendees of that same festival—a move that was characterized in the international media as explicitly exacerbating any danger to attendees.[158] As of 2016, the 1033 Program had distributed $7.4 billion of equipment (and military-grade surveillance technologies[159]) to more than eight thousand law enforcement agencies.[160] Meanwhile, evidence has shown that police militarization neither reduces crime nor makes the police safer.[161]

The accelerating acquisition of military equipment by the police is fueled not only by the same racist undercurrent that motivated the War on Drugs more generally but also by the changes in the psychology of policing that have emerged over the past few decades. Today's police are increasingly trained to adopt a "warrior" mindset[162] in which they are told to "have a plan to kill everyone you meet"[163] due to a shift in both public and police conceptualization of "real" police work as masculine, dangerous, and for the purpose of "fighting crime."[164] Under this increasingly dominant police imaginary, policing is masculine due to increasing valorization of a "cult of masculinity" in which officers denigrate anything viewed as feminine, such as compassion, caring for victims, or dealing with children.[165] Policing is also viewed as dangerous, despite policing being safer than ever[166] and rarely involving responses to violent crime,[167] as "crime rates" have fallen to historic lows.[168] In fact, policing is less dangerous than professions such as truck driving (in terms of workplace deaths[169]) and nursing (in terms of frequency of workplace violence[170]). Finally, policing is increasingly perceived as "fighting crime," despite evidence that crime is not reduced by policing,[171] including community policing.[172] As David Bayley writes in *Police Reform from the Bottom Up*:

> The police do not prevent crime. This is one of the best kept secrets of modern life. Experts know it, the police know it, but the public does not know it..Yet the police pretend that they are society's best defense against crime and continually argue that if they are given more resources, especially personnel, they will be able to protect communities against crime. This is a myth.[173]

This overarching conceptualization of policing as a crime-fighting warrior brotherhood (despite the reality of policing) has produced an overarching ethos: "You're not a police officer serving community, you're a soldier at war."[174] This ethos has made the police increasingly willing to embrace—and seek out—the military equipment made available to them, including robots.

As of 2016, more than seven hundred robots had already been provided to the police by the Department of Defense alone,[175] and the police are increasingly finding new means of acquiring robots. Simple robots such as drones have long been used by US law enforcement, with nearly 1,600 such drones in use as of 2020,[176] and well-funded police departments are now beginning to invest in more expensive and complex robots. The NYPD, for example, has recently acquired a variety of new robots, such as Knightscope's K5 (which costs around $70,000 a year to lease) and Boston Dynamics' robot "dog" Spot (the cost of which starts at $75,000).[177] This decision comes two years after the NYPD last tried to deploy Spot—an effort that was met with massive public outcry, forcing the "retirement" of the robot before it could see full deployment.[178] Although former New York City Mayor Bill de Blasio's office was ostensibly "glad the Digidog was put down," Mayor Eric Adams, a former cop, has no such reservations.

In both academia and industry, many of the robots now being used by the police are not being purchased off the shelf or inherited from the military but instead being explicitly built for use in policing. Although robots such as drones are "general purpose" and can be used across a multitude of contexts, policing-specific robots feature policing-specific perceptual capabilities (such as automated license plate reading[179]), policing-specific actuation and manipulation capabilities (e.g., the ability to lay down "spike strips" during traffic stops[180] or deploy smoke[181]), and capabilities that meet the human–robot teaming demands specific to organizations such as SWAT teams.[182]

It is easy for non-roboticists to underestimate the amount of close collaboration with roboticists that is needed for organizations to use even off-the-shelf robots. Robots deployed in any real-world context must typically be augmented with additional sensors and integrated with other software programs, and they may even need extensive custom software built by roboticists to facilitate deployment. In her keynote address at the 2023 International Conference on Human–Robot Interaction, for example, roboticist Andrea Thomaz, the CEO of hospital robotics company Diligent Robotics, described the intense efforts that Diligent has had to undertake when deploying robots in different hospitals. One of the biggest challenges faced by hospital robots is doors. To travel through a building such as a hospital, robots need to be able to perceive, locate, and open a variety of different doors. As such, Diligent has worked hand in hand with the hospitals into which its robots are deployed to make sure that those robots have the tools they need to make it through each troublesome door. Robots acquired by the police may need similar cooperation from roboticists to be used effectively. As such, even if companies such as Boston Dynamics claim their robots are general purpose and not explicitly designed for policing, the sale of their robots to the police likely requires specific collaborative actions by roboticists to make their robots work as desired in policing contexts.

What Are the Police Doing with Robots?

Some news outlets have cast police acquisition of robots as merely a PR stunt or as the police flaunting their ability to purchase expensive toys.[183] Although this certainly plays a role in why the police might acquire robots, it would be a mistake to assume that this is all these robots are truly being used for. Indeed, since the police have begun to acquire robots, they have used them to intensify the activities that otherwise define policing: surveillance and violence.

Robot-Assisted Police Surveillance Robot companies are quick to claim that the robots they provide to the police are not designed to be used as tools of surveillance. A representative of Boston Dynamics, for example, pushed back on such concerns by stating, "It's noisy and has flashing lights . . . It's not something that is discreet."[184] Yet, a lack of discreetness may actually make some robots ideally suited for certain types of police surveillance. One of the defining characteristics of panopticon-like systems of surveillance is that while those surveilled cannot see *who* is surveilling them, they can see the *mechanisms* of surveillance. Specifically, the knowledge that one is being surveilled under such a system is what allows the overall surveillance architecture to exert disciplinary power, steering behavior regardless of whether anyone is actually watching.

Police use of overt surveillance can be seen outside of robotics through their use of helicopters. The police have a long history of using aircraft in their operations, with police use of fixed-wing aircraft dating to 1929 and police use of helicopters dating to 1947.[185] Police helicopters have proved particularly effective for wielding disciplinary power, discouraging undesirable activities through their hypervisibility. This is especially effective when searchlights are used to transform helicopters into "nightsuns" of sound and light.[186] Today, the LAPD uses helicopters in this fashion to an almost ridiculous extent. As Brayne reports,[187] the LAPD has been shown to fly their expensive helicopters—which they call "ghetto birds"—*thirteen times a day* over software-defined hot spots, a rate justified by evidence that a neighborhood needs to be flown over fifty times per week in order to induce any meaningful "reduction in crime."

When the police use drones for surveillance, they often rely on this same hypervisibility in order to make their surveillance intimidating.[188] And because drones are so cheap and easy to use, they allow a much wider range of police departments (who cannot afford helicopters) to engage in such surveillance. For example, journalists Dhruv Mehrotra and Jesse Marx have reported that the Chula Vista Police Department has made twenty thousand drone flights over the past six years alone. These flights have been made predominantly over Latinx and working-class neighborhoods and on "flight paths [that] routinely take them over backyards and above public pools, high schools, hospitals, churches, mosques, immigration law firms, and even the city's Planned Parenthood facility."[189] The stress induced by this atmosphere of mass surveillance has led to the hospitalization of some Chula Vista residents. As one resident despaired, "It feels like our home is not ours anymore. It's like it belongs to the Chula Vista Police Department."[190]

On the other hand, the police are also using less overt drones for more intentionally covert surveillance. When Texas DPS chief pilot Bill C. Nabors Jr. was interviewed about the use of Drones in Texas SWAT operations, he stated, "The nice thing is it's covert . . . You don't hear it, and unless you know what you're looking for, you can't see it."[191] Accordingly, the police have used drones to enact covert surveillance in a variety of contexts, with targets ranging from those attending music festivals[192] to those living in homeless encampments[193] to more overtly racialized contexts of surveillance.[194]

Whether aerial police robots are hypervisible or invisible, their use in the big data dragnet will accelerate as robots become more autonomous. Unlike security cameras that watch from static positions, the autonomous aerial robots of the future will seek out and follow individuals, targeting them for enhanced surveillance on the basis of policies set by the police or automated recommendations made by predictive policing software[195] and recording what they see in databases where it can be automatically mined and used to feed those same predictive policing systems. From their new vantage point, these mobile surveillance robots will see more, will be harder to avoid, and will further dehumanize those being surveilled.

Whereas police drones increasingly take to the sky, ground-based police robots are taking to the streets—and are being designed in a way that represents a perfect storm of ethical concerns. Many ground-based police robots are intentionally anthropomorphic in design, appearing humanlike in order to increase their acceptance[196] and to enhance their persuasive power so that they can better encourage users to comply with robot-given commands. Moreover, many of these anthropomorphic policing robots are designed to evoke the image of a White male police officer. An example of this is the White masculine REEM robot from PAL robotics discussed in previous chapters. As previously discussed, one place where REEM has been deployed has been on the streets of Dubai as a robocop, as Dubai officials reportedly want these robots to eventually comprise 25 percent of their police force.[197] According to reporting in *Futurism*, these robots are being used to perform facial recognition on users, detect crimes, and report those crimes back to the police. Most of these robots are, of course, still teleoperated rather than autonomous, and in these cases, anthropomorphic robot design is often intended to remind those surveilled that there is a human behind the robot.[198] By extension, we can assume that the White masculine design of police robots is likely intended to remind the people of color being surveilled[199] that although the identities of those watching behind robots' cameras may be obscured, they—or the state power they represent—should be understood as White and male. Robots such as REEM, however, are simply the most obvious and visible form of robocop. Although the police robots built for or acquired by the police allow law enforcement to surveil people in public spaces, consumer social robots such as Amazon's Astro may allow police surveillance efforts to move inside the home.

Although many consumers think about Astro as a new kind of Echo smart speaker, and despite the team of careful, thoughtful, privacy-sensitive, and genuinely brilliant human–robot interaction researchers who have worked on the robot, Astro may be better thought of as a new frontier for police surveillance. Investigation of Astro's product page reveals that although Astro is, in part, marketed as a fun household robot that protects user privacy, at its core, it is a mobile Ring camera that tracks users inside their homes. Astro is advertised with an optional bundled subscription to Ring and highlights Astro's ability to "proactively patrol, investigate activity, save videos in Ring's cloud storage for up to 180 days, and more." As we have seen, Ring's cloud storage represents a surveillance clearinghouse to which the police have priority access, and Amazon regularly accedes to police requests for voice data from their smart speakers.[200]

Due to this casting by Amazon of Astro as a Ring device, it is reasonable to suspect that there may eventually be a set of policeman's eyes watching what is recorded by this social robot, and as such, it is reasonable to suspect that Astro may ultimately enable the encroachment of racialized police surveillance into the private sphere. In the commercial that introduced the Astro robot, a Black woman is shown asking her husband, "A robot? What are we going to do with a robot?" Although this is a reasonable question for her to ask her husband, who apparently purchased a $1,600 robot without consulting her, perhaps the more important question is what Amazon and their police collaborators, rather than she and her husband, intend to do with the police surveillance camera on wheels that they welcomed into their home.

This is not to say that Astro is uniquely at risk of being used to extend police networks of racialized surveillance. As introduced in chapter 4, classroom robots, for example, might well present similar risks. Robots for tutoring and education may represent one of the greatest success stories and opportunities of social robotics (assuming, at least, that robots are publicly funded and distributed equitably to schools). But as the police increasingly infiltrate schools in the role of school resource officers, roboticists should become more concerned about whether the data collected by their robots will be used as an opportunity or excuse for these school-stationed police to accelerate the rate at which they funnel students into the punitive carceral system through the school-to-prison pipeline.[201] On the other hand, police (mis)use of robots—in schools, homes, and city streets—goes far beyond passive surveillance. Just as the police have been using robots as their newest tool of (racialized) surveillance, so too have they been using robots as their newest tool of (racial) violence.

Robot-Assisted Police Violence One casualty of robot-assisted policing is Jose Guerena, a young Marine veteran killed by robot-equipped and heavily militarized police forces in Tucson, Arizona. After two tours in Iraq, the twenty-six-year-old veteran was shot with twenty-two bullets in his own home, leaving behind his wife and two children, in an ostensible drug raid. No drugs were ever found at the home. The somber conclusion of author David Axe reads: "One thing is clear. With military-grade vehicles, armor, assault weapons and robots, the raid on Guerena's home was all but indistinguishable from the kind of house-clearing operations U.S. forces perform every day in Iraq and Afghanistan. Guerena survived two tours in the desert only to perish in a military-style action in his own home."[202]

Since Guerena's death, the police have ramped up the violent capabilities of their robots by combining those robots with other tools of violence, such as explosives and chemical weapons. In 2014, for example, Albuquerque police used a bomb disposal robot (a Northrop Grumman Remotec ANDROS) to (in their own words) "deploy chemical munitions"[203] in order to induce the surrender of Stephen Fought, a mentally ill man who had barricaded himself into a hotel room. As Fought recounted in his interview with VICE News, "They sent that robot over there, and I thought, 'Man, what is that?' You know? . . . They threw a grenade in there, it started smoking, I threw it back out. They threw another one, I threw it back out."[204] Although the use of a bomb *disposal* robot being used to inflict harm may seem ironic, this use case had already been envisioned by roboticists for nearly a decade. In 2007, for example, roboticists Carl Lundberg and Henrik Christensen deployed a bomb disposal robot with Swedish SWAT teams for five months, an effort that was so successful that the roboticists projected the robot could be similarly deployed by the Stockholm SWAT unit "at least 20 times a year" in the future. How did these roboticists foresee their bomb disposal robot being used? In part, for "long-term surveillances and deploying non-lethal weapons."[205]

Two years after the Albuquerque Police Department's attempted gassing of Steven Fought, Dallas police used their own bomb disposal robot (a Northrop Grumman Remotec Androx Mark V A-1) to kill Micah Xavier Johnson, a Black veteran with post-traumatic stress disorder[206] who had killed five White police officers in an apparent response to anti-Black police violence.[207] To kill Johnson, the police sent a robot down a hallway toward him, telling him they were bringing him a phone.[208] Instead of bringing a phone, however, the robot instead delivered a pound of C4 to the opposite side of a wall from Johnson, releasing a blast that turned the rubble of the brick wall into "little missiles" passing through Johnson's body "like butter" so that he would bleed out while awake and confused.[209] Although Johnson's killing of the police officers was clearly unjustified and rooted in his own mental illness, the police response to his violence disconcertingly echoes the original rationale for robots envisioned by Ellis in *The Steam Man of the Prairies* discussed in chapter 2, in which a young White boy assaults Native Americans and then, when they fight back, sends his robot to blow itself—and them—up, leveraging his technological advantage to violently assert his racial dominance.

The police used a similar tactic two years later in Bangor, Maine, deploying a robot to blow up the home of sixty-one-year-old Michael Grendell, a mentally ill man living alone.[210] After Grendell had shot at (but not hit) a friend of his, the friend called the police in hopes of getting Grendell help for his mental illness. The police responded by driving an armored vehicle into Grendell's home and smashing his windows. "Not surprisingly, Mr. Grendell was terrified and began firing."[211] The police then navigated an explosive-laden robot next to Grendell's home and used it to destroy part of the house. Grendell emerged from the house, and the police shot him in the face, wounding but not killing him.[212] Here again, we see the police using explosive-laden robots to (attempt to) kill individuals with mental illness. And in Grendell's case, it is particularly obvious how unnecessary the police's actions were: no immediate danger required the police to destroy Grendell's home or to attempt to kill him.

Thankfully, the police have not used robots to deliver explosives since 2018, perhaps due to the blowback this tactic received and perhaps due to the shifting nature of policing during the COVID-19 pandemic. But recent rumblings, such as San Francisco's expressed intention to explicitly allow for this type of robot-assisted police violence[213]—and roboticists' continued development of White masculine robots intended to brutally enact physical violence[214]—suggest that we may be in the eye of the storm rather than beyond its borders. As such, it is worth considering more carefully the unique risks posed by police robots, especially the ways in which robots may exacerbate police violence.

Robots Exacerbate Police Violence

Robots stand to exacerbate police violence due to the influence they have on police psychology. Police robots lead policing to be perceived as physically and psychologically safer while making police violence more cognitively accessible. First, let's consider the argument sometimes made in favor of giving explicitly violent robots to the police because it might "make them safer." As we have established above, the danger of policing is largely a myth. Policing is rarely dangerous, and when the police do encounter danger, it is often due to situations that they themselves created. Moreover, evidence shows that police killings don't correlate with the amount of danger they encounter or even with crime rates[215] and are instead more accurately predicted by differences in police department culture.[216] A second argument made in favor of police robots is that although robots don't actually enhance police safety, they might decrease police violence by making the police *feel* safer, since police shootings are typically committed in response to claimed concerns about their own safety.[217] Yet, it is well established that dishonesty is a core facet of police work. The police learn nuanced tactics of dishonesty in order to coerce compliance from those they target,[218] manipulate the public,[219] and sway legal proceedings when under oath.[220] The police are rewarded and incentivized for lying and are told it is just part of good police work.[221] In the context of police violence, the police regularly kill people where it is obvious that there is no possible source of threat yet claim that they felt threatened in order to exonerate themselves,[222] despite it being exceptionally rare for the police to be held accountable for violence, even in cases where there they lack any convincing excuse. As such, it seems unlikely that robots would play any meaningful role in decreasing police violence by increasing police safety.

Even if robots *did* increase police safety or *did* increase police perception of safety, the use of robots as a technological buffer between the police and their victims would itself lead to increased violence.[223] It has been well established in military contexts that the physical distance robots put between operators and targets exacerbates the use of extrajudicial lethal violence. In part, this is due to the "moral buffering" enabled by technology.[224] As human–robot interaction researcher and former fighter pilot Missy Cummings describes, "A moral buffer adds an additional layer of ambiguity and possible diminishment of accountability and responsibility through an artifact or process, such as a computer interface or automated recommendations. Moral buffers can be the conduits for moral disengagement."[225] Accordingly, drones used in Iraq provided a moral buffer that dehumanized Iraqis and reinforced their perception as physically, psychologically, socially, and culturally distant, facilitating their killing by US drone pilots. Similarly, the use of robots will make it psychologically easier for the police to pursue violence.

Police use of robots may also make it psychologically easier for the police to enact violence due to the reduced threat of embarrassment if their efforts fail. For example, we can consider the case of manhunts—an activity that allows the police to reassert their identities using asymmetric force to justify their existence[226] and in which violence is wielded to "track, search, seize, and accumulate the property and bodies of those deemed illegal."[227] Due to the racial character of policing, and because those deemed illegal are, in general, those racialized as non-White, manhunts allow the police to reassert their original role as trackers, searchers, seizers, and thieves of the property and bodies of people of color. But today's manhunts also present a source of risk for the police, as failure to execute manhunts successfully is a public sign of police weakness.[228] In 2022, for example, the LAPD was publicly humiliated by news footage showing a suspect hiding to one side of a parked car while the police obliviously walked by on its other side. In the video, the aerial reporter watching the scene notes, "He's trying to hide but doing a very poor job of it" before shouting at the officers as one might shout at their TV, "C'mon guys! He's right there!" The LAPD must have been particularly unhappy when this footage went viral.[229] Robots provide a unique opportunity for the police to avoid such situations, projecting power without projecting vulnerability.[230] Officers flying a drone may be afforded the same view as the media, and footage of a drone flying near a suspect would not present the same visual effect as that of a bumbling and visibly confused officer. As such, when robots are used to conduct manhunts, they simultaneously enhance (and serve as symbols of) police power and prevent opportunities for humiliation, providing the state a position of unassailable dominance.[231]

The final way that robots make police violence easier is by making physical violence more cognitively accessible. Dutch philosopher Peter-Paul Verbeek has argued, through his theory of technological mediation, that technologies play a key role in shaping how humans perceive their world.[232] Just as glasses allow perception at a more precise and refined level of detail and encourage focusing on otherwise obscured details, teleoperated robots encourage the police to focus on the details visually perceivable through the field of view of, and from the perspective of, their robots rather than from the perspective they would have if they were "on the ground." And just as magnetic resonance imaging scans provide access to certain otherwise imperceptible types of information and encourage radiologists to view things in term of what such scans make visible, computer vision–assisted robot perception that perceives and calls attention to individuals, or that automatically classifies those perceived in terms of particular attributes (such as race and gender, as discussed in chapter 5), shapes police perceptions and encourages them to make sense of the world in terms of those categorizations.

Verbeek's theory of technological mediation also explains how technologies shape the ways that humans act in their world by presenting or obstructing opportunities for action and enforcing scripts used to guide sequences of actions. A clear example of this is the design of social media platforms, each of which encourages different forms of interaction between users by providing (and incentivizing) different types of actions, such as commenting, liking, and reposting. Similarly, designing police robots to safely transport and manipulate explosives encourages the police to use "transporting and manipulating explosives" to resolve their problems, and outfitting police robots with shotguns, tasers, and teargas encourages the police to use those weapons to "interact" with citizens.

In the case of policing, these new affordances align with existing police motivations in dangerous ways. LAPD officers have previously commented that they would like to fly drones into Black communities to wipe them "off the map."[233] So, when these police are provided with data science technologies that encourage them to see the social world as a network of racialized suspects and are provided with militarized robots that encourage them to interact with that world by disorienting, demolishing, and dispatching, it is unsurprising that they readily adopt those technologies and their affordances for viewing and acting on the world.

Robots Exacerbate Cycles of Oppression

Although all police robots stand to exacerbate violence, anthropomorphic police robots may acutely increase the state's ability to reinforce White patriarchal power across all four domains of power. Police robots designed as White and masculine in their appearance and voice serve as tools for wielding cultural power (as discussed in chapter 3), reinforcing the associations between race, technology, and power that first emerged with the envisioning of *The Steam Man of the Prairies* (as discussed in chapter 2). The humanlike appearance and language capability of these robots also allows them to serve as tools for wielding interpersonal and disciplinary power by leveraging the persuasive power of humanlikeness and language capability to enforce compliance with the norms, laws, and power structures of White patriarchy (as discussed in chapter 4). The use of police robots as obvious mobile surveillance systems—especially when they employ face recognition and identity classification systems (as discussed in chapter 5)—enhances their ability to serve as tools for wielding disciplinary power because the knowledge of their surveillance (and the threat of their violence) encourages compliance with the norms, laws, and power structures of White patriarchy and because the recognition and categorization of individuals according to the logics of White patriarchy serve as key mechanisms for reinforcing and reinscribing those logics.

But perhaps most fundamentally, these police robots, however they look and however they speak, serve as tools for wielding structural power through the ways that robotic surveillance feeds the cycles of oppression in which the police operate as central cogs. Earlier in this chapter, I described Michelle Alexander's cycle of the New Jim Crow,[234] in which the police (1) disproportionately surveil Black communities, (2) disproportionately stop members of those communities, (3) assess those stopped against laws disproportionately designed to target members of those communities, and (4) disproportionately arrest members of those communities, after which other elements of the criminal justice system convict and sentence members of those communities at disproportionate rates. This cycle leads to disproportionate mass incarceration of Black Americans and their disproportionate condemnation to a racialized under-caste—an effect that should come as no surprise, given that this was the original purpose of policing. Based on what we've seen in this chapter, police robots have a clear potential to accelerate this cycle. Robots enable the police to more effectively blanket Black neighborhoods with surveillance, marking them for looting by the police[235] and exacerbating the numbers of Black Americans being subjected to every other part of this pipeline. As such, the use of robots for surveillance—especially when paired with predictive policing technologies and facial recognition systems that give the police an excuse to stop, accuse, and arrest members of those surveilled communities—may dramatically accelerate mass incarceration and social inequality.[236]

Although the White patriarchal robot imaginary may have originally envisioned robots as slaves, and although Asimov's laws may have envisioned robot ethics as a matter of keeping robotic slaves in line, we have seen in this book how robots may just as easily serve as the enforcers of White patriarchy by applying those slave codes to humans. And as we are seeing in this chapter, when robots are placed in the hands of the police, they may complete this transformation by directly—and indirectly—facilitating the (re-)enslavement of people of color. This should raise a serious question for the field of robotics: Are roboticists who collaborate with the police necessarily and inherently complicit in the White supremacist order and system of slavery that policing facilitates?

Roboticist–Police Collaborations Legitimize Policing

Due to the ways that surveillance and even bomb disposal robots can exacerbate police violence, some roboticists have restricted their collaborations with the police to only those domains where the potential for police surveillance and violence is minimized. The creep of police power has led to the police being put in charge of a variety of important tasks that they have no business leading, such as responding to mental health crises and coordinating search-and-rescue efforts. Some roboticists have thus argued in favor of pursuing police collaborations in these domains on the basis that the objectively prosocial outcomes possible in those domains cannot be pursued without those collaborations. But as I have argued with my coauthor Kerstin Haring, when the police use robots in service of these prosocial activities, those robots help to legitimize policing as a system.[237]

Police legitimization is the process whereby the perceived credibility of the police—and of police use of technology—is bolstered through their association with higher-credibility institutions, such as companies and universities. This heightens the public's view of the police as a natural and legitimate feature of society rather than as an anachronistic artifact of our racist past that we actively empower to do harm. Police legitimization is often achieved through academics or companies collaborating with the police and disseminating the results of those efforts through academic publications. When those publications are accepted and published by a scientific society (e.g., the IEEE or ACM), this gives the police–roboticist collaboration the tacit approval of the researcher's institution and the society who published the work. This, in turn, not only leads to the development of technologies that (typically) increase societal inequality and exacerbate police violence but also lends the collaborating police department an air of *scientific* legitimacy, thus reinforcing the overall legitimacy of that department and of the institution of policing as a whole. Indeed, the use of technology has historically played a central role in the legitimization of policing, as the police have used technology, along with statistics, to signal efficacy and accountability[238] where efficacy and accountability do not exist.

To see how police legitimization works, I'll provide two examples. First, we can consider the case of IBM described by Charles McIlwain in *Black Software*. In the 1960s, IBM collaborated with the Kansas City Police Department on a variety of predictive policing efforts, including racial profiling systems that equated poor Black masculinity with criminal intent.[239] As a result, IBM provided the Kansas City Police Department with more effective tools for enacting racialized surveillance and racial violence.[240] But moreover, because IBM published papers on this work in scientific venues, it essentially served as a mule, smuggling this societally harmful work into venues sponsored by organizations whose professional mission is to do societal good and casting the Kansas City Police Department as a participant in scientific endeavors with internationally renowned research institutions.

Second, let us consider a more recent—and subtler—example. On June 30, 2023, Peoria Illinois news station WMBD reported a collaboration between the Peoria police and local company Pringle Robotics, in which robots were to be deployed at Peoria's "Red White and Boom" Fourth of July celebration. As reported by WMBD:

> The bots will have ads and celebratory messages on their front screens and the bottles of water distributed by them will be donated by the American Red Cross, Central Illinois Chapter. The bots will have special wrap on them indicating the partnership with the police department. Pringle Robotics stated that students and those local to Bradley Univeristy [sic] will recognize the bots as they are in collaboration with Bradley to pilot food delivery robots.[241]

Handing out bottles of water to prevent heatstroke seems like an obviously prosocial use of robots. But it is also unclear why it would be the responsibility of the police to provide this public health service. As such, both the event (where police involvement was literally plastered across the robots) and the reporting on that event (with headlines such as "Police, Pringle Robotics collaborate for Red White and Boom safety") primarily served to paint the police as the beneficent providers of a public health service. Moreover, both the event and its reporting highlighted the associations between the police, the American Red Cross, the local university, and the local tech industry, thus leveraging the reputations of those institutions to provide PR support for the Peoria Police Department (a department in the top [worst] 25 percent of police departments in terms of overfunding[242]). Were it not for the collaboration with roboticists to add a sense of excitement, perhaps Peorians would have been more likely to question why men with guns were being put in charge of bottled water distribution.

Although debating who hands out bottled water may seem like a minor issue, this is just one small example of how the funneling of money, resources, power, and responsibility to police departments depends on continued legitimization of the police. Just as the militarization of the police relies on the public accepting a variety of myths (that policing is uniquely dangerous, that the world is dangerous without policing, and that police officers have a unique specialized professional authority[243]), the overall legitimacy of the police depends on the public accepting myths surrounding the role of the police as an ultimately benevolent institution.

Roboticists also legitimize the police when their research paints policing as socially conscious and ethically grounded. Technology and society researchers have raised a number of concerns recently about tech companies avoiding regulation by engaging in "ethics washing," in which they use "ethics boards" or other mechanisms to fend off concerns about the impacts of their technologies or engage in "ethics shopping" to cherry-pick ethical analyses that justify their behaviors.[244] We are now beginning to see similar phenomena emerge in the context of police robots, with some researchers beginning to analyze the privacy and ethics risks surrounding police robots based on interviews with the police themselves rather than with the communities impacted by policing. Unsurprisingly, the conclusions of such efforts tend to emphasize the need to encourage public acceptance and trust in police robots, thus taking police robots, police surveillance, and policing itself as foregone conclusions.

Not only do prosocial uses of police robots and ethics-washed police–robot collaborations serve to legitimize the police, but they also risk *de*-legitimization of the companies, universities, and professional organizations whose reputations are exploited to bolster the police. Specifically, police–roboticist collaborations may harm the reputations of the institutions involved, especially in the communities most likely to be harmed by the police. To see how reputational harm to academic institutions might result from collaboration, we can consider the harm done to MIT through their relationship with notorious pedophile and sex trafficker Jeffrey Epstein. As Ronan Farrow reported in 2019 for *The New Yorker*, the MIT Media Lab (a massive creative technologies research center) established a deep relationship with Epstein, with more than $7,000,000 sent to the Media Lab through Epstein. Despite Epstein being placed on a donor blacklist by MIT, Media Lab director Joi Ito nevertheless took donations from Epstein, despite knowledge of his role in child sex trafficking, and attempted to obscure the source of those donations.[245] Moreover, Ito took further funds from Epstein for tech investment funds that Ito controlled and stayed at Epstein's properties multiple times.[246] Around the same time, it was revealed that one of Epstein's victims had testified to being forced to have sex with Media Lab cofounder and AI pioneer Marvin Minsky at one of Epstein's properties.[247]

When the relationship between Epstein and the Media Lab was revealed, it did immediate harm to MIT: Ito was forced to resign; high-profile Media Lab researcher Ethan Zuckerman voluntarily resigned in protest[248]; and star researcher Richard Stallman, a pioneer of the open source software movement, was forced to resign after *defending* Minsky by arguing that "sex with a child isn't always assault."[249] Moreover, the association between MIT and Epstein became seared into the public consciousness.

Although the revelation of collaborations between roboticists and the police might not harm roboticists' reputations among the large percentage of (predominantly White) Americans who actively support the police, they may well do harm to themselves, their institutions, and their academic fields in the eyes of Black, Latinx, and Indigenous communities harmed by the police. As a result, students from those communities may avoid the laboratories, departments, universities, and entire fields that they perceive as complicit. This is especially concerning, given the existing demographic disparities that plague the fields of robotics and computer science[250] and due to the explicit need for a demographic shift to address the White supremacist status quo of robotics across the cultural, disciplinary, and structural domains of power that have been explored in this book.

III Subverting Robots' Default Reinforcement of White Patriarchy in the Structural Domain

As we have discussed in this chapter, roboticists wield power in the structural domain through their decisions as to whom to collaborate with. When American roboticists collaborate with the police, they collaborate with the institution that is responsible for using force to ensure that the domestic hierarchies of race, class, and gender that benefit White men with a high socioeconomic status are maintained. This means not only that roboticists collaborating with the police risk enhancing the ability of the police to reinforce those hierarchies in the cultural and disciplinary domains, but also that they risk wielding their power in the structural domain in a way that legitimizes the police as an institution and normalizes the use of the police as a means of addressing societal problems and exerting social control.

In this section, we will once again, and for the last time, consider how we might address this problem if we (1) revised the ways that roboticists approached collaborations between robotics and the police, (2) rethought the nature of police–roboticist relationships, or (3) ripped up the status quo by more directly addressing the sociohistorically determined root causes of the problems identified in this chapter.

Revising Our Approach to Police–Robot Collaboration

First, let us consider how we might simply revise how roboticists approach collaborations with the police. If providing robots to the police risks enabling the police to more effectively pursue their founding purpose of enforcing White heteropatriarchal capitalism, then perhaps roboticists should consider not building robots for the police that help them achieve this purpose. Refusing to build robots intended to help the police surveil or impose violence on already oppressed communities is a good start toward revising the nature of police–roboticist collaborations. However, there are a number of reasons to suspect this approach might not be effective. As described above, even when robots are provided to the police for purposes such as bomb disposal, those robots too often end up misused for exactly the opposite of that intended purpose. Moreover, as described above, the real power roboticists wield in the structural domain is their ability to legitimize or normalize institutions such as policing.

As Matthew Guariglia argues, academics who collaborate with the police have long served as violence workers. Writing about early twentieth-century proponents of policing research, Guariglia argues, "Although [they] never personally pulled a trigger or swung a club, they helped to build, justify, and legitimize a system of policing and power relationships that caused violence."[251] If roboticists wish to revise the status quo of police–roboticist relationships to avoid similar classification as violence workers, they may start by boycotting police collaborations of any form.

To understand how the refusal of individual roboticists to collaborate with the police could represent an effective means of wielding structural power, let's consider an example of how acts of individual refusal can be seeds of collective action. Two days after the outbreak of police violence described at the start of this chapter, I sent an email to robotics colleagues in Colorado in which I argued that it was "important for us to make clear to the public that we do not support, and refuse to actively facilitate, the use of robots to further militarize our police," and invited them to join me in crafting a statement to that effect. Several colleagues responded positively, and together we wrote an open letter to the robotics community entitled "#NoJusticeNoRobots." The letter (which I broadcasted to the robotics community through the roboticsworldwide mailing list) made three major commitments:

1. We refuse to facilitate the execution or publication of research in robotics or Artificial Intelligence (AI) that is performed in collaboration with State and Local Law Enforcement agencies (or national agencies such as ICE and CBP).
2. We refuse to pursue or accept funding for research projects whose express goal is development of robotic technologies or AI technologies (e.g., nonconsensual face recognition technologies, predictive policing models, etc.) intended for use by such agencies.
3. For those of us who are affiliated with colleges and universities, we pledge to petition our universities to divest themselves of all ties with law enforcement agencies that have demonstrated brutality, prejudice, or White supremacist tendencies.

The #NoJusticeNoRobots campaign was ultimately signed by around one hundred professors, postdocs, industry members, and PhD students.[252] This was an important step to take as a community and showed how we could collectively redirect our structural power. Yet, the limited reach of this campaign, and its ultimate failure to stem the tide of police robotics, also showed the inherent limitations of merely revising the status quo. This suggests a need to consider how roboticists might go farther to more fundamentally rethink the relationship between roboticists and the police.

Rethinking the Relationship Between Roboticists and the Police

What would it mean to more fundamentally rethink the relationship between roboticists and the police? If roboticists wish to go beyond mere refusal to collaborate with the police, perhaps they should work to actively oppose or disrupt the police use of robots. If our primary social justice concerns surrounding Amazon's Astro, for example, stem from privacy concerns surrounding the videos stored by Astro and the ways the police might access those videos, given Astro's use as a Ring platform, then these concerns might be addressed by adopting a field-wide standard that robots be private by design,[253] incapable of recording, or perhaps of even perceiving, information that could be used by the police or fed into police databases. This could be achieved either by community-wide movements to assume basic robot privacy standards or by pursuing legislation that makes it illegal for robots to store recordings of individuals without their consent.

Alternatively, roboticists might take even more overtly oppositional approaches to police robots by exploring ways that robots could be used to counter police surveillance and violence—an approach that has historically been used to oppose policing while simultaneously sowing the seeds for the next generation of technological innovation.[254] This might be achieved, for example, by developing drones and drone control interfaces that meet the specific needs of activists hoping to use such drones for sousveillance. As defined by Steve Mann, sousveillance inverts the typical panoptic gaze of surveillance, enabling those typically watched by the police to themselves watch and monitor the police or others with asymmetric power over them.[255] As an example, we can consider the ultra-cheap "occucopter" drones used by Occupy Wall Street protesters in New York City in 2011.[256] Due to the low cost of these drones, they became an effective way to skirt police cordons intended to limit the gaze of protesters and media alike. Similarly, DIY camera-equipped kites have been used to document the Israeli oppression of Palestinians in East Jerusalem.[257] And as Dennis Zueb and Gary Bratchford describe, low-cost drones were used by Sioux natives at the Standing Rock protests of 2016 to expose the environmental devastation of the Dakota Access Pipeline project and the police violence wrought against those protesting it.[258]

Roboticists hoping to pursue sousveillance applications might also take inspiration from recent projects in neighboring fields, such as Matt Franchi and colleagues' use of rideshare drivers' dashcam footage to precisely quantify disparities in police presence throughout New York City, including the overconcentration of the police in high-income centers of White commerce and in lower-income centers of Black and Latinx life.[259] There are clear ways that this type of approach could be leveraged in robotics, especially with the rise of robots in intracity grocery, restaurant, and package deliveries.

Ripping Up the Status Quo: Toward an Abolitionist Robotics

Finally, let us consider what it might mean to rip up the White patriarchal status quo of social robot design within the domain of structural power as it pertains to the recognition and categorization of humans.

If the reason why the use of police robots stands to reinforce White patriarchal power in the structural domain (regardless of what those robots are actually designed to do) is because policing began, and has historically functioned, as a way to enforce (wealthy) White male interests, then perhaps we should attack that root cause more directly through an abolitionist approach that works to move our society beyond the need for policing.

Why Try to Move Beyond Policing Rather Than Reform Policing? If roboticists' collaborations with the police lead to unavoidable harms, regardless of the nature of such collaboration, then shouldn't we be pushing for police reform? After all, there are plenty of socially beneficial applications of social robots that require working with the police. Bomb disposal, search-and-rescue-efforts, and public health initiatives present clear opportunities for social good.

Since at least the 1960s, organizers have pushed for a number of major reforms that have had a significant impact on policing.[260] Unfortunately (as foreshadowed earlier in this chapter), many of these reform efforts (such as COMPSTAT, Hot-Spot policing, and Signs-of-Crime or Broken Window policing) have demonstrably exacerbated police violence and mass incarceration.[261] Moreover, as criminologist Alex Vitale has shown, even community policing approaches and efforts to increase police accountability have been ultimately (and perhaps unavoidably) ineffective.[262] As Vitale discusses in his book *The End of Policing*, police reform efforts tend to backfire for at least three key reasons. First, police reform efforts tend to divert money to the police from government programs that actually prevent crime, such as affordable housing, income support, and community health initiatives. Second, these efforts integrate the police into even *more* elements of society,[263] not only leading to the mission creep that saw police-affiliated robots distributing water at Peoria's Fourth of July celebration but, more importantly, presenting new opportunities for corruption, discrimination, and abuse[264] without yielding any demonstrable improvements. And third, these efforts, especially accountability efforts, are often rendered ineffective by governmental incentive structures and organizational challenges. One example examined by Vitale is that of marijuana decriminalization efforts. In New York, for example, marijuana possession was reclassified as a violation rather than a felony, ostensibly reducing the risk of overly harsh sentences being meted out for drug crime.[265] Unfortunately, the NYPD nevertheless used this to target minorities, *ramping up* non-felony drug arrests through stop-and-frisk policies.[266] Furthermore, by increasing the use of summonses to appear in court for these minor violations, those targeted by the police were increasingly forced to choose between losing their jobs and facing criminal charges for failure to appear (even though it is the police themselves who are more likely to fail to appear when required[267]). Ultimately, the evidence suggests that regardless of the motives of individual police departments, the origins, history, and nature of policing makes reform unlikely to be successful as a means for roboticists to rip up the status quo and thus an ineffective antidote to the harms wrought by current police–roboticist collaborations.

What Would It Look Like for Roboticists to Move Beyond Policing? If roboticists wish to rip up the historical root causes that lead to current police–roboticist collaborations reinforcing White patriarchal power in the structural domain, and if advocating for police reforms might, if history serves as a guide, only serve to entrench police power and provide new opportunities for police violence and corruption, then roboticists might be best served adopting an abolitionist perspective that pushes for non-police solutions to the societal problems roboticists wish to address.

Abolitionist movements are perhaps most well-known for their attempts to shift resources away from police departments or to do away with police departments altogether,[268] as called for by abolitionists such as Angela Davis,[269] Ruth Wilson Gilmore,[270] and Mariame Kaba.[271] But focusing solely on the desire of such movements to minimize or do away with policing ignores those movements' *constructive* nature.[272]

As Lisa Lowe writes in her analysis[273] of the work of Ruth Wilson Gilmore[274]:

> Abolition . . . must mean forging robust alternative solidarities and social relations that do not yet exist . . . abolition is not simply the absence of prisons, policing, and criminalization; it is the presence of social relations of care, mutuality, and interdependency that make punishment and incarceration unnecessary. Abolition is aimed at the elimination of the racial capitalist social order that produces vast wealth accumulation for the few and vulnerability for the many, and which violently enforces and deepens those conditions.[275]

The need to create new institutions has been a critical facet of abolitionist efforts for centuries. Even W. E. B. Du Bois's *Black Reconstruction* emphasized that emancipation was both about ending slavery *and* about establishing citizenship for those enslaved.[276] And today, as Juan Lucena and Jon Leydens argue, a constructive approach is a critical facet of just engineering, as engineering for social justice requires both avoiding harms *and* finding ways to novelly enhance the human capabilities of those whose capabilities are structurally constrained.[277]

Roboticists seeking to adopt an abolitionist robotics would thus not only refuse to collaborate with the police and work to oppose them, but also look beyond policing-oriented solutions to societal problems by working to understand how robots can (1) meet communities' health and safety needs in new ways that do not involve or require policing (e.g., as part of non-police mobile mental health teams[278]), *especially* in domains where policing is currently the required or normalized solution; (2) provide new opportunities for forging solidarities, connections, and social relations between otherwise isolated community members (e.g., through robots that encourage community members to support the needs of their unhoused neighbors[279]); (3) be developed and deployed in ways that neither require nor support capitalist modes of production and accumulation; and (4) afford communities new opportunities to create, encode, and make manifest their shared values and their alternative imaginaries for what our future could hold. Finally, roboticists seeking to adopt an abolitionist robotics could work not only to pursue these goals in the research laboratory but also to push for the policy and regulatory changes needed to achieve them[280] and to teach new students about the importance of these goals through pedagogical frameworks such as Tiera Tanksley's abolitionist pedagogy.[281]

Conclusion and Chapter Summary

In this chapter, we have seen how roboticists reinforce White patriarchal power in the structural domain by designing robots in collaboration with, or intended for use by, institutions that are primarily used to enforce White patriarchal power structures, such as the police.

Specifically, we have seen two key ways that roboticists' collaboration with or facilitation of police use of robots might reinforce oppression. First, if roboticists build robots for use by (or that are likely to be used by) the police, this could exacerbate and legitimize the use of the police as an institution for surveilling, controlling, and enacting violence upon people of color and the poor, thus reinforcing White and patriarchal power structures. Second, any collaboration between roboticists and the police may serve to legitimize and normalize the use of the police to surveil, control, and enact violence upon people of color and the poor and the use of the police as the appropriate means of addressing other societal problems.[282]

These structural power dynamics may be reflected in the design of robots such as the REEM robot highlighted in chapter 1. In the context of REEM's deployment as a robocop in Dubai, REEM serves to legitimize policing in several key ways. First, REEM casts policing as a scientific and high-tech endeavor, bolstering the image of the police and serving as a display of power. But moreover, in the context of its Dubai deployment, REEM serves to reinforce policing as an institution in very specific ways. In the UAE, policing is an even more recent phenomenon than it is in Western contexts, with the UAE's police force serving as a legacy of the previous British colonialist military presence. As Al Shaali and Kibble write, "The UAE inherited a law enforcement system that was established by a colonial power. During the colonial era, the police formed part of the military organisation. At this time, the principal role of the police was to achieve stability in the region by serving as an extension to the military forces that guarded security between the various Emirates."[283] By deploying robotic police officers (and uniformed, saluting, White, and masculine robotic police officers at that), Dubai stands not only to further confuse the already murky relationship between policing and military occupation within the UAE[284] but, moreover, to further entrench the specific vision of law enforcement they inherited from British rule. This is an especially troubling parallel, given the UAE's reliance on "modern-day slavery" to construct its massive infrastructural projects.[285]

Overall, in this chapter, we have identified several historical and social factors that played a key role in producing the technological status quo by which roboticists (especially American roboticists) wield not only disciplinary but also structural power. Specifically, we have traced the history of policing to show how it originated—and continues to function—as a tool for violently maintaining White heteropatriarchal capitalism. And we saw how American police militarization efforts have been historically correlated with political movements to criminalize Black and working-class protests and unrest. If robot design continues along its current path (with roboticists either collaborating with the police or working on products that stand to be uniquely leveraged by the police), roboticists will continue to be complicit in the use of policing as an institution for structuring oppression and maintaining White male hegemonic power—and students from historically oppressed groups may be further dissuaded from joining the field of robotics, thus perpetuating the current White masculinity of robotics and exacerbating all of the previous problems discussed in this book.

Based on this understanding of our technological status quo, the way it reinforces our sociological status quo, and the historical and social root causes that explain this dynamic, we have identified key strategies that roboticists can use to build toward a more equitable robot future by revising, rethinking, and ripping up the status quo in the structural domain.

First, roboticists might revise the status quo by saying, "We won't design robots for use by the police, especially in areas where the police have a history of misusing robots." This is the approach taken by signatories of the #NoJusticeNoRobots campaign,[286] who publicly "refuse[d] to facilitate the execution or publication of research in robotics or Artificial Intelligence (AI) that is performed in collaboration with State and Local Law Enforcement agencies (or national agencies such as ICE and CBP)." Although these types of petition efforts may be a necessary first step, they may be insufficient to address police misuse of more general purpose robots.

Second, roboticists might rethink the status quo by saying, "If the problem is with how robots are being used, let's just design them so they cannot be misused in that way or so that they will primarily be used in ways that run counter to that misuse." This might include designing robots that are private by design[287] and thus avoid storing sensor data in ways that could later be accessed by the police. Alternatively, this could include working with activists to design robots for sousveillance,[288] explicitly working to maintain communities' safety by allowing them to monitor and document police activity.

Finally, roboticists might rip up the status quo by saying, "Let's attack the root causes that explain how we have arrived at the status quo." If the reason why roboticists who collaborate with the police tend by default to reinforce White patriarchal power is because American society is designed under the assumption that race- and class-directed violence is the solution to societal health and safety concerns, then perhaps we should be working to negate this assumption by dismantling society's reliance on policing. As such, this book calls on roboticists who seek to subvert White patriarchal power (or, more generally, White capitalist heteropatriarchal power) within the structural domain to adopt an abolitionist robotics in which robots are designed for use by non-policing institutions to address problems society has currently abdicated to the police and to push for societal support of those alternative institutions.

7 Future Tense

Over the course of this book, I have used the guiding framework of matrix-guided technology power analysis[1] to provide a new understanding of the White patriarchal origins of robotics—the ways that today's robots continue to reinforce White supremacy, capitalism, and heteropatriarchy. I have introduced new terminology that helps us reason and communicate about these dynamics; identifying the White patriarchal robot imaginary as the dominant lens through which robots are conceived and designed, social identity design cues as the design elements through which this imaginary is reinforced or subverted in the cultural domain, New Jim Cobots as the robots that enforce the New Jim Crow and New Jim Code in the disciplinary domain, social identity trait perception as the set of computer vision technologies that invisibly exacerbate this enforcement, and abolitionist robotics as the overarching design strategy that roboticists must follow if they wish to truly subvert oppression in the structural domain. And overall, I have highlighted a variety of design strategies, tools, and techniques that roboticists can use to subvert rather than reinforce oppression in different ways, culminating in this vision of an abolitionist robotics.

To conclude, then, let us consider the degrees of freedom[2] available to roboticists who wish to subvert robotics' dominant modes of oppression. In robotics, "degrees of freedom" refers to the axes of motion available to a robot. A robot arm with a single joint that pivots back and forth along one axis has a single degree of freedom, affording an extremely limited range of motion. Meanwhile, the human arm has seven degrees of freedom (three for the shoulder, three for the wrist, and one for the elbow). In this chapter, we will play off this concept to consider the ethical degrees of freedom available to roboticists—the factors that are within their power to change and the types of tools available to effect those changes—and to look at the phrase another way: the different degrees of freedom and liberation that might be achieved through the use of those tools.

Where We've Been

First, let us briefly summarize the different types of power we've seen roboticists wield across this book and the default ways that power is wielded today.

When roboticists design and deploy robots, they bend the world to align with a set of motivations and values: a combination of their own motivations and values, those of their employers, those of the communities they work with, and those baked into robotics through its unique history. As we saw in chapter 2, these default motivations and values are overwhelmingly those of White men due to the emergence of the White patriarchal robot imaginary from the postbellum aspirations of White men in the American South patriarchal future. As demonstrated across the chapters of this book, roboticists cannot afford to ignore this history or view it as a "thing of the past."[3]

These historical antecedents have continued to shape robotics in both the cultural and techno-scientific imaginaries: American and European robots in both science fiction and science fact are typically depicted and designed in the image of White men or to suit the gaze of White men, both in terms of how those robots are depicted and in terms of the purposes for which those robots are envisioned. This has cast robotics as the domain of White men, leading to a population of robot designers who, at least in the US and Europe, are overwhelmingly White men operating within a White patriarchal society—a trend that has reinscribed the White patriarchal robot imaginary and created a vicious cycle that will take a concerted effort to escape.

As we saw across the course of this book, the White patriarchal robot imaginary, if left unchallenged, will continue to propel the field along a course where its default effect is to reinforce White patriarchal power, not only through the wielding of "mere" interpersonal power to persuade and influence specific courses of immediate action but also across each of the other domains of the matrix of domination, where subtler and longer-lasting impacts might be felt. This gives roboticists substantial power across each of these domains to either reinforce or subvert that status quo.

As we saw in chapter 3, the White patriarchal motivations and values of robotics have led to today's humanlike, interactive, and language-capable robots being designed using social identity design cues that feed off and reinforce a host of racialized and gendered biases, stereotypes, and norms, thus reinforcing White patriarchal power. Roboticists thus hold considerable power within the cultural domain to either reinforce or subvert White patriarchy through the ways they design robots to look and speak.

As we saw in chapter 4, machine morality efforts are having a default effect of encoding and enforcing the informal moral and social rules that already structure our societies. Given the existing power systems in the US and Europe, this means that American and European roboticists' machine morality efforts tend to enforce, perpetuate, and globalize White patriarchal and colonialist norms, categorization systems, and logics, turning otherwise helpful cobots into New Jim Cobots that serve as enforcers of the New Jim Crow and the New Jim Code. Roboticists thus hold considerable power within the disciplinary domain to either reinforce or subvert White patriarchy through the ways they design robots to morally reason, act, and communicate.

As we saw in chapter 5, robots are increasingly being designed in ways that render women and people of color simultaneously invisible when they would wish to be seen and hypervisible when they would wish to be ignored. Moreover, social identity trait perception systems are having the default effect of encoding and enforcing the very systems of race and gender that are used to socially structure those societies. Given the existing power systems in the US and Europe, this means that American and European roboticists' machine perception efforts may similarly tend to enforce, perpetuate, and globalize White patriarchal and colonialist norms, categorization systems, and logics. This accentuates roboticists' power within the disciplinary domain to either reinforce or subvert White patriarchy through the ways they design robots to enable—or force—others to be seen.

And as we saw in chapter 6, today's robots are increasingly being used by the police to violently enforce White patriarchy, consistent with the original purposes of both robots and the police. Given the existing power systems in the US and Europe, this means that merely collaborating with the police, even in ostensibly prosocial use cases, may have the ultimate effect of legitimizing and normalizing the use of the police as an ineffective cudgel for addressing serious societal problems. Roboticists thus hold considerable power within the structural domain to either legitimize or contest White patriarchy through their choices of whom to collaborate with and their decisions to accept or subvert the dominant imaginaries of public health and safety—a power whose acknowledgement motivates a call for an abolitionist robotics.

Overall, then, we have seen why it is critical for roboticists to recognize the power they wield across these domains and to use that power to subvert rather than reinforce inequitable power dynamics. But moreover, we have seen concrete ways that roboticists can work toward subverting rather than reinforcing oppression—both through specific strategies to revise, rethink, or rip up the status quo and through the use of recurring patterns and design frameworks they can uptake and combine in different ways to achieve those goals.

Where We're Going

As we have seen in this book, we are at an inflection point where the fundamental history and assumptions of robotics are being questioned, presenting unique opportunities to shift the landscape of power so that robots, by default, elevate the aesthetics, motivations, and values of otherwise-marginalized communities rather than those that robotics currently elevates by default. Today's roboticists have the power not only to rethink what robots look like, how robots make decisions, how robots see the world, and where robots are used but, moreover, to envision and pursue a fundamentally new robotic imaginary in which robots serve as tools for subverting rather than reinforcing White patriarchy.

Over the course of this book, we have seen a wide array of tools that roboticists who are dedicated to a vision of robots for social justice[4] could use to pursue this new robotic imaginary. We have seen key analytic frameworks (such as Black feminist thought, feminist human–robot interaction, and Confucian robot ethics) that can help us to more precisely identify the type of power that robots wield and the risks this presents. And we have seen key design frameworks (such as engineering for social justice and design justice) that show us how, once roboticists have effected this analysis and come to this understanding, they can wield positive moral influence with respect to the needs, values, and priorities of minoritized communities rather than continuing to reinforce White patriarchy.

To conclude, then, let us take a step back and walk through how each of these different analytic and design frameworks can help roboticists. That is, let us consider the specific tools they have for moving robotics along the distinct degrees of freedom that define the matrix of domination in order to enable some greater degree of freedom within their society.

1. *Black feminist thought* helps us understand the types of power roboticists wield. As we saw in chapter 1, when human–robot interaction researchers talk about robots' persuasive power, they are typically talking simply about the power that roboticists wield in the interpersonal domain through the ways they leverage robots' ability to persuade others *to* pursue some course of action or *of* some state of affairs. But Black feminist thought reminds us that the interpersonal domain is just the most visible domain of power and encourages us to consider other domains of power in which roboticists hold sway. Within this intellectual tradition, the specific framework used to guide the structure of this book (matrix-guided technology power analysis[5]) further illuminates the specific ways that roboticists wield power in each of those other domains That is, it illuminates robotics' degrees of freedom.
2. *Feminist human–robot interaction* helps us to more acutely understand the ways that roboticists wield power in the specific social-relational context of human–robot interaction, especially at the macroscale that goes beyond the interaction. Specifically, feminist human–robot interaction[6] helps us understand the importance of this careful consideration of the types of power relations and hierarchies we wish to support or subvert. Moreover, feminist human–robot interaction encourages us to take a morally pluralist view, not only by combining the insights of different moral and ethical theories but also by encouraging us to resist robot design approaches that mandate, reinforce, and spread a single "universal" set of designer-chosen moral principles or virtues. Furthermore, feminist human–robot interaction suggests the use of participatory design processes in order to embrace pluralism.
3. *Confucian robot ethics* further helps us to understand the ways that roboticists wield power in the specific context of human–robot interaction, especially at the microscale of interactions. Specifically, Confucian robot ethics[7] helps us to understand not simply that humans form relational attachments with robots or that robots wield persuasive power and shape moral norms but, moreover, the *way* that robots wield different types of power and encourage different types of virtues based on the types of relationships they participate in with humans. By extension, Confucian role ethics encourages us to be explicit in our consideration of the virtue-bearing relational roles we want to encourage in the contexts into which robots are embedded. But most importantly, Confucian role ethics

encourages us to shift our focus from how robots can avoid harm by following enumerated rules to how robots can wield their power to help others become good by remonstrating with interactants to encourage them to fulfill the virtue-laden relational roles that roboticists wish to encourage.

4. *Engineering for social justice* shows how these types of power analyses may be embedded into a practice of socially just robotics research, engineering, and design. Specifically, engineering for social justice[8] encourages roboticists to work with communities to understand the economic, social, and political factors that constrain them and the priorities that motivate them, and to use this knowledge to design technologies that provide more genuine benefit to those communities.
5. *Design justice*, finally, further encourages roboticists to develop engineering solutions that can be extended and hacked by communities to meet their own needs without the involvement of roboticists. And more broadly, design justice[9] encourages us to support design work led by Black, Latinx, and Indigenous communities; by women, non-binary, and LGBTQIA+ communities; by disabled communities; and by other communities underrepresented in and excluded from today's robot design teams but best positioned to successfully carry out the design processes that will build a more equitable future for robotics.

These different theories and methods represent key tools in the toolbox of roboticists hoping to join the movement for robots for social justice.[10] But moreover, the utility of these theories and methods shows the ways that we need to fundamentally rethink the entire project of robot ethics. If we wish to identify the harms caused by robots, we need to move beyond the philosopher's armchair. Although robot ethics is typically viewed and taught through the lens of moral philosophy, I've shown in this book that the ethical design of robots requires attention to sociological issues of power, and that understanding those power dynamics requires us to draw on history and politics; law, criminology, and sociology; feminist, ethnic, and Black studies; literary and media studies; and social, moral, and cognitive psychology. Similarly, if we wish to address the harms we have identified, we need to move beyond the academy entirely, as the most meaningful actions we can take toward a more equitable robotic future, especially those that may be taken through the lens of an abolitionist robotics, may fall entirely outside the realm of robotics research. First, we must expand education[11] and tech literacy efforts[12] to promote public awareness of the ways robots reinforce inequitable power structures and the harms this causes to society. Second, we must identify and pursue more effective means of auditing tech companies, both to understand the harms those companies are causing and to hold those companies to account.[13] Third, we must enact effective policies and legislation that prevent tech companies from developing robots and AI systems that reinforce inequitable power structures,[14] building on preliminary legislative efforts such as the EU and Colorado AI Acts.[15] And finally, we must divest from and abolish the institutions of White patriarchy and replace them with institutions that genuinely and equitably serve the needs of our communities.

In other words, rather than debating whether to pull the levers of the trolley tracks of robotics, roboticists and non-roboticists should perhaps instead be working together to rethink where those tracks are laid, who is deciding where we are headed, and what makes a trolley a reasonable means of travel in the first place.

Notes

Acknowledgments

1. Tom Williams and Kerstin Haring, "No Justice, No robots: From the Dispositions of Policing to an Abolitionist Robotics," in *Proceedings of the AAAI/ACM Conference on Artificial Intelligence, Ethics, and Society (AIES)* (ACM, 2023), 566–575.

2. Tom Williams, "The Eye of the Robot Beholder: Ethical Risks of Representation, Recognition, and Reasoning over Identity Characteristics in Human-Robot Interaction," in *Companion of the ACM/IEEE International Conference on Human-Robot Interaction (alt.HRI)* (ACM, 2023), 1–10.

3. Tom Williams, "Understanding Roboticists' Power Through Matrix-Guided Technology Power Analysis," in *Companion of the 2024 ACM/IEEE International Conference on Human–Robot Interaction (alt.HRI)* (ACM, 2024), 46–56.

Chapter 1

1. The exact phrasing of the engineer's comment may have differed; the interaction was recorded only in my memory.

2. Isaac Asimov, *I, Robot* (Gnome Press, 1950).

3. Association for Advancing Automation, "Unimate // The First Industrial Robot: A Tribute to Joseph Engelberger," 2015, https://www.automate.org/a3-content/joseph-engelberger-unimate; and Yakov Fet, "Norbert Wiener in Moscow," in *Proceedings of the 3rd International Conference on Computer Technology in Russia and in the Former Soviet Union (SoRuCom)* (IEEE Computer Society, 2014), 194–195.

4. Karel Čapek, *R.U.R. (Rossum's Universal Robots)* (Aventinum, 1920).

5. Kara Weisman, "Extraordinary Entities: Insights into Folk Ontology from Studies of Lay People's Beliefs about Robots," in *Proceedings of the Annual Meeting of the Cognitive Science Society* (Cognitive Science Society, 2022): 3493–3499.

6. Belinda J. Dunstan and Guy Hoffman, "Social Robot Morphology: Cultural Histories of Robot Design," in *Cultural Social Robots and Their Emergent Cultural Ecologies,*

ed. Belinda J. Dunstan, Jeffrey T. K. V. Koh, Deborah Turnbull Tillman, and Scott Andrew Brown (Springer, 2023).

7. Tom Standage, *The Turk: The Life and Times of the Famous Eighteenth-Century Chess-Playing Machine* (Walker & Company, 2002).

8. There is a story of one particular observer who did view the Turk as having agency, but only because they thought it was haunted—a view that only underscores their view of the machine itself as non-agentic; Karl Gottlieb von Windisch, *Inanimate Reason; or, A Circumstantial Account of That Astonishing Piece of Mechanism, M. de Kempelen's Chess-Player: Now Exhibiting at No. 8 Saville-Row, Burlington Gardens* (Bladon, 1784).

9. Edward S. Ellis, *The Steam Man of the Prairies* (Beadle and Adams, 1868).

10. However, see Abnet's account of the "magical automatons," in Fitz-James O'Brien's *The Wondersmith* (The Atlantic Monthly, 1859), which O'Brien used to stoke fears of "gypsies"; and Dustin A. Abnet, *The American Robot: A Cultural History* (University of Chicago Press, 2020), 51. O'Brien's work did not, however, obtain the same commercial success as Ellis's and thus did not have the same impact on the public's robotic proto-concept. Moreover, the automata in *The Wondersmith* were magical in nature rather than premised in engineering.

11. J. Stanley Lemons, "Black Stereotypes as Reflected in Popular Culture, 1880–1920," *American Quarterly* 29, no. 1 (1977): 102–116.

12. Taylor Scott Evans, "The Race of Machines: A Prehistory of the Posthuman" (PhD thesis, University of California, Riverside, 2018). See also the coverage by Abnet in *The American robot.*

13. Dorothy Roberts, "Race," in *The 1619 Project: A New American Origin Story*, ed. Nikole Hannah-Jones (Random House, 2021): 45–62.

14. Evans, "The Race of Machines," 2018.

15. Charles W. Mills, *Black Rights/White Wrongs: The Critique of Racial Liberalism* (Oxford University Press, 2017); and Charles W. Mills, *The Racial Contract* (Cornell University Press, 2014).

16. Isaac Asimov, "Runaround," *Astounding Science Fiction* 29, no. 1 (1942): 94–103.

17. Roger Lind, "The Robots Are Coming, the Robots Are Coming," *Oui* 3, no. 2, 1974: 40, as cited in Salam Elzway, "Technoliberal Machines: Robotic Work(ers) from Science Fiction to the Assembly Line," *Work: The Labors of Language, Culture, and History in North America*, ed. J. Jesse Ramírez and Sixta Quassdorf (2021), 159–186.

18. Salam Elzway, "Arms of the State: A History of the Industrial Robot in Postwar America" (PhD thesis, University of Michigan, 2024).

19. Michelle Alexander, *The New Jim Crow: Mass Incarceration in the Age of Colorblindness* (New Press, 2020).

20. Yoyo Tsung-Yu Hou et al., "Power in Human–Robot Interaction," in *Proceedings of the ACM/IEEE International Conference on Human–Robot Interaction* (ACM, 2024), 269–282.

21. Gordon Briggs and Matthias Scheutz, "How Robots Can Affect Human Behavior: Investigating the Effects of Robotic Displays of Protest and Distress," *International Journal of Social Robotics* 6 (2014): 343–355; and Vijay Chidambaram et al., "Designing Persuasive Robots: How Robots Might Persuade People Using Vocal and Nonverbal Cues," in *Proceedings of the ACM/IEEE International Conference on Human–Robot Interaction* (ACM, 2012), 293–300.

22. Ilaria Baroni et al., "Designing Motivational Robot: How Robots Might Motivate Children to Eat Fruits and Vegetables," in *Proceedings of the 23rd IEEE International Conference on Robot and Human Interactive Communication* (IEEE, 2014), 796–801.

23. Solace Shen et al., "'Stop. I See a Conflict Happening.' A Robot Mediator for Young Children's Interpersonal Conflict Resolution," in *Proceedings of the ACM/IEEE International Conference on Human–Robot Interaction* (ACM, 2018), 69–77.

24. Jason Borenstein and Ronald C. Arkin, "Nudging for Good: Robots and the Ethical Appropriateness of Nurturing Empathy and Charitable Behavior," *AI & Society* 32 (2017): 499–507; and Boyoung Kim et al., "The Impact of Different Ethical Frameworks Underlying a Robot's Advice on Charitable Donations," in *Proceedings of the IEEE International Conference on Robot and Human Interactive Communication (RO-MAN)* (IEEE, 2023), 2433–2438.

25. Cass R. Sunstein, *Why Nudge? The Politics of Libertarian Paternalism* (Yale University Press, 2014).

26. Katie Winkle et al., "Effective Persuasion Strategies for Socially Assistive Robots," in *Proceedings of the 14th ACM/IEEE International Conference on Human–Robot Interaction (HRI)* (IEEE, 2019), 277–285.

27. Katie Winkle et al., "In-Situ Learning from a Domain Expert for Real-World Socially Assistive Robot Deployment," in *Proceedings of Robotics: Science and Systems* (RSS Foundation, 2020).

28. Chidambaram et al., "Designing Persuasive Robots"; Jaap Ham et al., "Making Robots Persuasive: The Influence of Combining Persuasive Strategies (Gazing and Gestures) by a Storytelling Robot on Its Persuasive Power," in *Proceedings of the International Conference on Social Robotics* (Springer, 2011), 71–83; and Peter A. Hancock et al., "A Meta-Analysis of Factors Affecting Trust in Human–Robot Interaction," *Human Factors* 53, no. 5 (2011): 517–527.

29. Paul Robinette et al., "Overtrust of Robots in Emergency Evacuation Scenarios," in *Proceedings of the 11th ACM/IEEE International Conference on Human–Robot Interaction (HRI)* (IEEE, 2016), 101–108.

30. The definitions of agency and moral agency used here are provided by Luciano Floridi and J. W. Sanders, "On the Morality of Artificial Agents," *Minds and Machines*

14 (2004): 349–379; explanation of the role of action in agency is provided by Lucy A. Suchman, *Plans and Situated Actions: The Problem of Human–Machine Communication* (Cambridge University Press, 1987); and compatible definitions of social agency are provided by Ryan Blake Jackson and Tom Williams, "*A Theory of Social Agency for Human–Robot Interaction*," *Frontiers in Robotics and AI* 8 (2021): 1–15.

31. Susanne Göckeritz et al., "Young Children's Creation and Transmission of Social Norms," *Cognitive Development* 30 (2014): 81–95.

32. Of course, robots' ability to sanction norm violations and enforce norm adherence depends on how the robot frames its moral communication. For instance, some attempts at robot-delivered sanction have proved ineffective; see Himavath Jois and Alan R. Wagner, "What Happens When Robots Punish? Evaluating Human Task Performance during Robot-Initiated Punishment," *ACM Transactions on Human–Robot Interaction (THRI)* 10, no. 4 (2021): 1–18. Some attempts have even backfired; see Boyoung Kim et al., "Can Robot Advisers Encourage Honesty?: Considering the Impact of Rule, Identity, and Role-Based Moral Advice," *International Journal of Human–Computer Studies* 184 (2024): 1–16. We will return to this topic in chapter 4.

33. Ryan Blake Jackson and Tom Williams, "Language-Capable Robots May Inadvertently Weaken Human Moral Norms," in *Proceedings of the 14th ACM/IEEE International Conference on Human–Robot Interaction (HRI)* (IEEE, 2019), 401–410; and Ryan Blake Jackson and Tom Williams, "Enabling Morally Sensitive Robotic Clarification Requests," *ACM Transactions on Human–Robot Interaction (THRI)* 11, no. 2 (2022): 1–18.

34. Lorin Dole, "The Influence of a Robot's Mere Presence on Human Communication" (PhD thesis, Stanford University, 2017).

35. Min Kyung Lee et al., "Ripple Effects of an Embedded Social Agent: A Field Study of a Social Robot in the Workplace," in *Proceedings of the SIGCHI Conference on Human Factors in Computing Systems* (ACM, 2012), 695–704.

36. See also Weisswange et al. for good coverage of the types of influence robots can have when mediating group interactions; and Thomas H. Weisswange et al., "What Could a Social Mediator Robot Do? Lessons from Real-World Mediation Scenarios," in *Proceedings of the ICRA Workshop "Towards a Balanced Cyberphysical Society: A Focus on Group Social Dynamics"* (2023).

37. Hamish Tennent et al., "MicBot: A Peripheral Robotic Object to Shape Conversational Dynamics and Team Performance," in *Proceedings of the 14th ACM/IEEE International Conference on Human–Robot Interaction (HRI)* (IEEE, 2019), 133–142.

38. Shen et al., "'Stop. I See a Conflict Happening.'"

39. Sarah Gillet et al., "A Social Robot Mediator to Foster Collaboration and Inclusion among Children," in *Proceedings of Robotics: Science and Systems* (RSS Foundation, 2020). See also Sarah Gillet et al., "Robot Gaze Can Mediate Participation Imbalance in

Groups with Different Skill Levels," in *Proceedings of the ACM/IEEE International Conference on Human–Robot Interaction* (ACM, 2021), 303–311; Yanghee Kim et al., "Designing for Robot-Mediated Interaction among Culturally and Linguistically Diverse Children," *Educational Technology Research and Development* 69 (2021): 3233–3254; and Isabel Neto et al., "The Robot Made Us Hear Each Other: Fostering Inclusive Conversations among Mixed-Visual-Ability Children," in *Proceedings of the ACM/IEEE International Conference on Human–Robot Interaction* (ACM, 2023), 13–23.

40. Tom Hitron et al., "Implications of AI Bias in HRI: Risks (and Opportunities) When Interacting with a Biased Robot," in *Proceedings of the ACM/IEEE International Conference on Human–Robot Interaction* (ACM, 2023), 83–92.

41. Jennifer Graham, "Are Alexa and Siri Making Your Kids Rude?" *Deseret News*, June 24, 2019; Hunter Walk, "Amazon Echo is Magical. It's Also Turning My Kid into an Asshole," *Hunter Walk*, April 6, 2016; Ken Gordon, "Alexa and the Age of Casual Rudeness," *The Atlantic*, April 23, 2018; and Edward C. Baig, "Kids Were Being Rude to Alexa, So Amazon Updated It," *USA Today*, April 25, 2018.

42. Elisa Giaccardi et al., "The Repertoire of Meaningful Voice Interactions: How to Design Good Smart Speakers," in *The State of Responsible IoT 2020* (ThingsCon e.V., 2020), 53–57.

43. BBC, "Amazon Alexa to Reward Kids Who Say: 'Please,'" 2018, https://www.bbc.com/news/technology-43897516. See also J. Vincent, "Google Assistant Adds Pretty Please Functionality, Updated Lists, and Other Features," *The Verge*, November 29, 2018; and Baig, "Kids Were Being Rude to Alexa," 2018.

44. Cristian Danescu-Niculescu-Mizil et al., "A Computational Approach to Politeness with Application to Social Factors," in *Proceedings of the 51st Annual Meeting of the Association for Computational Linguistics* (ACL, 2013), 250–259.

45. Ruchen Wen et al., "Fresh Start: Encouraging Politeness in Wakeword-Driven Human–Robot Interaction," in *Proceedings of the ACM/IEEE International Conference on Human–Robot Interaction* (ACM, 2023), 112–121, XXX; Ruchen Wen et al., "Unpretty Please: Ostensibly Polite Wakewords Discourage Politeness in Both Robot-Directed and Human-Directed Communication," in *Proceedings of the ACM International Conference on Multimodal Interaction* (ACM, 2022), 181–190; and Tom Williams et al., "'Excuse Me, Robot': Impact of Polite Robot Wakewords on Human–Robot Politeness," in *Proceedings of the International Conference on Social Robotics* (Springer, 2020), 404–415.

46. Astrid M. Rosenthal-von der Pütten et al., "An Experimental Study on Emotional Reactions towards a Robot," *International Journal of Social Robotics* 5 (2013): 17–34.

47. Hideki Garcia Goo et al., "Victims and Observers: How Identity, Experience, and Biases Shape Perceptions of Robot Abuse," in *Proceedings of the IEEE International Conference on Robot and Human Interactive Communication (RO-MAN)* (IEEE, 2023): 2439–2446.

48. Henry Shevlin, "How Could We Know When a Robot Was a Moral Patient?" *Cambridge Quarterly of Healthcare Ethics* 30, no. 3 (2021): 459–471.

49. Briggs and Scheutz, "How Robots Can Affect Human Behavior."

50. Julie Carpenter, *Culture and Human–Robot Interaction in Militarized Spaces: A War Story* (Routledge, 2016).

51. Cherie Lacey and Catherine Caudwell, "Cuteness as a 'Dark Pattern' in Home Robots," in *Proceedings of the 14th ACM/IEEE International Conference on Human–Robot Interaction (HRI)* (ACM, 2019), 374–381.

52. Harry Brignull, "Dark Patterns," https://darkpatterns.org/. Accessed 2023.

53. Chad Edwards et al., "Initial Interaction Expectations with Robots: Testing the Human-to-Human Interaction Script," *Communication Studies* 67, no. 2 (2016): 227–238.

54. Peter H. Kahn Jr. et al., "Will Do People Keep the Secret of a Humanoid Robot? Psychological Intimacy in HRI," in *Proceedings of the Tenth Annual ACM/IEEE International Conference on Human–Robot Interaction* (ACM, 2015), 173–180.

55. Shoshana Zuboff, *The Age of Surveillance Capitalism: The Fight for a Human Future at the New Frontier of Power* (Profile Books, 2019), 293.

56. Cf. Zuboff, *The Age of Surveillance Capitalism*; and Nick Couldry and Ulises A. Mejias, *The Costs of Connection: How Data Are Colonizing Human Life and Appropriating It for Capitalism* (Oxford University Press, 2020).

57. Patricia Hill Collins, *Black Feminist Thought: Knowledge, Consciousness, and the Politics of Empowerment* (Routledge, 2022).

58. Patricia Hill Collins, "Black Feminist Thought in the Matrix of Domination," in *Black Feminist Thought: Knowledge, Consciousness, and the Politics of Empowerment* (Hyman, 1990), 221–238.

59. Collins, *Black Feminist Thought*, 350–354.

60. Collins, *Black Feminist Thought*, 355–358.

61. Collins, *Black Feminist Thought*, 359–362.

62. Collins, *Black Feminist Thought*, 363–365.

63. Collins, *Black Feminist Thought*, 363–365.

64. Collins, "Black Feminist Thought in the Matrix of Domination"; and bell hooks, *Talking Back: Thinking Feminist, Thinking Black* (South End Press, 1989), 175.

65. Williams, "Understanding Roboticists' Power."

66. The more comprehensive set of questions guiding this work can be found in my work on matrix-guided technology power analysis; Williams, "Understanding Roboticists' Power."

67. Alexander, *The New Jim Crow*.

68. Ruha Benjamin, *Race after Technology: Abolitionist Tools for the New Jim Code* (Social Forces, 2019).

69. D'Ignazio and Klein, *Data Feminism*.

70. Benjamin, *Race after Technology*.

71. I like to think that a "joint PhD" means I did two PhDs worth of work but only got one diploma and, more importantly, only one funny hat to wear at graduation.

72. Randy Connolly, "Why Computing Belongs within the Social Science," *Communications of the ACM* 63, no. 8 (2020): 54–59.

73. K. E. Hayes and M. Kaba, *Let This Radicalize You: Organizing and the Revolution of Reciprocal Care* (Haymarket Books, 2023).

74. Hayes and Kaba, *Let This Radicalize You*.

75. Collins, *Black Feminist Thought*, 49.

Chapter 2

1. A brief note for non-roboticists: The word "servo" in this chapter's title is short for "servomotor," a common type of actuator used in robotics.

2. Carl Remick, "Robots: New Faces on the Production Line," *Management Review* 68, no. 5 (1979):24–39: 3, as quoted in Elzway, "Arms of the State."

3. Audrey Smedley and Brian D. Smedley, "Race as Biology Is Fiction, Racism as a Social Problem Is Real: Anthropological and Historical Perspectives on the Social Construction of Race," *American Psychologist* 60, no. 1 (2005): 16–26.

4. Bruce R. Dain, *A Hideous Monster of the Mind* (Harvard University Press, 2021), as cited in Wendy Hui Kyong Chun, "Race and/as Technology, or How to Do Things to Race," in *Race after the Internet*, ed. Lisa Nakamura and Peter A. Chow-White (Routledge, 2013), 44–66.

5. Roger Ascham, *The Scholemaster* (1570).

6. John Trenchard and Thomas Gordon, *The Independent Whig*, vol. 1, 7th ed. (J. Peele, 1743).

7. Charles Harrington Elliot, *The Republican Refuted: In a Series of Biographical, Critical and Political Strictures on Thomas Paine's Rights of Man* (London, 1791).

8. Audrey Smedley and Brian D. Smedley, *Race in North America: Origin and Evolution of a Worldview* (Westview Press, 2012).

9. Vilna Bashi Treitler, *The Ethnic Project* (Stanford University Press 2020), 45–57.

10. Bashi Treitler, *The Ethnic Project*, 45–57.

11. Bashi Treitler, "Racialization: Paradigmatic Frames From British Colonization to Today, and Beyond," (DesiguALdades.net, 2015).

12. Bashi Treitler, *The Ethnic Project.*

13. Michael Omi and Howard Winant, "On the Theoretical Status of the Concept of Race," in *Race, Identity, and Representation in Education*, ed. Cameron McCarthy and Warren Crichlow (Routledge, 1993), 3–10.

14. Anthony S. Parent, *Foul Means: The Formation of a Slave Society in Virginia, 1660–1740* (University of North Carolina Press, 2003).

15. Smedley and Smedley, *Race in North America*; and Audrey Smedley, "Antecedents of the Racial Worldview," in *Race and Racialization: Essential Readings*, ed. Tania Das Gupta, Carl E. James, Roger C. A. Maaka, Grace-Edward Galabuzi, and Chris Andersen (Canadian Scholars Press, 2007), 31–44.

16. Zeus Leonardo, "The Color of Supremacy: Beyond the Discourse of 'White Privilege,'" *Educational Philosophy and Theory* 36, no. 2 (2004): 137–152.

17. Moreover, the elevation of these virtues may have been necessary, given the magnitude of the oppression wrought by America's "founding fathers." As Nick Couldry and Ulises A. Mejias note, oppressors need to cast themselves in terms of virtuous principles in order to rehabilitate themselves; Couldry and Mejias, *The costs of connection*, 70.

18. Mills, *Black Rights/White Wrongs*, 62.

19. David Baker, "Systemic White Racism and the Brutalization of Executed Black Women in the United States," in *It's a Crime: Women and Justice*, ed. Roslyn Muraskin (Pearson Prentice Hall, 2007), 406, as quoted in Andrea J. Ritchie, *Invisible No More: Police Violence Against Black Women and Women of Color* (Beacon Press, 2017), 25.

20. Lawrence M. Friedman, *Crime and Punishment in American History* (Basic Books, 1994), 86, as cited in Ritchie, *Invisible No More*, 26.

21. Sally E. Hadden, *Slave Patrols: Law and Violence in Virginia and the Carolinas* (Harvard University Press, 2003).

22. Douglas A. Blackmon, *Slavery By Another Name: The Re-Enslavement of Black Americans from the Civil War to World War II* (Anchor, 2009).

23. Bashi Treitler, "Racialization."

24. Eduardo Bonilla-Silva, "Rethinking Racism: Toward a Structural Interpretation," *American Sociological Review* 62, no. 3 (1997): 465–480.

25. Smedley and Smedley, *Race in North America.*

26. Bashi Treitler, *The Ethnic Project.*

27. Chun, "Race and/as Technology"; and Beth Coleman, "Race as Technology," *Camera Obscura: Feminism, Culture, and Media Studies* 24, no. 1 (2009): 177–207.

28. Smedley and Smedley, "Race as Biology Is Fiction."

29. There is some evidence, however, that slavery may have continued to linger in Massachusetts into the 1800s.

30. Kevin Coffee, "Industrial Lowell and the Dawn of the Anthropocene," *Industrial Archaeology Review* 43, no. 1 (2021): 20–33.

31. Thomas Dublin, *Lowell: The Story of an Industrial City: A Guide to Lowell National Historical Park and Lowell Heritage State Park, Lowell, Massachusetts*, vol. 140 (Government Printing Office, 1992).

32. Edward Lillie Pierce, *Memoir and Letters of Charles Sumner*, vol. 4 (Roberts Brothers, 1893), as quoted in Carla Bosco, "Harvard University and the Fugitive Slave Act," *The New England Quarterly* 79, no. 2 (2006): 227–247.

33. Dublin, *Lowell*, 20.

34. Note, however, that the underground railroad appears to have detoured around Lowell to avoid "that cotton-mill city"; Wilbur H. Siebert, "The Underground Railroad in Massachusetts," *The New England Quarterly* 9, no. 3 (1936): 447–467.

35. A map of antislavery locations is maintained by the University of Massachusetts Lowell, available at https://libguides.uml.edu/c.php?g=1125577.

36. William Lloyd Garrison and James Brown Yerrinton, "Convention in Lowell," *The Liberator* 14, no. 9 (1844): 35.

37. Thomas Dublin, "Women, Work, and Protest in the Early Lowell Mills: 'The Oppressing Hand of Avarice Would Enslave US,'" *Labor History* 16, no. 1 (1975): 99–116.

38. Julie Husband, "'The White Slave of the North': Lowell Mill Women and the Reproduction of Free Labor," *Legacy* 16, no. 1 (1999): 11–21.

39. Jocelyn M. Boryczka, "The Virtues of Vice: The Lowell Mill Girl Debate and Contemporary Feminist Ethics," *Feminist Theory* 7, no. 1 (2006): 49–67.

40. Ralph Waldo Emerson, *Journals of Ralph Waldo Emerson with Annotations*, vol. 7 (Houghton Mifflin, 1913), 300, as quoted in John F. Kasson, *Civilizing the Machine: Technology and Republican Values in America, 1776–1900* (Macmillan, 1999), 124, as cited in Evans, "The Race of Machines," 68.

41. Trevor J. Saunders, and T. A. Sinclair, transl., *Aristotle: The Politics* (Penguin Classics, 1981), 8, as quoted in Elzway, "Arms of the State."

42. Evans, "The Race of Machines," 124–127.

43. Laurence F. Gross, *The Course of Industrial Decline: The Boott Cotton Mills of Lowell, Massachusetts, 1835–1955* (Johns Hopkins University Press, 1993), 26.

44. Charles Cowley, *Illustrated History of Lowell* (Lee & Shepard, 1868), as quoted in Gross, *The Course of Industrial Decline*, 35.

45. Gross, *The Course of Industrial Decline*, 36.

46. Blackmon, *Slavery by Another Name*, 378. See also Luis C. deBaca and Griffin Thomas Black, "The Development of US Anti-Slavery Law: A Historical Review," in *The Historical Roots of Human Trafficking: Informing Primary Prevention of Commercialized Violence*, ed. Makini Chisolm-Straker and Kathrine Chon (Springer, 2021), 127–168.

47. Michele Goodwin, "The Thirteenth Amendment: Modern Slavery, Capitalism, and Mass Incarceration," *Cornell Law Review* 104 (2018), 899–990.

48. Belinda J. Dunstan and Guy Hoffman, "Social Robot Morphology: Cultural Histories of Robot Design," in *Cultural Robotics: Social Robots and Their Emergent Cultural Ecologies*, ed. Belinda J. Dunstan, Jeffrey TKV Koh, Deborah Turnbull Tillman, and Scott Andrew Brown (Springer, 2023), 13–34; and Lucy Suchman, *Human–Machine Reconfigurations: Plans and Situated Actions* (Cambridge University Press, 2007), 35.

49. Alfred Chapuis et al., *Automata: A Historical and Technological Study*. Neuchatel / Editions Du Griffon (Central Book Company, 1958), 67–155.

50. Chapuis et al., *Automata*, 156–222.

51. Chapuis et al., *Automata*, 236.

52. Chapuis et al., *Automata*, 245–314.

53. Scientific American, "Automaton Extraordinary," *Scientific American* 3, no. 34 (1848), 272.

54. The Racine Advocate, "A Singular Automaton," *The Racine Advocate*, May 10, 1848.

55. Louis Joseph Gras, *Histoire de la rubanerie* (Société de l'imprimerie Théolier, 1906), 595, as cited in Ronald Aminzade, "Reinterpreting Capitalist Industrialization: A Study of Nineteenth-Century France: Proletarianization, Small-Scale Industry, and Capitalist Industrialization," *Social History* 9, no. 3 (1984): 329–350.

56. Aminzade, "Reinterpreting Capitalist Industrialization."

57. D. A. Tompkins, "The Cotton Industry," *Publications of the American Economic Association* 5, no. 1 (1904): 144–153.

58. Everett F. Bleiler, "From the Newark Steam Man to Tom Swift," *Extrapolation* 30, no. 2 (1989): 101–116.

59. Nathanial Williams, "War Machines and Child Geniuses: American Edisonades," in *The Cambridge History of Science Fiction*, ed. Gerry Canavan and Eric Carl Link (Cambridge University Press, 2019), 86–100.

60. Ellis, *The Steam Man of the Prairies*, 89.

61. Ellis, *The Steam Man of the Prairies*, 96.

62. Evans, "The Race of Machines," 120.

63. Lemons, "Black Stereotypes."

64. Alan W. C. Green, "'Jim Crow,' 'Zip Coon': The Northern Origins of Negro Minstrelsy," *The Massachusetts Review* 11, no. 2 (1970): 385–397.

65. Annemarie Bean, James V. Hatch and Brooks McNamara. *Inside the Minstrel Mask: Readings in Nineteenth-Century Blackface Minstrelsy* (Wesleyan University Press, 1996).

66. Evans, "The Race of Machines," 129.

67. Evans, "The Race of Machines," 136.

68. Neda Atanasoski and Kalinda Vora, *Surrogate Humanity: Race, Robots, and the Politics of Technological Futures* (Duke University Press, 2019), 34.

69. Evans, "The Race of Machines."

70. Evans, "The Race of Machines," 123.

71. Suchman, *Human–Machine Reconfigurations*.

72. Helen Verran, "Re-Imagining Land Ownership in Australia," *Postcolonial Studies: Culture, Politics, Economy* 1, no. 2 (1998): 237–254.

73. Cornelius Castoriadis, *The Imaginary Institution of Society* (MIT Press, 1997).

74. Castoriadis, *The Imaginary Institution of Society*, as cited in Claudia Strauss, "The Imaginary," *Anthropological Theory* 6, no. 3 (2006): 322–344.

75. Selma Šabanović, "Robots in Society, Society in Robots: Mutual Shaping of Society and Technology as a Framework for Social Robot Design," *International Journal of Social Robotics* 2, no. 4 (2010): 439–450.

76. Jennifer Rhee, *The Robotic Imaginary: The Human and the Price of Dehumanized Labor* (University of Minnesota Press, 2018).

77. Louis Chude-Sokei, "Race and Robotics," in *Cyborg Futures: Cross-Disciplinary Perspectives on Artificial Intelligence and Robotics*, ed. Teresa Heffernan (Palgrave Macmillan, 2019), 161.

78. Simone Browne, "Thinking AI through Rastus Robot, the Westinghouse Mechanical Slave," AI Now Experts Workshop, 2017, https://www.youtube.com/watch?v=HvC6kvrwsN4.

79. Scott Schaut, *Robots of Westinghouse, 1924–Today* (Mansfield Memorial Museum, 2006), as cited in Bob Johnson, "Energy Slaves: Carbon Technologies, Climate Change, and the Stratified History of the Fossil Economy," *American Quarterly* 68, no. 4 (2016):

955–979, and further discussed in Robert Sparrow, "Do Robots Have Race? Race, Social Construction, and HRI," *IEEE Robotics & Automation Magazine* 27, no. 3 (2019): 144–150.

80. Otto O. Binder, "You'll Own Slaves By 1965," *Mechanix Illustrated*, 1957, 62–65, as cited in Benjamin, *Race After Technology*.

81. One exception to this is Hanson Robotics' Bina48, although the motivations beyond its creation are less fraught; Hanson Robotics, "Bina48," 2010, https://www.hansonrobotics.com/bina48-9/.

82. Auguste Villiers de l'Isle-Adam, *L'Ève future* (Paris, 1886).

83. Samuel Butler, *Erewhon: Or, Over the Range* (Trübner & Co, 1872).

84. Butler's characterization of Erewhon's robots as a "race" has led Chude-Sokei to call Butler "the first to explicitly racialize machines"; Chude-Sokei, "Race and Robotics," 169.

85. See also Samuel Butler's essay "Darwin Among the Machines" (The Press, 1863).

86. Cristiano Turbil, "Memory, Heredity and Machines: From Darwinism to Lamarckism in Samuel Butler's *Erewhon*," *Journal of Victorian Culture* 25, no. 1 (2020): 31–46.

87. Čapek, *R.U.R.*

88. Rhee, *The Robotic Imaginary*, 18.

89. Chude-Sokei, "Race and Robotics," 167–168.

90. Chude-Sokei, "Race and Robotics."

91. Butler, *Erewhon*.

92. Nelson Maldonado-Torres, "Race, Religion and Ethics in the Modern/Colonial World," *Journal of Religious Ethics* 42, no. 4 (2014): 691–711. See also William Edward Burghardt Du Bois's famous *The Souls of Black Folk* (Routledge, 2015), written in part as a strong response to such questions.

93. Chude-Sokei, "Race and Robotics," 168.

94. Chude-Sokei, "Race and Robotics," 168.

95. Louis Chude-Sokei, *The Sound of Culture: Diaspora and Black Technopoetics* (Wesleyan University Press, 2015); and Taras Romanenko and Polina Shcherbinina, "Robot vs Worker," *Technology and Language* 3, no. 1 (2022): 17–28.

96. Nathanial Williams, *Gears and God: Technocratic Fiction, Faith, and Empire in Mark Twain's America* (University of Alabama Press, 2018).

97. Ondřej Pilný et al., *Cultural Convergence: The Dublin Gate Theatre, 1928–1960* (Springer Nature, 2021), 47.

98. Kara Reilly, "From Automata to Automation: The Birth of the Robot in *RUR (Rossum's Universal Robots),"* in *Automata and Mimesis on the Stage of Theatre History* (Springer, 2011), 148–176.

99. Derek Johnston, "Genre, Taste and the BBC: The Origins of British Television Science Fiction" (PhD thesis, University of East Anglia, 2009).

100. Jonathan Nolan and Lisa Joy, "Westworld" (2016). For commentary, see Bella Poynton, "From RUR to Westworld: Personal Revolt, Digital Technology, and the Making of a New Robot Ur-Text," *Comparative Drama* 56, no. 4 (2022): 363–388.

101. Joss Whedon, "Dollhouse" (2009). For commentary, see Kristen Noone, "Rossum's Universal Robots: Karel Capek meets Joss Whedon in the Dollhouse," in *Inside Joss' Dollhouse: From Alpha to Rossum*, ed. Jane Espenson (Smart Pop, 2011), 21–32.

102. Fritz Lang, "Metropolis" (1973). For commentary, see Kat Sinclair, "The Feminised Robot: Labour and Harawayan Afterlives," in *Artificial Intelligence and Its Discontents: Critiques from the Social Sciences and Humanities*, ed. Ariane Hanemaayer (Springer, 2022), 43–60.

103. Ridley Scott, "Blade Runner" (1982).

104. Ronald D. Moore, "Battlestar Galactica" (2004). For commentary, see Tobias Higbie, "Why Do Robots Rebel? The Labor History of a Cultural Icon," *Labor: Studies in Working-Class History of the Americas* 10, no. 1 (2013): 99–121.

105. Chun, "Race and/as Technology."

106. Yulia Frumer, "The Short, Strange Life of the First Friendly Robot: Japan's Gakutensoku Was a Giant Pneumatic Automaton That Toured through Asia—Until It Mysteriously Disappeared," *IEEE Spectrum* 57, no. 6 (2020): 42–48.

107. Christopher Bolton et al., *Robot Ghosts and Wired Dreams: Japanese Science Fiction from Origins to Anime* (University of Minnesota Press, 2007); and Tadashi Nagasawa, "The Reception of American Science Fiction in Japan," in *Oxford Research Encyclopedia of Literature* (2016). https://doi.org/10.1093/acrefore/9780190201098.013.166.

108. Note, however, that some typical assumptions about the ways that robots are viewed in Japan, and why, have come under recent critique; Cosima Wagner, "'The Japanese Way of Robotics': Interacting 'Naturally' with Robots as a National Character?" in *Proceedings of the IEEE International Conference on Robot and Human Interactive Communication (RO-MAN)* (IEEE, 2009), 510–515.

109. Asimov, *I, Robot.*

110. Isaac Asimov, "Liar!" *Astounding Science Fiction* 28, no. 3 (1941), 43–87.

111. Asimov, "Runaround."

112. Isaac Asimov, *In Memory Yet Green: The Autobiography of Isaac Asimov, 1920–1954* (Doubleday, 1979); Bill Schelly, *Otto Binder: The Life and Work of a Comic Book and Science Fiction Visionary* (North Atlantic Books, 2016), as cited in Elzway, "Arms of the state."

113. Earl Binder and Otto Binder, "I, Robot," in *Amazing Stories,* edited by B. G. Davis, 13, no. 1, (Experimenter Publishing, 1939), 8–21.

114. Isaac Asimov, *The Rest of the Robots* (Doubleday, 1964), as quoted in Gorman Beauchamp, "The Frankenstein Complex and Asimov's Robots," *Mosaic: A Journal for the Interdisciplinary Study of Literature* 13, no. 3–4 (1980): 83–94.

115. Simon Ings, "Clones Are Not Us," *New Scientist* 244, no. 3254 (2019): 30.

116. Asimov, "Runaround."

117. Wesley L. Stone, "The History of Robotics," in *Robotics and Automation Handbook,* ed. Thomas R. Kurfess (CRC Press, 2018), 8–19.

118. Peter Libbey, "John W. Campbell Award Is Renamed," *New York Times,* August 28, 2019.

119. Isaac Asimov, *Robot Visions* (Penguin, 1991), 453, as cited in Elzway, "Techno-liberal Machines."

120. Fet, "Norbert Wiener in Moscow."

121. Ronald R. Kline, *The Cybernetics Moment: Or Why We Call Our Age the Information Age* (Johns Hopkins University Press, 2015).

122. Kline, *The Cybernetics Moment,* as cited in Thomas Swann, "Information, Cybernetics and the Second Industrial Revolution," *Ephemera* 17, no. 2 (2017): 457–465.

123. Norbert Wiener, *Cybernetics: Control and Communication in the Animal and the Machine* (Hermann, 1948).

124. Chude-Sokei, "Race and Robotics," 161.

125. Norbert Wiener, *I Am a Mathematician: The Later Life of a Prodigy* (MIT Press, 1964).

126. Norbert Wiener, *The Human Use of Human Beings: Cybernetics and Society* (Da Capo Press, 1988), 152, as cited in Chude-Sokei, "Race and Robotics," 163.

127. Elzway, "Arms of the State."

128. Jeremy Pearce, "George C. Devol, Inventor of Robot Arm, Dies at 99," *New York Times,* August 15, 2011.

129. Elzway, "Arms of the State."

130. Association for Advancing Automation, "Unimate // the First Industrial Robot."

131. J. F. Engelberger, "Economic and Sociological Impact of Industrial Robots," *Transactions of the American Nuclear Society* 14, no. 2, 1971, 873, as quoted in Elzway, "Arms of the State."

132. Lind, "The Robots Are Coming," as cited in Elzway, "Technoliberal Machines."

133. Kevin Anderson, "Marcuse's and Fromm's Correspondence with the Socialist Feminist Raya Dunayevskaya: A New Window on Critical Theory," *Logos: A Journal of Modern Society and Culture* 11, no. 2 (2012), as cited in Jason Resnikoff, *Labor's End: How the Promise of Automation Degraded Work* (University of Illinois Press, 2022), 76.

134. Robert Theobald, *Free Men and Free Markets* (Anchor Books, 1965), as cited in Resnikoff, *Labor's End*, 76.

135. Waymon R. Hinson and Edward Robinson, "'We Didn't Get Nothing': The Plight of Black Farmers," *Journal of African American Studies* 12 (2008): 283–302; and Valerie Grim, "The Impact of Mechanized Farming on Black Farm Families in the Rural South: A Study of Farm Life in the Brooks Farm Community, 1940–1970," *Agricultural History* 68, no. 2 (1994): 169–184.

136. Hinson and Robinson, "'We Didn't Get Nothing.'"

137. Jeremy Rifkin, *The End of Work: The Decline of the Global Labor Force and the Dawn of the Post-Market Era* (Putnam's Sons, 1995), 82.

138. Martin Luther King Jr., "Illinois Wesleyan University Student Senate Convocation," 1966. https://www.iwu.edu/mlk/page-3.html.

139. Hak-Chong Lee, "The Lordstown Plant of General Motors," in *Folder: Histories—1974–1975—Lordstown historical documents, box 3, Lordstown collection* (General Motors Heritage Center, 1974), as cited in Elzway, "Technoliberal Machines."

140. David F. Moberg, "Rattling the Golden Chains: Conflict and Consciousness of Auto Workers" (PhD thesis, University of Chicago, 1978), as cited in Salem Elzway and Jason Resnikoff, "Whence Automation? The History (and Possible Futures) of a Concept," *Labor* 21, no. 1 (2024): 27–41.

141. Elzway, "Arms of the State."

142. Elzway, "Technoliberal Machines."

143. Moberg, "Rattling the Golden Chains," as cited in Elzway and Resnikoff, "Whence Automation?"

144. Elzway and Resnikoff, "Whence Automation?"

145. I place "automation" in quotes here due to the ways it served as a philosophy of labor more than an actual technically meaningful term (see Resnikoff, *Labor's End*), leading the *Baltimore Sun* to term "automation" their 1955 "Cliche of the Year"; Elzway, "Arms of the State."

146. Resnikoff, *Labor's End*, 16, 19.

147. Elzway and Resnikoff, "Whence Automation?"

148. Gordon Briggs et al., "Why and How Robots Should Say 'No,'" *International Journal of Social Robotics* 14, no. 2 (2022): 323–339.

149. Alan M. Turing, "Computing Machinery and Intelligence," *Mind* 49 (1950): 433–460.

150. Alan M. Turing, "On Computable Numbers, With an Application to the Entscheidungsproblem," *Proceedings of the London Mathematical Society* s2–42 (1937): 230–265.

151. Warren S. McCulloch and Walter Pitts, "A Logical Calculus of the Ideas Immanent in Nervous Activity," *The Bulletin of Mathematical Biophysics* 5 (1943): 115–133.

152. Nils J. Nilsson, *The quest for Artificial Intelligence* (Cambridge University Press, 2009), as cited in Yarden Katz, *Artificial Whiteness: Politics and Ideology in Artificial Intelligence* (Columbia University Press, 2020).

153. Katz, *Artificial Whiteness.*

154. Rhee, *The Robotic Imaginary*, 74.

155. Patricia Simons, "(Check) Mating The Grand Masters: The Gendered, Sexualized Politics of Chess in Renaissance Italy," *Oxford Art Journal* 16, no. 1 (1993): 59–74.

156. Allen Newell et al., "Report on a General Problem Solving Program, in *Proceedings of the International Conference on Information Processing* (RAND Corporation, 1959), 256–264, as cited in Katz, *Artificial Whiteness.*

157. Charles Spearman, "'General Intelligence' Objectively Determined and Measured," *American Journal of Psychology* 15, no. 2 (1904): 201–293.

158. Charles Spearman, "The Heredity of Abilities," *The Eugenics Review* 6, no. 3 (1914): 219–237; and Charles Spearman, "The Measurement of Intelligence," *The Eugenics Review* 6, no. 4 (1915): 577–578.

159. Derek William Forrest, *Francis Galton: The Life and Work of a Victorian Genius* (Taplinger, 1974), as cited in Roy Lowe, "The Educational Impact of the Eugenics Movement," *International Journal of Educational Research* 27, no. 8 (1998): 647–660.

160. Alfred Binet and Theophile Simon, "The Development of the Binet–Simon Scale, 1905–1908," in *Readings in the History of Psychology*, ed. Wayne Dennis (Appleton-Century-Crofts, 1948), 412–424.

161. L. T. Benjamin, *Monitor on Psychology* 40, no. 1 (2009), 20–21.

162. Leila Zenderland, *Measuring Minds: Henry Herbert Goddard and the Origins of American Intelligence Testing* (Cambridge University Press, 2001), 182.

163. Carl Campbell Brigham, *A Study of American Intelligence* (Princeton University Press, 1922), 194.

164. Jean-Claude Croizet, "The Racism of Intelligence: How Mental Testing Practices Have Constituted an Institutionalized Form of Group Domination," in *The Oxford Handbook of African American citizenship, 1865–Present*, ed. Lawrence D. Bobo, Lisa Crooms-Robinson, Linda Darling-Hammond, Michael C. Dawson, Henry Louis Gates, Gerald Jaynes, and Claude Steele (Oxford Academic, 2012), 806.

165. Croizet, "The Racism of Intelligence."

166. Jean Belkhir, "Race, Sex, Class and 'Intelligence' Scientific Racism, Sexism and Classism," *Race, Sex & Class* 1, no. 2 (1994): 53–83.

167. Cynthia Solomon et al., "History of Logo," in *Proceedings of the ACM on Programming Languages*, 4(HOPL) (ACM, 2020): 1–66.

168. Sheldon Teitelbaum, "Scientists Say Asimov Put the Stars in Their Eyes," *Los Angeles Times*, April 8, 1992, as cited in Elzway, "Arms of the State."

169. Nils J. Nilsson, "Shakey the Robot," Technical report (SRI International Menlo Park, California, 1984).

170. Robert Sparrow, "Robotics Has a Race Problem," *Science, Technology, & Human Values* 45, no. 3 (2020): 538–560.

171. Joseph F. Engelberger, "Sociological Impact of Robots," in *Robotics in Practice: Management and Applications of Industrial Robots* (Springer, 1980), ed. Dennis Lock and Kenneth Willis, 111–116, as quoted in Elzway, "Technoliberal Machines."

172. Wiener, *The Human Use of Human Beings*, as cited in Chude-Sokei, "Race and Robotics," 164.

173. Marvin Minsky and Doug Riecken, "A Conversation with Marvin Minsky About Agents," *Communications of the ACM* 37, no. 7 (1994): 22–29, as quoted in Suchman, *Human–Machine Reconfigurations*, 220.

174. Mark Walker, "A Moral Paradox in the Creation of Artificial Intelligence: Mary Poppins 3000s of the World Unite!," in *Proceedings of the AAAI Workshop on Human Implications of Human–Robot Interaction* (AAAI, 2006), 23–28.

175. Bartek Chomanski, "What's Wrong with Designing People to Serve?" *Ethical Theory and Moral Practice* 22, no. 4 (2019): 993–1015.

176. Steve Petersen, "Designing People to Serve," in *Robot Ethics: The Ethical and Social Implications of Robotics*, ed. Patrick Linn, Keith Abney, and George A. Bekey (MIT Press, 2011), 283–298.

177. Joanna J. Bryson, "Robots Should Be Slaves," in *Close Engagements with Artificial Companions: Key Social, Psychological, Ethical and Design Issues*, ed. Yorik Wilks, Natural Language Processing 8 (John Benjamins, 2010), 63–74.

178. According to Zippia (zippia.com), 82 percent of robot designers, 93 percent of robot engineers, and an astounding 96 percent of robot system engineers are male.

179. John Danaher, "Robotic Rape and Robotic Child Sexual Abuse: Should They Be Criminalised?" *Criminal Law and Philosophy* 11, no. 1 (2017): 71–95.

180. Yolande Strengers and Jenny Kennedy, *The Smart Wife: Why Siri, Alexa, and Other Smart Home Devices Need a Feminist Reboot* (MIT Press, 2021), 122.

181. Norman Makoto Su et al., "Of dolls and Men: Anticipating Sexual Intimacy with Robots," *ACM Transactions on Computer–Human Interaction (TOCHI)* 26, no. 3 (2019): 1–35.

182. Kimberlé Crenshaw, *Demarginalizing the Intersection of Race and Sex: A Black Feminist Critique of Antidiscrimination Doctrine, Feminist Theory and Antiracist Politics* (University of Chicago Legal Forum, 1989).

183. As a rich literature of intersectional analysis has shown, Black women are indeed subject to unique harms under White patriarchy not faced by Black men or White women, including confinement to unique corners of the labor market (Collins, *Black Feminist Thought*, 38); sexual exploitation both within the media (Collins, *Black Feminist Thought*, 90, 160) and at the hands of specific members of society, historically including slaveowners (Thelma Jennings, "'Us Colored Women Had to go Through a Plenty': Sexual Exploitation of African-American Slave Women," *Journal of Women's History* 1, no. 3 [1990]: 45–74) and the police (Ritchie, *Invisible No More*); and unique treatment within the United States' punitive carceral system and prison industrial complex (Crenshaw, *Demarginalizing the Intersection of Race and Sex*). This motivates a need both to address the harms inflicted on Black women and to elevate the standpoints and epistemologies that Black women have developed in collective response to those harms; Collins, *Black Feminist Thought*, 38.

184. Isis H. Settles et al., "Scrutinized But Not Recognized: (In)visibility and Hypervisibility Experiences of Faculty of Color," *Journal of Vocational Behavior* 113 (2019): 62–74, citing Andrea Brighenti, "Visibility: A Category for the Social Sciences," *Current Sociology* 55, no. 3 (2007): 323–342.

185. Saidiya V. Hartman, *Scenes of Subjection: Terror, Slavery, and Self-Making in Nineteenth-Century America* (Oxford University Press, 1997).

186. Collins, *Black Feminist Thought.*

187. Rasul A. Mowatt et al., "Black/Female/Body Hypervisibility and Invisibility: A Black Feminist Augmentation of Feminist Leisure Research," *Journal of Leisure Research* 45, no. 5 (2013): 644–660.

188. Joy Buolamwini and Timnit Gebru, "Gender Shades: Intersectional Accuracy Disparities in Commercial Gender Classification," in *Proceedings of the ACM Conference on Fairness, Accountability and Transparency (FAccT)* (ACM, 2018), 77–91.

189. M. Ryland, "Hypervisibility: How Scrutiny and Surveillance Makes You Watched, But Not Seen," *The Body Is Not an Apology* (2013) https://www.tumblr.com/thebodyisnotanapology/57763238146/hypervisibility-how-scrutiny-and-surveillance, as cited in Settles et al., "Scrutinized But Not Recognized."

190. Francine J. Lipman, "Tax Audits, Economics, and Racism," in *Oxford Research Encyclopedia of Economics and Finance* (2022.) https://doi.org/10.1093/acrefore/9780190625979.013.759.

191. Patricia Lewis and Ruth Simpson, "Kanter Revisited: Gender, Power and (In)visibility," *International Journal of Management Reviews* 14, no. 2 (2012): 141–158, as cited in Settles et al., "Scrutinized But Not Recognized."

192. Patricia Lewis and Ruth Simpson, "Introduction: Theoretical Insights into the Practices of Revealing and Concealing Gender Within Organizations," in *Revealing and Concealing Gender: Issues of Visibility in Organizations*, ed. Patricia Lewis and Ruth Simpson (Springer, 2010), 1–22; and Ruth Simpson and Patricia Lewis, "An Investigation of Silence and a Scrutiny of Transparency: Re-Examining Gender in Organization Literature Through the Concepts of Voice and Visibility," *Human Relations* 58, no. 10 (2005): 1253–1275, as cited in Settles et al., "Scrutinized But Not Recognized."

193. Valerie Stead, "Learning to Deploy (In)visibility: An Examination of Women Leaders' Lived Experiences, *Management Learning* 44, no. 1 (2013): 63–79, as cited in Settles et al., "Scrutinized But Not Recognized."

194. Veronica A. Newton, "Hypervisibility and Invisibility: Black Women's Experiences with Gendered Racial Microaggressions on a White Campus," *Sociology of Race and Ethnicity* 9, no. 2 (2023): 164–178.

195. Benjamin, *Race after Technology*, 105.

196. bell hooks, *Feminist Theory: From Margin to Center Boston: South End* (South End Press, 1984), as cited in Leonardo, "The Color of Supremacy."

197. Melissa E. Wooten and Enobong H. Branch, "Defining Appropriate Labor: Race, Gender, and Idealization of Black Women in Domestic Service," *Race, Gender & Class* 19 no. ¾ (2012): 292–308.

198. Collins, *Black Feminist Thought*, 74.

199. Karen Dugger, "Social Location and Gender-Role Attitudes: A Comparison of Black and White Women," *Gender & Society* 2, no. 4 (1988): 425–448.

200. Julia Wrigley, "Feminists and Domestic Workers," *Feminist Studies* 17, no. 2 (1991): 317–329, as cited in Wooten and Branch, "Defining Appropriate Labor." Note also that although Black women have largely shifted out of domestic service positions, (1) those positions are now filled primarily by other women of color,

especially undocumented women; and (2) the positions into which Black women have moved are characterized by similar dynamics and still characterized as "second-class"; Collins, *Black Feminist Thought*, 52.

201. Verónica Caridad Rabelo and Ramaswami Mahalingam, "'They Really Don't Want to See Us': How Cleaners Experience Invisible 'Dirty' Work," *Journal of Vocational Behavior* 113 (2019): 103–114.

202. Natalee Kehaulani Bauer, *(En)Gendering Whiteness: A Historical Analysis of White Womanhood, Colonial Anxieties, and "Tender Violence" in US Schools* (University of California, Berkeley, 2017), 3.

203. Anne Witz and Mike Savage, "The Gender of Organizations," *The Sociological Review* 39 (1991): 3–62.

204. Rosabeth Kanter, *Men and Women of the Corporation* (Basic Books, 1977), as cited in Witz and Savage, "The Gender of Organizations."

205. Rhee, *The Robotic Imaginary*, 78.

206. Spike Jonze, "Her" (2014).

207. Bryan Forbes, "The Stepford Wives" (1975).

208. Alex Garland, "Ex Machina" (2015).

209. Rhee, *The Robotic Imaginary*, 88.

210. Rhee, *The Robotic Imaginary*, 36, 55; and Heather Suzanne Woods, "Asking More of Siri and Alexa: Feminine Persona in Service of Surveillance Capitalism," *Critical Studies in Media Communication* 35, no. 4 (2018): 334–349.

211. Evelyn Nakano Glenn, "From Servitude to Service Work: Historical Continuities in the Racial Division of Paid Reproductive Labor," *Signs: Journal of Women in Culture and Society* 18, no. 1 (1992): 1–44, as cited in Rhee, *The Robotic Imaginary*.

212. Erico Guizzo, "Cynthia Breazeal Unveils Jibo, a Social Robot for the Home," *IEEE Spectrum*, July 16, 2014.

213. Atanasoski and Vora, *Surrogate Humanity*, 89.

214. Teresa Heffernan, "Fiction Meets Science: Ex Machina, Artificial Intelligence, and the Robotics Industry," in *Cyborg Futures: Cross-Disciplinary Perspectives on Artificial Intelligence and Robotics*, ed. Teresa Heffernan (Springer, 2019), 127–140.

215. Atanasoski and Vora, *Surrogate Humanity*, 89.

216. Strengers and Kennedy, *The Smart Wife*.

217. Note that contemporary industrial and field roboticists' explanations that robots are primarily used in domains that are "dull, dirty, and dangerous" similarly devalues and demeans those whose jobs they are replacing.

218. Resnikoff, *Labor's End*, 142.

219. Resnikoff, *Labor's End*, 143.

220. The differential valuation of labor by robotics companies is demonstrated by Rhee through her case study of the Roomba. As Rhee argues, the Roomba is designed to be mindless. After all, the first versions of the Roomba operated through an intentionally unintelligent navigation policy, turning random amounts when running into obstacles. Rhee insightfully juxtaposes this mindlessness with the advertisements for the Roomba, which emphasize the liberation of White women from domestic labor. As this juxtaposition shows, the Roomba's guiding narrative feeds off and into these discourses of racialized and gendered labor, casting White women's time as valuable and the labor typically invisibly conducted by women of color as mindless and replaceable; Rhee, *The Robotic Imaginary*, 94.

221. Here, I use the sense of "generative" in the current zeitgeist, even though this is not the traditional meaning of generative within the machine learning literature.

222. Jeremias Prassl, *Humans as a Service: The Promise and Perils of Work in the Gig Economy* (Oxford University Press, 2018).

223. Bill Perrigo, "OpenAI used Kenyan Workers on Less Than $2 Per Hour to Make ChatGPT Less Toxic," *Time*, January 18, 2023.

224. Couldry and Mejias, *The Costs of Connection*.

225. Sparrow, "Robotics Has a Race Problem."

226. Standage, *The Turk*.

227. Prassl, *Humans as a Service*.

228. Jason Pontin, "Artificial Intelligence, with Help from the Humans," *New York Times*, March 25, 2007.

229. Jonathan Grudin and Richard Jacques, "Chatbots, Humbots, and the Quest for Artificial General Intelligence," in *Proceedings of the CHI Conference on Human Factors in Computing Systems* (ACM, 2019), 1–11.

230. With some exceptions. At least one onlooker thought the machine to be possessed by a spirit (von Windisch, *Inanimate Reason*, 15)—an intriguing application of another interstitial category (cf. Weisman, "Extraordinary Entities") to make sense of the machine in the absence of the robot imaginary that would have otherwise facilitated this sense making.

231. Sangseok Yun et al., "Engkey: Tele-Education Robot," in *Proceedings of the International Conference on Social Robotics* (Springer, 2011), 142–152.

232. Anna Romina Guevarra, "Techno-Modeling Care: Racial Branding, Dis/Embodied Labor, and 'Cybraceros' in South Korea," *Frontiers: A Journal of Women Studies* 36, no. 3 (2015): 139–159.

233. Guevarra, "Techno-Modeling Care."

234. Elizabeth Gillespie McRae, *Mothers of Massive Resistance: White Women and the Politics of White Supremacy* (Oxford University Press, 2018).

235. Guevarra, "Techno-Modeling Care."

236. Guevarra, "Techno-Modeling Care."

237. Guevarra, "Techno-Modeling Care."

238. See Cave and Dihal's Conception of the White Utopian Imaginary; Stephen Cave and Kanta Dihal, "The Whiteness of AI," *Philosophy & Technology* 33, no. 4 (2020): 685–703.

239. Susan Leigh Star and Anselm Strauss, "Layers of Silence, Arenas of Voice: The Ecology of Visible and Invisible Work," in *Proceedings of the ACM on Human–Computer Interaction*, (CSCW), vol. 8 (ACM, 1999), 9–30.

240. Catherine Knight Steele, *Digital Black Feminism* (New York University Press, 2021).

241. Lisa Nakamura, "Race in/for Cyberspace: Identity Tourism and Racial Passing on the Internet," *Works and Days* 13, no. 1–2 (1995): 181–193.

242. Saad Elbeleidy et al., "Practical, Ethical, and Overlooked: Tele-Operated Socially Assistive Robots in the Quest for Autonomy," in *Companion of the 17th ACM/IEEE International Conference on Human–Robot Interaction (alt.HRI)* (ACM, 2022), 577–587.

243. Saad Elbeleidy et al., "The Invisible Labor of Authoring Dialogue for Teleoperated Socially Assistive Robots," in *Proceedings of the IEEE International Conference on Robot and Human Interactive Communication (RO-MAN)* (IEEE, 2023), 2462–2469; and Saad Elbeleidy et al., "Beyond the Session: Centering Teleoperators in Robot-Assisted Therapy Reveals the Bigger Picture," in *Proceedings of the ACM on Human–Computer Interaction*, (CSCW) (ACM, 2023), 1–33.

244. Elbeleidy et al., "The Invisible Labor," citing Marion Crain et al. *Invisible Labor: Hidden Work in the Contemporary World* (University of California Press, 2016).

245. Elbeleidy et al., "The Invisible Labor"; and Elbeleidy et al., "Beyond the Session."

246. Shuichi Nishio et al., "Geminoid: Teleoperated Android of an Existing Person," *Humanoid Robots: New Developments* 14, no. 343–352 (2007): 10–1109.

247. Scientific American, "Automaton Extraordinary"; and The Racine Advocate, "A Singular Automaton."

248. Justin McCurry, "Erica, the 'Most Beautiful and Intelligent' Android, Leads Japan's Robot Revolution," *The Guardian*, December 31, 2015.

249. Takayuki Todo and Takanari Miisho, "Interactions on Eyeballs of Humanoid-Robots," in *Proceedings of the 2nd International Conference on Human–Agent Interaction* (ACM, 2014), 281–283.

250. James Wright, *Robots Won't Save Japan: An Ethnography of Eldercare Automation* (Cornell University Press, 2023).

251. Jennifer Robertson, "Robo Sapiens Japanicus: Humanoid Robots and the Posthuman Family," *Critical Asian Studies* 39, no. 3 (2007): 369–398, as cited in Wright, *Robots Won't Save Japan*, 116.

252. Wright, *Robots Won't Save Japan*, 32.

253. Aziz Ansari and Alan Yang, "Master of None: Season 1" (2015).

254. Garland, "Ex Machina."

Chapter 3

1. Johannes Lemburg et al., "AILA-Design of an Autonomous Mobile Dual-Arm Robot," in *Proceedings of the IEEE International Conference on Robotics and Automation (ICRA)* (IEEE, 2011), 5147–5153.

2. Robert Kratz et al., "Control Approach for a Novel High Power-to-Weight Robotic Actuator Scalable in Force and Length," in *2007 IEEE/ASME International Conference on Advanced Intelligent Mechatronics* (IEEE, 2007).

3. David M. Higgins and Matthew Iung, "Comic Books 1," in *The Routledge Companion to Cyberpunk Culture*, ed. Anna McFarlane, Graham J. Murphy, and Lars Schmeink (Routledge, 2019), 91–100.

4. Chris Cunningham, "All is Full of Love," *Leonardo* 35, no. 5 (2002): 558–558.

5. Masahiro Mori et al., "The Uncanny Valley [From the Field]," *IEEE Robotics & Automation Magazine* 19, no. 2 (2012): 98–100.

6. Jasmin Bernotat et al., "The (Fe)male Robot: How Robot Body Shape Impacts First Impressions and Trust Towards Robots," *International Journal of Social Robotics* 13 (2021): 477–489.

7. Joe Feagin, *Systemic Racism: A Theory of Oppression* (Routledge, 2013) and Joe R. Feagin, *The White Racial Frame: Centuries of Racial Framing and Counter-Framing* (Routledge, 2020), as cited in Cave and Dihal, "The Whiteness of AI."

8. Feagin, *Racist America*.

9. Cave and Dihal, "The Whiteness of AI."

10. Elizabeth Phillips et al., "What is Human-Like?: Decomposing Robots' Human-Like Appearance Using the Anthropomorphic Robot (ABOT) Database," in *Proceedings of the 13th ACM/IEEE International Conference on Human–Robot Interaction (HRI)* (ACM, 2018), 105–113.

11. Sparrow, "Do Robots Have Race?"

12. Giulia Perugia et al., "The Shape of Our Bias: Perceived Age and Gender in the Humanoid Robots of the ABOT Database," in *Proceedings of the ACM/IEEE International Conference on Human–Robot Interaction* (IEEE, 2022), 110–119.

13. Jennifer Robertson, "Gendering Humanoid Robots: Robo-Sexism in Japan," *Body & Society* 16, no. 2 (2010): 1–36, as cited in Simone Alesich and Michael Rigby, "Gendered Robots: Implications for Our Humanoid Future," *IEEE Technology and Society Magazine* 36, no. 2 (2017): 50–59.

14. Dennler, Nathaniel S., Mina Kian, Stefanos Nikolaidis, and Maja Matarić. "Designing Robot Identity: The Role of Voice, Clothing, and Task on Robot Gender Perception," arXiv preprint arXiv:2404.00494 (2024).

15. Technically, "android" refers to a highly anthropomorphic robot modeled after *male* humans, whereas those modeled after female humans are called "gynoids." However, in practice, the term "android" is used to capture both types of robots, perhaps due to the overwhelming perception of robotics as a masculine field.

16. Aldebaran Robotics, "Nao," 2008, https://www.aldebaran.com/en/nao.

17. Christoph Bartneck et al., "Robots and Racism," in *Proceedings of the ACM/IEEE International Conference on Human–Robot Interaction* (ACM, 2018), 196–204; Arifah Addison et al., "Robots Can Be More Than Black and White: Examining Racial Bias Towards Robots," in *Proceedings of the AAAI/ACM Conference on AI, Ethics, and Society* (ACM, 2019), 493–498.

18. Sparrow, "Do Robots Have Race?"

19. Sparrow, "Do Robots Have Race?"

20. Sparrow, "Robotics Has a Race Problem."

21. Mark West et al., "I'd Blush if I Could: Closing Gender Divides in Digital Skills Through Education," Technical report (UNESCO, 2019).

22. West et al., "I'd Blush if I Could," as cited in Jasper Feine et al., "Gender Bias in Chatbot Design," in *Proceedings of the Third International Workshop on Chatbot Research and Design* (Springer, 2020), 79–93.

23. Perugia et al., "The Shape of Our Bias"; and Katie Seaborn and Alexa Frank, "What Pronouns for Pepper? A Critical Review of Gender/ing in Research," in *CHI Conference on Human Factors in Computing Systems* (ACM, 2022), 1–15.

24. Amit Kumar Pandey and Rodolphe Gelin, "A Mass-Produced Sociable Humanoid Robot: Pepper: The First Machine of its Kind," *IEEE Robotics & Automation Magazine* 25, no. 3 (2018): 40–48; Roger Andre Søraa, "Mechanical Genders: How do Humans Gender Robots?" *Gender, Technology and Development* 21 (2017): 99–115.

25. In the wake of UNESCO's report, some companies announced they would no longer have a default voice and would instead ask users to choose one; Matthew

Panzarino, "Apple Adds Two Brand New Siri Voices and Will no Longer Default to a Female or Male Voice in iOS," *Tech Crunch*, March 31, 2021. Even so, users continue to overwhelmingly gender voice assistants as feminine, as evidenced through the pronouns people use to refer to them; Gavin Abercrombie et al., "Alexa, Google, Siri: What Are Your Pronouns? Gender and Anthropomorphism in the Design and Perception of Conversational Assistants," in *Proceedings of the 3rd Workshop on Gender Bias in Natural Language Processing* (ACL, 2021), 24–33.

26. Sinduja Rangarajan, "Hey Siri—Why Don't You Understand More People Like Me?" *Mother Jones*, February 23, 2021.

27. Kaveh Waddell, "Hey Siri, Is That You? Apple's New Voices Resonate with Some Black iPhone Users," *Consumer Reports*, April 23, 2021.

28. Benjamin, *Race after Technology*.

29. Ilaria Torre and Sébastien Le Maguer, "Should Robots Have Accents?," in *Proceedings of the 29th IEEE International Conference on Robot and Human Interactive Communication (RO-MAN)* (IEEE, 2020), 208–214.

30. Jessica Barfield, "Designing Social Robots to Accommodate Diversity, Equity, and Inclusion in Human–Robot Interaction," in *Proceedings of the ACM SIGIR Conference on Human Information Interaction and Retrieval* (ACM, 2023), 463–466.

31. West et al., "I'd Blush if I Could."

32. See also Katie Winkle et al., "Boosting Robot Credibility and Challenging Gender Norms in Responding to Abusive Behaviour: A Case for Feminist Robots," in *Companion of the ACM/IEEE International Conference on Human–Robot Interaction (alt.HRI)* (ACM, 2021), 29–37.

33. Michael Stolp-Smith and Tom Williams, "More Than Binary: Transgender and Nonbinary Perspectives on Human Robot Interaction," in *Proceedings of the ACM/IEEE International Conference on Human–Robot Interaction* (ACM, 2024), 697–705.

34. Geoffrey K. Pullum, "African American Vernacular English Is Not Standard English with Mistakes," in *The workings of Language: From Prescriptions to Perspectives*, ed. Rebecca S. Wheeler (Praeger, 1999), 59–66.

35. Geneva Smitherman, "Word From the Hood: The Lexicon of African-American Vernacular English," in *African-American English* (Routledge, 2021), 223–248.

36. Jenny Cheshire et al., "Contact, the Feature Pool and the Speech Community: The Emergence of Multicultural London English," *Journal of Sociolinguistics* 15, no. 2 (2011): 151–196.

37. Frantz Fanon, *Black Skin, White Masks* (Grove Press, 2008), 2.

38. Some sociologists have recently called for increased work within Whiteness studies (Woody Doane, "Rethinking Whiteness Studies," in *White Out: The Continuing Significance*

of Racism, ed. Ashley W. Doane and Eduardo Bonilla-Silva (Routledge, 2013, 11–26) to make Whiteness a more explicit topic of critique rather than an unspoken default.

39. AI systems more generally have reflected a default Whiteness through their linguistic choices as far back as Weizenbaum's computerized psychotherapist ELIZA; Joseph Weizenbaum, "Eliza—A Computer Program for the Study of Natural Language Communication Between Man and Machine," *Communications of the ACM* 9, no. 1 (1966): 36–45. As Mark Marino writes in "The Racial Formation of Chatbots," "If ELIZA presented a bot that tried to imitate language, it was performing standard White middle-class English, without a specific identifying cultural inflection . . . language without culture, disembodied, hegemonic, and, in a word, White"; Mark Marino, "The racial Formation of Chatbots," *CLCWeb: Comparative Literature and Culture* 16, no. 5 (2014):1481–4374, as cited in Cave and Dihal, "The Whiteness of AI."

40. Cave and Dihal, "The Whiteness of AI."

41. bell hooks, *Representing Whiteness in the Black imagination* (Routledge, 1992), as cited in Cave and Dihal, "The Whiteness of AI."

42. Common Crawl, "Common Crawl," 2023, https://commoncrawl.org/ 2023.

43. Wikipedia, "Community Insights 2021 Report," https://meta.wikimedia.org/wiki/Community Insights/Community_Insights_2021_Report.

44. Robert W. Fairlie, "Have We Finally Bridged the Digital Divide? Smart Phone and Internet Use Patterns by Race and Ethnicity," *First Monday*, September 1, 2017.

45. Mark Graham, "Inequitable Distributions in Internet Geographies: The Global South is Gaining Access, but Lags in Local Content," *Innovations: Technology, Governance, Globalization* 9, no. 3–4 (2014): 3–19.

46. Ryan Blake Jackson et al., "Tact in Noncompliance: The Need for Pragmatically Apt Responses to Unethical Commands," in *Proceedings of the AAAI/ACM Conference on AI, Ethics, and Society* (ACM, 2019), 499–505.

47. Ryan Blake Jackson et al., "Exploring the Role of Gender in Perceptions of Robotic Noncompliance," in *Proceedings of the ACM/IEEE International Conference on Human–Robot Interaction* (ACM, 2020), 559–567.

48. The results are slightly more nuanced and complicated than described here. For more details, please refer to Jackson et al., "Exploring the Role of Gender."

49. Sara Mills, *Gender and Politeness* (Cambridge University Press, 2003); Robin Tolmach Lakoff, *Language and Woman's Place: Text and Commentaries*, vol. 3 (Oxford University Press, 2004); and Sara Mills, "Gender and Impoliteness," *Journal of Politeness Research* 1, no. 2 (2005): 263–280.

50. Manuela Barreto and Naomi Ellemers, "The Burden of Benevolent Sexism: How it Contributes to the Maintenance of Gender Inequalities," *European Journal of Social Psychology* 35, no. 5 (2005): 633–642.

51. Another possibility is that women may simply have recognized that, despite the existence of apparent norms of benevolent sexism, men often *are* rude to women and may thus have expected that masculine robots were probably *going to* be ruder. Yet another possibility is that women may have been more sensitive to in-group dynamics, viewing humans and robots with similar gender expressions as more likely to be part of the same in-groups and thus more forgiving of perceived social norm violations within those in-groups than in the case of perceived social norm violations toward out-group members.

52. Addison et al., "Robots Can Be More Than Black and White."

53. Joshua Correll et al., "The Police Officer's Dilemma: Using Ethnicity to Disambiguate Potentially Threatening Individuals," *Journal of Personality and Social Psychology* 83, no. 6 (2002): 1314–1329.

54. This manipulation has also been used in other studies of racial perception in human–robot interaction; Friederike Eyssel and Steve Loughnan, "It Don't Matter if You're Black or White?" in *Proceedings of the International Conference on Social Robotics* (Springer, 2013), 422–431.

55. Barfield, "Designing Social Robots," 2023.

56. Megan Strait et al., "Robots Racialized in the Likeness of Marginalized Social Identities are Subject to Greater Dehumanization Than Those Racialized as White," in *Proceedings of the 27th IEEE International Conference on Robot and Human Interactive Communication (RO-MAN)* (IEEE, 2018), 452–457.

57. Tobi Ogunyale, De'Aira Bryant, and Ayanna Howard, "Does Removing Stereotype Priming Remove Bias? A Pilot Human–Robot Interaction Study," in *5th Workshop on Fairness, Accountability, and Transparency in Machine Learning*, 2018.

58. Jaime Banks and Kevin Koban, "A Kind Apart: The Limited Application of Human Race and Sex Stereotypes to a Humanoid Social Robot," *International Journal of Social Robotics* 15, no. 11 (2022): 1949–1961.

59. Andrea Keay, "The Naming of Robots: Biomorphism, Gender and Identity (PhD thesis, master thesis in digital cultures, University of Sydney, 2012), as cited in Kate Darling, "'Who's Johnny?' Anthropomorphic Framing in Human–Robot Interaction, Integration, and Policy," in *Robot Ethics 2.0*, ed. Patrick Lin, Ryan Jenkins, and Keith Abney (Oxford University Press, 2015), 173–192.

60. Friederike Eyssel and Dieta Kuchenbrandt, "Social Categorization of Social Robots: Anthropomorphism as a Function of Robot Group Membership," *British Journal of Social Psychology* 51, no. 4 (2012): 724–731.

61. Kylie L. Goodman and Christopher B. Mayhorn, "It's Not What You Say but How You Say It: Examining the Influence of Perceived Voice Assistant Gender and Pitch on Trust and Reliance," *Applied Ergonomics*, 106, 2023, 1–8.

62. Yuting Liao and Jiangen He, "Racial Mirroring Effects on Human–Agent Interaction in Psychotherapeutic Conversations," in *Proceedings of the 25th International Conference on Intelligent User Interfaces* (ACM, 2020), 430–442.

63. Woods, "Asking More of Siri and Alexa."

64. Zuboff, *The Age of Surveillance Capitalism.*

65. Leopoldina Fortunati et al., "Is Alexa Female, Male, or Neutral? A Cross-National and Cross-Gender Comparison of Perceptions of Alexa's Gender and Status as a Communicator," *Computers in Human Behavior*, 137, 2022, 1–9.

66. Hilary Bergen, "'I'd Blush if I Could': Digital Assistants, Disembodied Cyborgs and the Problem of Gender," *Word and Text, a Journal of Literary Studies and Linguistics* 6, no. 1 (2016): 95–113.

67. Reynolds 2016, as cited in Bergen, "'I'd Blush if I Could.'"

68. John Danaher and Neil McArthur, eds., *Robot Sex: Social and Ethical Implications* (MIT Press, 2017).

69. Tomasz Oleksy and Anna Wnuk, "Do Women Perceive Sex Robots as Threatening? The Role of Political Views and Presenting the Robot as a Female-vs Male-Friendly Product," *Computers in Human Behavior* 117 (2021): 1–9.

70. Nicola Döring and Sandra Pöschl, "Sex Toys, Sex Dolls, Sex Robots: Our Under-Researched Bed-Fellows," *Sexologies* 27, no. 3 (2018), 51–55.

71. Nicola Döring, "Vom internetsex zum robotersex: Forschungsstand und herausforderungen für die sexualwissenschaft," *Zeitschrift für Sexualforschung* 30, no. 01 (2017): 35–57.

72. Oleksy and Wnuk, "Do Women Perceive Sex Robots as Threatening?"

73. Unrealistic Beauty Standards are Already Being Reinforced and Spread in Similar Ways by AI-Generated Social Media Influencers; Pamela C. Scorzin, "> More Human Than Human <. Digital Dolls on Social Media," *Denkste: Puppe-multidisziplinäre zeitschrift für mensch-puppen-diskurse* 4, no. 1 (2021): 157–166.

74. This argument is taken up by the Campaign Against Sex Robots (Kathleen Richardson, "The Asymmetrical 'Relationship' Parallels Between Prostitution and the Development of Sex Robots," *Computers and Society* 45, no. 3 [2016]: 290–293), although cp. John Danaher et al., "Should We Campaign Against Sex Robots?," in *Robot Sex: Social and Ethical Implications*, ed. John Danaher and Neil McArthur (MIT Press, 2017), 47–72.

75. Chantel Cox-George and Susan Bewley, "I, Sex Robot: The Health Implications of the Sex Robot Industry," *BMJ Sexual & Reproductive Health* 44, no. 3 (2018): 161–164.

76. Roc Morin, "Can Child Dolls Keep Pedophiles from Offending," *The Atlantic*, January 11, 2016.

77. Li Lucy and David Bamman, "Gender and Representation Bias in GPT-3 Generated Stories," in *Proceedings of the Third Workshop on Narrative Understanding* (ACL, 2021), 48–55.

78. Sparrow, "Do Robots Have Race?"

79. Syed Iylia Hariz Al-Qadri and Hafzi Hasani Mohamed, "Meet ADAM, Malaysia's Own Version of Sophia the Humanoid Robot," 2019, http://energy.bernama.com/news.php?id=17480972019.

80. Terrence Fong and Illah Nourbakhsh, "Interaction Challenges in Human–Robot Space Exploration," *Interactions* 12, no. 2 (2005): 42–45.

81. Aditi Ramachandran et al., "Thinking Aloud with a Tutoring Robot to Enhance Learning," in *Proceedings of the ACM/IEEE International Conference on Human–Robot Interaction* (ACM, 2018), 59–68.

82. Abubakar Abid et al., "Persistent Anti-Muslim Bias in Large Language Models," in *Proceedings of the AAAI/ACM Conference on AI, Ethics, and Society* (ACM, 2021), 298–306.

83. Jesutofunmi A. Omiye et al., "Large Language Models Propagate Race-Based Medicine". *npj Digital Medicine*, 6 (195), 2023.

84. Kelly M. Hoffman et al., "Racial Bias in Pain Assessment and Treatment Recommendations, and False Beliefs About Biological Differences Between Blacks and Whites," *Proceedings of the National Academy of Sciences* 113, no. 16 (2016): 4296–4301.

85. Heather H. Burris et al., "Black–White Disparities in Maternal In-Hospital Mortality According to Teaching and Black-Serving Hospital Status," *American Journal of Obstetrics and Gynecology* 225, no. 1 (2021), 83e1–83e9.

86. Tom Williams et al., "Scarecrows in Oz: The Use of Large Language Models in HRI," *ACM Transactions on Human–Robot Interaction (T-HRI)* 13, no. 1 (2023): 1–11.

87. Harry G. Frankfurt, *On bullshit* (Princeton University Press, 2009).

88. Large language models are not unique in their tendency toward deception. This is, in many ways, an inevitable side effect of the field's traditional obsession with fluency. It can be seen in the more sensationalist corners of the field of artificial intelligence, such as Turing test competitions. The Turing test is the colloquial name for the thought experiment devised by Alan Turing in 1950 (Turing, "Computing Machinery and Intelligence"), in which the performance of an intelligent system is adjudicated on the basis of how well a human judge could distinguish the system from a competing human, merely by asking questions over a text interface. Although the Turing test is not taken seriously by most AI researchers, it has attracted a lot of recent public attention due to its use in popular culture pieces such as *Ex Machina* (Garland, "Ex Machina"). Although Turing test advocates present it as a test of intelligence, at its

core, the Turing test is truly a test of deceptive capability, as "winning" systems need not actually demonstrate intelligent capabilities but rather only fool people into thinking they are human (Simone Natale, *Deceitful Media: Artificial Intelligence and Social Life After the Turing Test* [Oxford University Press, 2021]). Take, for example, the case of Eugene Goostman, a chatbot that won a 2014 Turing test competition and was declared to have "passed" after fooling one-third of its interactants into thinking it was human. What soaring feats of intelligence did Eugene display in order to achieve this feat? It pretended to be a thirteen-year-old Eastern European teenager, deftly avoiding questions through jokes and ignorance (Gary Marcus et al., "Beyond the Turing Test," *AI Magazine* 37, no. 1 [2016]: 3–4). When our measures of AI quality are grounded merely in how much their outputs "sound like" those of real people, we inherently optimize for exactly this sort of deception—a dangerous trait when embodied into a highly persuasive and humanlike agent.

89. Benjamin, *Race after Technology*; and Casey R. Lynch, "Glitch Epistemology and the Question of (Artificial) Intelligence: Perceptions, Encounters, Subjectivities," *Dialogues in Human Geography* 12, no. 3 (2022): 379–383.

90. Kathleen Richardson, *An Anthropology of Robots and AI: Annihilation Anxiety and Machines* (Routledge, 2015), as cited in Alesich and Rigby, "Gendered Robots."

91. Lemburg et al., "AILA-design of an Autonomous Mobile Dual-Arm Robot."

92. Cave and Dihal, "The Whiteness of AI."

93. Laurel Riek and Don Howard, "A Code of Ethics for the Human–Robot Interaction Profession," in *Proceedings of We Robot* (2014); and Inês Hipólito et al., "Enactive Artificial Intelligence: Subverting Gender Norms in Robot and Human Interaction," *Frontiers in Neurorobotics* 17 (2023) :1–11. See also recent work from social roboticists Lux Miranda, Ginevra Castellano, and Katie Winkle, who have shown that more *groups* of robots that exhibit diverse gender presentations are viewed more positively by children; Lux Miranda et al., "A Case for Diverse Social Robot Identity Performance in Education," in *Companion of the 2024 ACM/IEEE International Conference on Human–Robot Interaction* (ACM, 2024), 28–35.

94. Sparrow, "Robotics Has a Race Problem."

95. Kallyn Song-Nichols and Andrew Young, "Gendered Robots Can Change Children's Gender Stereotyping," in *Proceedings of the Annual Meeting of the Cognitive Science Society (CogSci)* (Cognitive Science Society, 2020), 2480–2485. Moreover, when this type of cross-stereotypic design is used in classrooms, it may have the added benefit of making children more engaged in learning; Natalia Reich-Stiebert and Friederike Eyssel, "(Ir)relevance of Gender? On the Influence of Gender Stereotypes on Learning with a Robot," in *Proceedings of the ACM/IEEE International Conference on Human–Robot Interaction* (ACM, 2017), 166–176.

96. Winkle et al., "Boosting Robot Credibility."

97. The utterances listed here are translations provided by Winkle for the Swedish language actually used in the experiment.

98. Here, "credibility" is defined as being more trustworthy, bearing more goodwill, and having more expertise.

99. Katie Winkle et al., "Norm-Breaking Responses to Sexist Abuse: A Cross-Cultural Human Robot Interaction Study," in *Proceedings of the 17th ACM/IEEE International Conference on Human–Robot Interaction (HRI)* (IEEE, 2022), 120–129.

100. Alessio Galatolo et al., "The Right (Wo)Man for the Job? Exploring the Role of Gender when Challenging Gender Stereotypes with a Social Robot," *International Journal of Social Robotics* 15 (2022): 1933–1947.

101. Atanasoski and Vora, *Surrogate Humanity*, 195.

102. Frédéric Delaunay et al., "Towards Retro-Projected Robot Faces: An Alternative to Mechatronic and Android Faces," in *Proceedings of the IEEE International Symposium on Robot and Human Interactive Communication (RO-MAN)* (IEEE, 2009), 306–311.

103. Michael Walker et al., "Virtual, Augmented, and Mixed Reality for Human–Robot Interaction: A Survey and Virtual Design Element Taxonomy," *ACM Transactions on Human–Robot Interaction (T-HRI)* 12, no. 4 (2023): 1–39.

104. Nhan Tran et al., "Now Look Here! ↓ Mixed Reality Improves Robot Communication Without Cognitive Overload," in *Proceedings of the International Conference on Virtual, Augmented, and Mixed Reality (VAMR),* held as part of the *International Conference on Human–Computer Interaction (HCI)* (Springer, 2023); Zhao Han et al., "Crossing Reality: Comparing Physical and Virtual Robot Deixis," in *Proceedings of the ACM/IEEE International Conference on Human–Robot Interaction* (ACM, 2023), 152–161; Landon Brown et al., "Best of Both Worlds? Combining Different Forms of Mixed Reality Deictic Gestures," *ACM Transactions on Human–Robot Interaction (T-HRI)* 12, no. 1 (2022): 1–23; and Jared Hamilton et al., "What's the Point? Tradeoffs Between Effectiveness and Social Perception when Using Mixed Reality to Enhance Gesturally Limited Robots," in *Proceedings of the 16th ACM/IEEE International Conference on Human–Robot Interaction (HRI)* (ACM, 2021), 177–186.

105. Yifei Zhu et al., "Designing Augmented Reality Robot Guidance Interactions Through the Metaphors of Re-Embodiment and Telepresence," in *Proceedings of the IEEE International Conference on Robot and Human Interactive Communication (RO-MAN)* (IEEE, 2024), 29–36.

106. Adam Poulsen et al., "Queering Machines," *Nature Machine Intelligence* 2, no. 3 (2020): 152–152.

107. Stolp-Smith and Williams, "More Than Binary." See also recent complementary work from Miranda et al., "A Case for Diverse Social Robot Identity Performance in Education."

108. Schaut, *Robots of Westinghouse.*

109. As defined by Karen Tong, as quoted in Sara Moniuszko, "'It Remains Purely Symbolic': Why Rainbow Capitalism Can Be Harmful to LGBTQ People," *USA Today*, June 7, 2021, as cited in Eric Smialek, "Who Needs to Calm Down? Taylor Swift and Rainbow Capitalism," *Contemporary Music Review* 40, no. 1 (2021): 99–119.

110. Marc Bruxelle, "The Problem with the 'Rainbow-Washing' of LGBTQ+ Pride," *Wired*, June 21, 2018.

111. Jaye Nias et al., "Decolonizing Technologies for Preserving Cultural and Societal Diversity," in *Proceedings of the CHI 2020 Workshop on Engaging in Race in HCI* (2020).

112. Ron Eglash et al., "Race-Positive Design: A Generative Approach to Decolonizing Computing," in *Proceedings of the CHI Workshop on Race in HCI* (2020). https://par.nsf.gov/servlets/purl/10182458.

113. Nias et al., "Decolonizing Technologies for Preserving Cultural and Societal Diversity."

114. Inyong Ha et al., "Development of Open Humanoid Platform DARwIn-OP," in *Proceedings of the SICE Annual Conference* (IEEE, 2011), 2178–2181.

115. Samer Al Moubayed et al., "Furhat: A Back-Projected Human-Like Robot Head for Multiparty Human-Machine Interaction," in *Cognitive Behavioural Systems: COST 2102 International Training School*, ed. Anna Esposito, Antoinetta M. Esposito, Alessandro Vinciarelli, Rüdiger Hoffmann, and Vincent C. Müller (Springer, 2012), 114–130.

116. Raymond S. Nickerson, "On Conversational Interaction with Computers," in *Proceedings of the ACM/SIGGRAPH Workshop on User-oriented Design of Interactive Graphics Systems* (ACM, 1976), 101–113, as cited in Suchman, *Human–Machine Reconfigurations*, 42.

117. Rhee, *The Robotic Imaginary*; Matthias Scheutz, "The Inherent Dangers of Unidirectional Emotional Bonds Between Humans and Social Robots," in *Robot Ethics: The Ethical and Social Implications of Robotics*, ed. Patrick Linn, Keith Abney, and George A. Bekey (MIT Press, 2011), 205–221; Sherry Turkle, *Alone Together: Why We Expect More from Technology and Less from Each Other* (Hachette UK, 2017); Neil M. Richards and William D. Smart, "How Should the Law Think About Robots?," in *Robot Law*, ed. Ryan Calo, A. Michael Froomkin, and Ian Kerr (Edward Elgar, 2016), 3–22; and Jean-Christophe Giger et al., "Humanization of Robots: Is It Really Such a Good Idea?," *Human Behavior and Emerging Technologies* 1, no. 2 (2019): 111–123.

118. Kate Darling, *The New Breed: What Our History with Animals Reveals About Our Future with Robots* (Henry Holt, 2021). See also Sparrow, "Robotics Has a Race Problem."

119. Jennifer R. Lambdin et al., "The Animal = Male Hypothesis: Children's and Adults' Beliefs About the Sex of Non-Sex-Specific Stuffed Animals," *Sex Roles* 48 (2003): 471–482.

120. Judy S. DeLoache et al., "The Three Bears Are All Boys: Mothers' Gender Labeling of Neutral Picture Book Characters," *Sex Roles* 17, no. 3–4 (1987): 163–178, as cited in Lambdin et al., "The Animal = Male Hypothesis."

121. Carole R. Beal et al., "Children's Moral Orientation: Does the Gender of Dilemma Character Make a Difference?" *Journal of Moral Education* 26, no. 1 (1997): 45–58, as cited in Lambdin et al., "The Animal = Male Hypothesis."

122. Lambdin et al., "The Animal = Male Hypothesis."

123. Sean A. McGlynn et al., "Paro's Stress-Reduction Potential for Older Adults," in *Proceedings of the Human Factors and Ergonomics Society Annual Meeting* (Sage, 2016), 1799–1803.

124. Elaine Short et al., "How to Train Your Dragonbot: Socially Assistive Robots for Teaching Children About Nutrition Through Play," in *Proceedings of the 23rd IEEE International Conference on Robot and Human Interactive Communication* (IEEE, 2014), 924–929.

125. Dexter Duckworth et al., "Therabot: The Initial Design of a Robotic Therapy Support System," in *Companion of the ACM/IEEE International Conference on Human–Robot Interaction (HRI Extended Abstracts)* (ACM, 2015), 13–14.

126. Katie Seaborn et al., "Exploring Gender-Expansive Categorization Options for Robots," in *Extended Abstracts of the CHI Conference on Human Factors in Computing Systems* (ACM, 2022), 1–6.

127. Anara Sandygulova and Gregory M. P. O'Hare, "Age-Related Differences in Children's Associations and Preferences for a Robot's Gender," in *Companion of the ACM/IEEE International Conference on Human–Robot Interaction* (ACM, 2018), 235–236, as cited in Katie Seaborn and Peter Pennefather, "Neither 'Hear' nor 'Their': Interrogating Gender Neutrality in Robots," in *Companion of the ACM/IEEE International Conference on Human–Robot Interaction (Extended Abstracts)* (IEEE, 2022), 1030–1034.

128. Julie Carpenter, "Why Project Q is More Than the World's First Nonbinary Voice for Technology," *Interactions* 26, no. 6 (2019): 56–59.

129. This was shortly before the public became cognizant of the "Q-Anon" group of conspiracy theorists.

130. Rhee, *The Robotic Imaginary*, 173.

131. Dunn, Anthony and Fiona Raby, "Speculative Everything: Design, Fiction, and Social Dreaming" (MIT Press, 2024).

132. Dunn, Anthony and Fiona Raby, "Speculative Everything: Design, Fiction, and Social Dreaming" (MIT Press, 2024), 80.

133. Pohflepp, Sascha. The Golden Institute. *Thresholds* 38 (2010): 70–75.

134. Katie Winkle, "Robots from Nowhere: A Case Study in Speculative Sociotechnical Design and Design Fiction for Human-Robot Interaction," in *Proceedings of the ACM/IEEE International Conference on Human–Robot Interaction (HRI)* (IEEE, 2024); and United Nations, "The 17 Goals" 2024, https://sdgs.un.org/goals.

135. Judith Gregory, "Scandinavian Approaches to Participatory Design," *International Journal of Engineering Education* 19, no. 1 (2003): 62–74.

136. Finn Kensing, *The Trade Unions Influence on Technological Change* (North Holland, 1983), 223, as quoted in Gregory, "Scandinavian Approaches to Participatory Design."

137. Jon A. Leydens and Juan C. Lucena, *Engineering Justice: Transforming Engineering Education and Practice* (John Wiley, 2017).

138. Yifei Zhu et al., "Robots for Social Justice (R4SJ): Toward a More Equitable Practice of Human–Robot Interaction," in *Proceedings of the ACM/IEEE International Conference on Human–Robot Interaction (HRI)* (IEEE, 2024), 850–859; and Tom Williams and Ruchen Wen, "Human Capabilities as Guiding Lights for the Field of AI-HRI: Insights from Engineering Education," in *Proceedings of the AAAI Fall Symposium on Artificial Intelligence for Human–Robot Interaction (AI-HRI)* (AAAI, 2021).

139. Nyomi Morris et al., "Trauma-Informed Insights from Co-Design of Self-Disclosure Robots with Domestic Abuse Survivors", in *Proceedings of the ACM/IEEE International Conference on Human–Robot Interaction (HRI)* (IEEE, 2025).

140. Katie Winkle et al., "LEADOR: A Method for End-To-End Participatory Design of Autonomous Social Robots," *Frontiers in Robotics and AI* 8 (2021): 1–18.

141. Sasha Costanza-Chock, *Design justice: Community-Led Practices to Build the Worlds We Need* (MIT Press, 2020). See also work on design justice in the human–robot interaction community (Anastasia K. Ostrowski et al., "Ethics, Equity, and Justice in Human–Robot Interaction: A Review and Future Directions," in *Proceedings of the 31st IEEE International Conference on Robot and Human Interactive Communication (RO-MAN)* [IEEE, 2022], 969–976), although that work places less emphasis on end user programming than does Costanza-Chock's.

142. Stolp-Smith and Williams, "More Than Binary."

143. Winkle et al., "Boosting Robot Credibility"; and Katie Winkle et al., "Norm-Breaking Responses to Sexist Abuse: A Cross-Cultural Human Robot Interaction Study," in *Proceedings of the 17th ACM/IEEE International Conference on Human–Robot Interaction (HRI)* (IEEE, 2022), 120–129.

144. Seaborn and Pennefather, "Neither 'Hear' Nor 'Their'"; and Sandygulova and O'Hare, "Age-Related Differences in Children's Associations and Preferences for a Robot's Gender."

145. Stolp-Smith and Williams, "More Than Binary."

Chapter 4

1. Asimov, "Runaround."

2. A brief note for non-roboticists: the phrase "control system" in this chapter's title refers to a system used to determine the motion of a robotic system in response to sensor data. For example, a control system might help to steer a manipulator (e.g., a robot gripper) in order to grasp an object. The phrase is also used here as a reference to the ways that machine morality systems exert control and, ironically, the ways they exert control over those who would otherwise serve as "manipulators" (i.e., humans).

3. This is not to say that everyone in the field of machine morality operates under this assumption. However, rule-based approaches remain a dominant force within the community.

4. Collins, *Black Feminist Thought.*

5. Alexander, *The New Jim Crow.*

6. Benjamin, *Race After Technology.*

7. James H. Moor, "The Nature, Importance, and Difficulty of Machine Ethics," *IEEE Intelligent Systems* 21, no. 4 (2006): 18–21.

8. Bertram F. Malle and Matthias Scheutz, "Moral Competence in Social Robots," in *Machine Ethics and Robot Ethics*, ed. Wendell Wallach and Peter Asaro (Routledge, 2020), 225–230.

9. Bertram Malle et al., "Networks of Social and Moral Norms in Human and Robot Agents," in *Proceedings of the International Conference on Robot Ethics* (CLAWAR, 2017), 3–17.

10. Vasanth Sarathy et al., "Mental Representations and Computational Modeling of Context-Specific Human Norm Systems," in *Proceedings of the Annual Meeting of the Cognitive Science Society (CogSci)* (Cognitive Science Society, 2017), 1035–1040.

11. Jackson and Williams, "Enabling Morally Sensitive Robotic Clarification Requests."

12. Ryan Blake Jackson et al., "An Integrated Approach to Context-Sensitive Moral Cognition in Robot Cognitive Architectures," in *Proceedings of the IEEE/RSJ International Conference on Intelligent Robots and Systems (IROS)* (IEEE, 2021), 1911–1918.

13. Jackson et al., "An Integrated Approach to Context-Sensitive Moral Cognition."

14. Karl Aschenbrenner, "Moral Judgment," in *The Concepts of Value: Foundations of Value Theory* (Springer, 1971), 95–110.

15. Nyx L. Ng et al., "Thinking about Reasons for One's Choices Increases Sensitivity to Moral Norms in Moral-Dilemma Judgments," *Personality and Social Psychology Bulletin* 51, no. 1 (2024): 33–48.

16. Although roboticists' predominant approach to enabling machine morality may be the formulation of slave codes, the motivation for enabling moral competence today is a desire more often to prevent kids, hackers, or other ne'er-do-wells from misusing robots than to keep robots in line. Despite much fearmongering to the contrary, few roboticists are worried about their robots "rising up" and are instead more worried about whether their robots will be able to walk across carpets, through doorways, and up steps without falling down.

17. Amanda Harmony Cooley, "Quiet Encroachments on School Prayer Jurisprudence," *West Virginia Law Review* 126, no. 2 (2024), 395–486.

18. Ed Kilgore, "Tax Dollars Are Now Funding Christian-Nationalist Schools," *Intelligencer*, June 2024.

19. Alyssa Hanson et al., "The Power of Advice: Differential Blame for Human and Robot Advisors and Deciders in a Moral Advising Context," in *Proceedings of the ACM/IEEE International Conference on Human–Robot Interaction* (ACM, 2024), 240–249.

20. Omar Mubin et al., "A Review of the Applicability of Robots in Education," *Journal of Technology in Education and Learning 1* (2013):1–7; and Saira Anwar et al., "A Systematic Review of Studies on Educational Robotics," *Journal of Pre-College Engineering Education Research (J-PEER)* 9, no. 2 (2019): 2.

21. Ben Williamson, "Calculating Children in the Dataveillance School: Personal and Learning Analytics," in *Surveillance futures* (Routledge, 2016), 50–66, as cited in Couldry and Mejias, *The Costs of Connection.*

22. Couldry and Mejias, *The Costs of Connection*, 175.

23. Khang Nhut Lam et al., "Monitoring Attendance and Checking School Uniforms Using Yolov8," in *International Conference on Intelligent Systems and Data Science* (Springer, 2023), 200–207.

24. Terran Mott and Tom Williams," Confrontation and Cultivation: Understanding Perspectives on Robot Responses to Norm Violations," in *2023 32nd IEEE International Conference on Robot and Human Interactive Communication (RO-MAN)* (IEEE, 2023), 2336–2343; and Winkle et al., "Norm-Breaking Responses to Sexist Abuse."

25. Deanna J. Glickman, "Fashioning Children: Gender Restrictive Dress Codes as an Entry Point for the Trans School to Prison Pipeline," *American University Journal of Gender, Social Policy, and the Law* 24 (2015): 263; and Nancy A. Heitzeg, *The School-to-Prison Pipeline: Education, Discipline, and Racialized Double Standards* (Bloomsbury, 2016).

26. Clea A. McNeely et al., "Exploring an Unexamined Source of Racial Disparities in Juvenile Court Involvement: Unexcused Absenteeism Policies in US Schools," *AERA Open*, 7 (2021).

27. ACLU (American Civil Liberties Union), "The School-to-Prison Pipeline in Black and White," Technical report (American Civil Liberties Union of Rhode Island, 2015).

28. ACLU, "The School-to-Prison Pipeline in Black and White."

29. T. Elon Dancy, "(Un)doing Hegemony in Education: Disrupting School-to-Prison Pipelines for Black Males," *Equity & Excellence in Education* 47, no. 4 (2014): 476–493, as cited in Jessica L. Yang et al., "'We Have to Educate Every Single Student, Not Just the Ones That Look Like Us': Support Service Providers' Beliefs About the Root Causes of the School-to-Prison Pipeline for Youth of Color," *Equity & Excellence in Education* 51, no. 3–4 (2018): 316–331.

30. Dan McQuillan, *Resisting AI: An Anti-Fascist Approach to Artificial Intelligence.* Bristol University Press, 2022), 71.

31. Benjamin, *Race After Technology*.

32. Alexander, *The New Jim Crow*.

33. Immanuel Kant, *Of the Different Human Races, 1777,* trans. Jon Mark Mikkelsen (Albany: SUNY Press, 2014).

34. Immanuel Kant, *Observations on the Feeling of the Beautiful and Sublime,* trans. John T. Goldthwait (University of California Press, 1960 [1764]), 111–113, as cited in Mills, *The Racial Contract*, 70.

35. Mills, *The Racial Contract*, 70.

36. Charles W. Mills, "Kant's Untermenschen," in *Race and Racism in Modern Philosophy*, ed. Andrew Valls (Cornell University Press, 2005), 169–193.

37. Ursula K. Le Guin, *The Ones Who Walk Away from Omelas* (Harper Perennial, 2017), as cited in Derrick Bell, *Faces at the Bottom of the Well: The Permanence of Racism* (Basic Books, 1992), 129.

38. Mirko Farina et al., "AI and Society: A Virtue Ethics Approach," *AI & Society* 39 (2022): 1127–1140.

39. Paul Bello, "Attention, Self-Control, and (Virtuous) Machine Ethics," in *Proceedings of the 52ND Annual Cal State Fullerton Philosophy Symposium* (XXX, 2024).

40. Martin Gibert, "The Case for Virtuous Robots," *AI and Ethics* 3, no. 1 (2023): 135–144.

41. That is, the pattern suggested by matrix-guided technology power analysis; Tom Williams, "Understanding Roboticists' Power Through Matrix Guided Technology Power Analysis," in *Companion of the 2024 ACM/IEEE International Conference on Human–Robot Interaction (alt.HRI)* (ACM, 2024), 46–56.

42. Much of the content in the following paragraphs is informed by my paper "The Confucian Matador" (Tom Williams et al., "The Confucian Matador: Three Defenses Against the Mechanical Bull," in *Companion of the ACM/IEEE International Conference on Human–Robot Interaction [alt.HRI]* [ACM, 2020], 25–33) or is directly informed by

the writings of my colleague Qin Zhu as part of our collaborative research efforts. This collaboration has been both fitting and ironic. My grandfather (Jay G. Williams) was a religious philosopher, who taught courses such as "Chinese Religion and Thought," "Time, Space, and Power," and "The Quest," and published articles such as "On Reading a Confucian Classic: The Rhetoric of the Lun Yu"; Jay G. Williams, "On Reading a Confucian Classic: The Rhetoric of the Lun Yu," *Journal of Chinese Religions* 19, no. 1 (1991): 105–111. Although I did not end up taking any classes from him while a student at Hamilton College, little did I know that my career as a computer scientist would ultimately lead to collaborations in his area of scholarship.

43. Mark Csikszentmihalyi, "Confucius," in *The Stanford encyclopedia of philosophy*, ed. Edward N. Zalta (Metaphysics Research Lab, Stanford University, Summer 2020).

44. David Wong, "Chinese Ethics," in *The Stanford Encyclopedia of Philosophy*, ed. Edward N. Zalta and Uri Nodelman (Metaphysics Research Lab, Stanford University, Fall 2023).

45. Tu Wei-Ming, "Self-Cultivation as Education Embodying Humanity," in *Proceedings of the Twentieth World Congress of Philosophy*, vol. 3 (Bowling Green State Univ Philosophy, 1999), 27–39.

46. Henry Rosemont and Donald J. Munro, "On Reappraising Ancient Chinese Philosophy: The Concept of Man in Early China," *Philosophy, East and West* 21, no. 2 (1971): 203–217.

47. Roger T. Ames and Henry Rosemont Jr., *The Analects of Confucius: A Philosophical Translation* (Ballantine Books, 2010).

48. Quin Zhu, "Engineering Ethics Education, Ethical Leadership, and Confucian Ethics," *International Journal of Ethics Education* 3, no. 2 (2018): 169–179.

49. Virgina Held, *The Ethics of Care: Personal, Political, and Global* (Oxford University Press, 2006).

50. John Santiago, "Confucian Ethics in the Analects as Virtue Ethics," in *Philosophical Ideas and Artistic Pursuits in the Traditions of Asia and the West: An NEH Faculty Humanities Workshop* (2008).

51. Henry Rosemont Jr. and Roger T. Ames, *Confucian Role Ethics: A Moral Vision for the 21st Century?* (V&R Academic, 2016).

52. E.g., the work of philosophers such as Epictetus 108, as discussed in Brian E. Johnson, *The Role Ethics of Epictetus: Stoicism in Ordinary Life* (Lexington Books, 2013); Christine M. Korsgaard, "The Reasons We Can Share: An Attack on the Distinction Between Agent-Relative and Agent-Neutral Values," *Social Philosophy and Policy* 10, no. 1 (1993): 24–51; and Thomas E. Randall, "Justifying Partiality in Care Ethics," *Res Publica* 26, no. 1 (2020): 67–87.

53. Quin Zhu et al., "Blame-Laden Moral Rebukes and the Morally Competent Robot: A Confucian Ethical Perspective," *Science and Engineering Ethics* 26 (2020): 2511–2526.

54. Massimiliano L. Cappuccio et al., "Can Robots Make Us Better Humans? Virtuous Robotics and the Good Life with Artificial Agents," *International Journal of Social Robotics* 13 (2021): 7–22.

55. Ruchen Wen et al., "Comparing Norm-Based and Role-Based Strategies for Robot Communication of Role-Grounded Moral Norms," *ACM Transactions on Human–Robot Interaction (T-HRI)* 12, no. 3 (2023): 1–25.

56. Malle and Scheutz, "Moral Competence in Social Robots."

57. Ruchen Wen et al., "Teacher, Teammate, Subordinate, Friend: Generating Norm Violation Responses Grounded in Role-Based Relational Norms," in *Proceedings of the 17th ACM/IEEE International Conference on Human–Robot Interaction (HRI)* (IEEE, 2022), 353–362.

58. JeeLoo Liu, "Confucian Robotic Ethics," in *Proceedings of the International Conference on the Relevance of the Classics under the Conditions of Modernity: Humanity and Science* (2017).

59. Williams et al., "The Confucian Matador."

60. Dim Cheuk Lau, *Mencius* (Chinese University Press, 2003).

61. Wen et al., "Teacher, Teammate, Subordinate, Friend."

62. Wen et al., "Comparing Norm-Based and Role-Based Strategies."

63. Zhu et al., "Blame-Laden Moral Rebukes."

64. Although, cf. Strengers and Kennedy's account of robots such as Pepper as "Smart Wives"; Strengers and Kennedy, *The Smart Wife*.

65. Katie Winkle et al., "Feminist Human–Robot Interaction: Disentangling Power, Principles and Practice for Better, More Ethical HRI," in *Proceedings of the ACM/IEEE International Conference on Human–Robot Interaction* (ACM, 2023), 72–82. See also earlier calls for a feminist robotics from Claudia Castañeda, "Robotic Skin: The Future of Touch?" in *Thinking Through the Skin*, ed. Sara Ahmed and Jackie Stacey (Routledge, 2003), 237–250, as discussed by Suchman, *Human–Machine Reconfigurations*, 274.

66. D'Ignazio and Klein, *Data Feminism*.

67. Star and Strauss, "Layers of Silence."

68. Liam J. Bannon, "The Politics of Design: Representing Work," *Communications of the ACM* 38, no. 9 (1995): 66–68; Lucy Suchman, "Making Work Visible," *Communications of the ACM* 38, no. 9 (1995): 56–64; Cathy O'Neil, *Weapons of Math*

Destruction: How Big Data Increases Inequality and Threatens Democracy (Crown, 2017); and D'Ignazio and Klein, *Data Feminism.*

69. Leydens and Lucena, *Engineering Justice.*

70. Zhu et al., "Robots for Social Justice (R4SJ)."

71. Winkle et al., "LEADOR: A Method for End-to-End Participatory Design."

72. Perrigo, "OpenAI used Kenyan Workers."

73. Niamh Rowe, "'It's Destroyed Me Completely': Kenyan Moderators Decry Toll of Training of AI Models," *The Guardian,* August 2, 2023.

74. Perrigo, "OpenAI Used Kenyan Workers."

75. Couldry and Mejias, *The Costs of Connection.*

76. Atanasoski and Vora, *Surrogate Humanity,* 90.

77. Gopika Ajaykumar et al., "A Survey on End-User Robot Programming," *ACM Computing Surveys (CSUR)* 54, no. 8 (2021): 1–36.

78. Ruth Wilson Gilmore, *Golden Gulag: Prisons, Surplus, Crisis, and Opposition in Globalizing California,* vol. 21 (University of California Press, 2007), 238.

79. John S. Mbiti, *African Religions and Philosophy* (Heinemann, 1990), as quoted in Mark Coeckelbergh, "The Ubuntu Robot: Towards a Relational Conceptual Framework for Intercultural Robotics," *Science and Engineering Ethics* 28, no. 2 (2022): 16.

80. Coeckelbergh, "The Ubuntu Robot."

81. Abeba Birhane, "Algorithmic Injustice: A Relational Ethics Approach," *Patterns* 2, no. 2 (2021), 1–9.

82. Joan Tronto, *Moral Boundaries: A Political Argument for an Ethic of Care* (Routledge, 2020).

83. Joan C. Tronto, "Creating Caring Institutions: Politics, Plurality, and Purpose," in *Ethics and Social Welfare 4, no. 2,* 2010*:158–171* as cited in Aimee Van Wynsberghe, "Service Robots, Care Ethics, and Design," *Ethics and Information Technology* 18, no. 4 (2016): 311–321.

84. Stephanie Collins, *The Core of Care Ethics* (Springer, 2015).

85. Tronto, *Moral Boundaries,* as cited in Madelaine Ley, "Care ethics and the future of work: A different voice," *Philosophy & Technology* 36, no. 1 (2023): 7.

86. Collins, *The Core of Care Ethics.*

87. Aimee Van Wynsberghe and Scott Robbins, "Critiquing the Reasons for Making Artificial Moral Agents," *Science and Engineering Ethics* 25, no. 3 (2019): 719–735.

88. Moor, "The Nature, Importance, and Difficulty of Machine Ethics."

89. Note that while a participatory approach to deontological machine morality might begin to question dominant *processes* and *logics* of Kantian Deontology, roboticists leveraging this approach might also need to careful reimagine other key assumptions made by Kantian deontology. As Mills writes in *"Kant's untermenschen"*, we would need to consider "[whether] crucial concepts such as 'autonomy' need to be given a different spin, if a case can be made that a tacitly masculinist experience has grounded its formation . . . [and whether] 'respect' [should] be reconceptualized in group rather than individualist terms, given that white supremacy has stigmatized entire races as less than worthy of respect."

90. Zuboff, *The Age of Surveillance Capitalism.*

91. Couldry and Mejias, *The Costs of Connection.*

Chapter 5

1. A brief note for non-roboticists: the phrase "bounding box" in this chapter's title refers to a box drawn by computer vision systems around an object of interest located in an image. The phrase is also used here to refer to the ways that computer vision systems place humans into boxes, and the ways that those systems shape the ways that the boundaries between human identity categories are drawn and negotiated.

2. Michel Foucault, *Discipline and Punish: The Birth of the Prison* (Vintage, 1979), per Patricia Hill Collins, "It's All in the Family: Intersections of Gender, Race, and Nation," *Hypatia* 13, no 3 (1998): 62–82.

3. Cf. Anja Kovacs, "Reading Surveillance Through a Gendered Lens: Some Theory," *Gendering Surveillance*, February 2017.

4. Tawfiq Ammari et al., "Music, Search, and IoT: How People (Really) Use Voice Assistants," *ACM Transactions on Computer–Human Interaction (TOCHI)* 26, no. 3 (2019): 1–28.

5. Yao Qian et al., "Bidirectional LSTM-RNN for Improving Automated Assessment of Non-Native Children's Speech," in *Proceedings of Interspeech* (Curran Associates, 2017), 1417–1421, as cited in Siyuan Feng, Olya Kudina, Bence Mark Halpern, and Odette Scharenborg, "Quantifying Bias in Automatic Speech Recognition," *arXiv*, preprint arXiv:2103.15122, 2021. However, also note recent evidence that this quality is improving with advances in transformer-based speech recognizers; Ruben Janssens et al., "Child Speech Recognition in Human–Robot Interaction: Problem Solved?," in *Proceedings of the International Symposium on Technological Advances in Human–Robot Interaction (TAHRI)* (2024).

6. Bence Mark Halpern et al., "Detecting and Analysing Spontaneous Oral Cancer Speech in the Wild," in *Proceedings of Interspeech* (Curran Associates, 2020), 4826–4830, as cited in Feng et al., "Quantifying Bias in Automatic Speech Recognition."

7. Rachel Tatman and Conner Kasten, "Effects of Talker Dialect, Gender and Race on Accuracy of Bing Speech and YouTube Automatic Captions," in *Proceedings of Interspeech* (Curren Associates, 2017), 934–938, as cited in Feng et al., "Quantifying Bias in Automatic Speech Recognition."

8. Yunhan Wu et al., "See What I'm Saying? Comparing Intelligent Personal Assistant Use for Native and Non-Native Language Speakers," in *Proceedings of the 22nd International Conference on Human–Computer Interaction with Mobile Devices and Services* (ACM, 2020) 1–9, as cited in Feng et al., "Quantifying Bias in Automatic Speech Recognition."

9. Adam Palanica et al., "Do You Understand the Words That Are Comin Outta My Mouth? Voice Assistant Comprehension of Medication Names," *NPJ Digital Medicine* 2, no. 1 (2019):1–6.

10. Allison Koenecke et al., "Racial Disparities in Automated Speech Recognition," *Proceedings of the National Academy of Sciences* 117, no. 14 (2020): 7684–7689.

11. Koenecke et al., "Racial Disparities in Automated Speech Recognition."

12. E.g., its use of the habitual "be"; Joshua L. Martin and Kevin Tang, "Understanding Racial Disparities in Automatic Speech Recognition: The case of Habitual 'Be,'" in *Proceedings of Interspeech* (Curran Associates, 2020), 626–630, as cited in Mikel K. Ngueajio and Gloria Washington, "Hey ASR system! Why Aren't You More Inclusive? Automatic Speech Recognition Systems' Bias and Proposed Bias Mitigation Techniques. A Literature Review," in *Proceedings of the International Conference on Human–Computer Interaction* (Springer, 2022), 421–440.

13. Zion Mengesha et al., "'I Don't Think These Devices Are Very Culturally Sensitive.'—Impact of Automated Speech Recognition Errors on African Americans," *Frontiers in Artificial Intelligence* 4 (2021):1–12.

14. Radkika Garg et al., "'Learn, Use, and (Intermittently) Abandon': Exploring the Practices of Early Smart Speaker Adopters in Urban India," in *Proceedings of the ACM on Human–Computer Interaction*, 5(CSCW2) (ACM, 2021), 1–28; and Linus Kendall et al., "Understanding Technology as Situated Practice: Everyday Use of Voice User Interfaces Among Diverse Groups of Users in Urban India," *Information Systems Frontiers* 22 (2020): 585–605, as cited in Christina N. Harrington et al. "'It's Kind of Like Code-Switching': Black Older Adults' Experiences with a Voice Assistant for Health Information Seeking," in *Proceedings of the CHI Conference on Human Factors in Computing Systems* (ACM, 2022), 1–15.

15. Gregory M. Walton et al., "Stereotype Threat in Organizations: Implications for Equity and Performance," *Annual Review of Organizational Psychology and Organizational Behavior* 2, no. 1 (2015): 523–550, as cited in Courtney L. McCluney et al., "The Costs of Code-Switching," *Harvard Business Review* 15 (2019).

16. Brendan F. Klare et al., "Face Recognition Performance: Role of Demographic Information," *IEEE Transactions on Information Forensics and Security* 7, no. 6 (2012): 1789–1801.

17. Patrick Grother et al., "Face Recognition Vendor Test (FVRT): Part 3, Demographic Effects," Technical report (National Institute of Standards and Technology Gaithersburg, 2019).

18. Samuel Dooley et al., "Robustness Disparities in Face Detection," *Advances in Neural Information Processing Systems* 35 (2022): 38245–38259.

19. Kyriakos Kyriakou et al., "Emotion-Based Stereotypes in Image Analysis Services," in *Adjunct Publication of the 28th ACM Conference on User Modeling, Adaptation and Personalization* (ACM, 2020), 252–259, as cited in Eugenia Kim et al., "Age Bias in Emotion Detection: An Analysis of Facial Emotion Recognition Performance on Young, Middle-Aged, and Older Adults," in *Proceedings of the AAAI/ACM Conference on AI, Ethics, and Society* (ACM, 2021), 638–644.

20. Kyriakou et al., "Emotion-Based Stereotypes," as cited in Kim et al., "Age Bias in Emotion Detection."

21. Buolamwini and Gebru, "Gender Shades."

22. Joy Buolamwini, "When the Robot Doesn't See Dark Skin," *New York Times*, June 21, 2018.

23. Shalini Kantayya and Joy Buolamwini, "Coded Bias" (2020).

24. Khari Johnson, "ExamSoft's Remote Bar Exam Sparks Privacy and Facial Recognition Concerns," *VentureBeat*, September 29, 2020.

25. Bobby Allyn, "'The Computer Got It Wrong': How Facial Recognition Led to False Arrest of Black Man," *NPR* (2020). https://www.npr.org/2020/06/24/882683463/the-computer-got-it-wrong-how-facial-recognition-led-to-a-false-arrest-in-michig.

26. Ninareh Mehrabi et al., "A Survey on Bias and Fairness in Machine Learning," *ACM Computing Surveys (CSUR)* 54, no. 6 (2021): 1–35.

27. O'Neil, *Weapons of Math Destruction*.

28. Richard Berk et al., "Fairness in Criminal Justice Risk Assessments: The State of the Art," *Sociological Methods & Research* 50, no. 1 (2021): 3–44.

29. Julia Angwin et al., "Machine Bias," in *Ethics of Data and Analytics: Concepts and Cases*, ed. Kirsten Martin (Auerbach, 2022), 254–264; and Jeff Larson et al., "How We Analyzed the COMPAS Recidivism Algorithm," *ProPublica*, May 23, 2016.

30. Mehrabi et al., "A Survey on Bias and Fairness."

31. Naman Goel et al., "Non-Discriminatory Machine Learning Through Convex Fairness Criteria," in *Proceedings of the 2018 AAAI/ACM Conference on AI, Ethics, and Society (AIES)* (AAAI, 2018), 116–116; Toshihiro Kamishima et al., "Fairness-Aware Classifier with Prejudice Remover Regularizer," in *Proceedings of the European Conference on Machine Learning and Knowledge Discovery in Databases* (Springer, 2012), 35–50; Emmanouil Krasanakis et al., "Adaptive Sensitive Reweighting to Mitigate Bias in

Fairness-Aware Classification," in *Proceedings of the 2018 World Wide Web Conference (WWW)* (ACM, 2018), 853–862; Aditya Krishna Menon and Robert C. Williamson, "The Cost of Fairness in Binary Classification," in *Proceedings of the ACM Conference on Fairness, Accountability and Transparency (FAccT)* (ACM, 2018), 107–118; and Kristian Lum et al., "De-Biasing 'Bias' Measurement," in *Proceedings of the ACM Conference on Fairness, Accountability, and Transparency (FAccT)* (ACM, 2022), 379–389. See also Solon Barocas et al., "Fairness and Machine Learning," 2019, https://fairmlbook.org/2019; and Shira Mitchell et al., "Algorithmic Fairness: Choices, Assumptions, and Definitions," *Annual Review of Statistics and Its Application* 8 (2021): 141–163.

32. Alexander Amini et al., "Uncovering and Mitigating Algorithmic Bias Through Learned Latent Structure," in *Proceedings of the 2019 AAAI/ACM Conference on AI, Ethics, and Society* (ACM, 2019), 289–295; Tolga Bolukbasi et al., "Man is to Computer Programmer as Woman is to Homemaker? Debiasing Word Embeddings," *Advances in Neural Information Processing Systems (NeurIPS)* 29 (2016), 4349–4357; Elliot Creager et al., "Flexibly Fair Representation Learning by Disentanglement," in *Proceedings of the International Conference on Machine Learning* (PMLR, 2019), 1436–1445; Christos Louizos et al., "The Variational Fair Autoencoder," in *Proceedings of the International Conference on Learning Representations* (Curran Associates, 2016); Jieyu Zhao et al., "Learning Gender-Neutral Word Embeddings," in *Proceedings of the Conference on Empirical Methods in Natural Language Processing (EMNLP)* (ACL, 2018): 4847–4853; and Ryan Steed and Aylin Caliskan, "Image Representations Learned with Unsupervised Pre-Training Contain Human-Like Biases," in *Proceedings of the ACM conference on Fairness, Accountability, and Transparency (FAccT)* (ACM, 2021), 701–713.

33. Atoosa Kasirzadeh, "Algorithmic Fairness and Structural Injustice: Insights from Feminist Political Philosophy," in *Proceedings of the 2022 AAAI/ACM Conference on AI, Ethics, and Society* (ACM, 2022), 349–356.

34. Sean Hollister, "Google Contractors Reportedly Targeted Homeless People for Pixel 4 facial Recognition," *The Verge*, October 3, 2019.

35. Ginger Adams Otis and Nancy Dillon, "Google Using Dubious Tactics to Target People with "Darker Skin" in Facial Recognition Project: Sources," *New York Daily News*, October 3, 2019.

36. Anna Lauren Hoffmann, "Terms of Inclusion: Data, Discourse, Violence," *New Media & Society* 23, no. 12 (2021): 3539–3556.

37. Birhane, "Algorithmic Injustice."

38. McQuillan, *Resisting AI*, 112.

39. Matthew Le Bui and Safiya Umoja Noble, "We're Missing a Moral Framework of Justice in Artificial Intelligence," in *The Oxford handbook of ethics of AI*, ed. Markus D. Dubber, Frank Pasquale, and Sunit Das (Oxford Academic, 2020), 537–553. See also similar concerns raised by Cynthia Bennett and Os Keyes, "What is the Point of Fairness?" *Interactions* 27, no. 3 (2020): 35–39.

40. Harvey Gee, "Surveillance state: Fourth Amendment Law, Big Data Policing, and Facial Recognition Technology," *Berkeley Journal of African-American Law & Policy* 21, no. 2 (2020), 43–83.

41. Zahra Mian, "'Black Identity Extremist' or Black Dissident? How United States v. Daniels Illustrates FBI Criminalization of Black Dissent of Law Enforcement, from COINTELPRO to Black Lives Matter," *Rutgers Race and Law Review* 21 (2020): 53.

42. Dave Maass and Matthew Guariglia, *San Francisco Police Accessed Business District Camera Network to Spy on Protestors* (Electronic Frontier Foundation, 2020), as cited in Gee, "Surveillance State."

43. Associated Press, "LAPD Used Facial Recognition Nearly 30,000 Times," *Associated Press*, September 22, 2020, as cited in Gee, "Surveillance State."

44. These are only what is documented: The full use of the technology was not fully tracked; Libor Jany, "LAPD Doesn't Fully Track its Use of Facial Recognition, Report Finds," *Los Angeles Times*, December 14, 2022, as cited in Gee, "Surveillance State."

45. Associated Press, "LAPD Used Facial Recognition Nearly 30,000 Times," as cited in Gee, "Surveillance State."

46. Laura C. Powell, "The Good, the Bad, and the Ugly: Black Lives Matter Protests, the January 6th Insurrection, and Facial Recognition Technology as Admissible Evidence," *American University Law Review* 72 (2022): 277.

47. Powell, "The Good, the Bad, and the Ugly"; and James Vincent, "NYPD Used Facial Recognition to Track Down Black Lives Matter Activist," *The Verge*, August 18, 2020.

48. Aldebaran Robotics, "ALFaceCharacteristics API," 2014, http://doc.aldebaran.com/2-4/naoqi/peopleperception/alfacecharacteristics-api.html 2014. Although this is described as detecting "gender," the module uses a sex-based (male/female) coding scheme. It is unclear whether this is best thought of as a sex classifier (using sex-based differences in facial structure) or a gender classifier (using perceivable gendered differences in grooming and presentation norms).

49. Wilma Bainbridge et al., "Establishing a Database for Studying Human Face Photograph Memory," in *Proceedings of the Annual Meeting of the Cognitive Science Society* (The Cognitive Science Society, 2012), 1302–1307.

50. Wilma Bainbridge et al., "The Intrinsic Memorability of Face Photographs," *Journal of Experimental Psychology: General* 142, no. 4 (2013), 1323–1334, as cited in Morgan Klaus Scheuerman et al., "How We've Taught Algorithms to See Identity: Constructing Race and Gender in Image Databases for Facial Analysis," in *Proceedings of the ACM on Human–Computer Interaction*, 4(CSCW1) (ACM, 2020):1–35.

51. Zhengyou Zhang, "Microsoft Kinect Sensor and its Effect," *IEEE Multimedia* 19, no. 2 (2012): 4–10.

52. Timm Linder et al., "Real-Time Full-Body Human Gender Recognition in (RGB)-D data," in *Proceedings of the IEEE International Conference on Robotics and Automation (ICRA)* (IEEE, 2015), 3039–3045.

53. Arnaud Ramey and Miguel A. Salichs, "Morphological Gender Recognition by a Social Robot and Privacy Concerns," in *Proceedings of the 9th ACM/IEEE International Conference on Human–Robot Interaction (HRI)* (ACM, 2014), 272–273.

54. Jaap Ham, "Influencing Robot Influence: Personalization of Persuasive Robots," *Interaction Studies* 22, no. 3 (2021): 464–487.

55. Alessia Saggese et al., "MIVIABot: A Cognitive Robot for Smart Museum," in *Proceedings of the International Conference on Computer Analysis of Images and Patterns* (Springer, 2019), 15–25.

56. Ramey and Salichs, "Morphological Gender Recognition by a Social Robot"; and Linder et al., "Real-Time Full-Body Human Gender Recognition," 2015.

57. Pasquale Foggia et al., "A System for Gender Recognition on Mobile Robots," in *Proceedings of the 2nd International Conference on Applications of Intelligent Systems* (ACM, 2019), 1–6.

58. Anara Sandygulova et al., "Real-Time Adaptive Child–Robot Interaction: Age and Gender Determination of Children Based on 3D Body Metrics," in *Proceedings of the 23rd IEEE International Conference on Robot and Human Interactive Communication* (IEEE, 2014), 826–831.

59. Derald Wing Sue et al., "Racial Microaggressions in Everyday Life: Implications for Clinical Practice," *American Psychologist* 62, no. 4 (2007): 271–286.

60. Eduardo Bonilla-Silva, *Racism Without Racists: Color-Blind Racism and the Persistence of Racial Inequality in the United States* (Rowman & Littlefield, 2006).

61. Sue et al., "Racial Microaggressions in Everyday Life"; and Derald Wing Sue et al., "Racial Microaggressions and the Power to Define Reality," *American Psychologist* 63, no. 4 (2008): 277–279, as cited in Edwards 2017.

62. Morgan Klaus Scheuerman et al., "How We've Taught Algorithms to See Identity: Constructing Race and Gender in Image Databases for Facial Analysis," in *Proceedings of the ACM on Human–Computer Interaction*, 4(CSCW1) (ACM, 2020): 277–279.

63. Yandong Guo et al., "MS-CELEB-1M: A Dataset and Benchmark for Large-Scale Face Recognition," in *Proceedings of the European Conference on Computer Vision (ECCV)* (Springer, 2016), 87–102, as cited in Scheuerman et al., "How We've Taught Algorithms to See Identity."

64. Smedley and Smedley, "Race as Biology is Fiction."

65. Smedley and Smedley, "Race as Biology is Fiction."

66. Scheuerman et al., "How We've Taught Algorithms to See Identity."

67. Os Keyes, "The Misgendering Machines: Trans/HCI Implications of Automatic Gender Recognition," in *Proceedings of the ACM on Human–Computer Interaction,* 2(CSCW) (ACM, 2018), 1–22.

68. Kevin A. McLemore, "Experiences with Misgendering: Identity Misclassification of Transgender Spectrum Individuals," *Self and Identity* 14, no. 1 (2015): 51–74; and Y. Gavriel Ansara, "Cisgenderism in Medical Settings: How Collaborative Partnerships Can Challenge Structural Violence," in *Out of the ordinary: Representations of LGBT lives*, ed. Ian Rivers and Richard Ward (Cambridge Scholars, 2012).

69. Bruce Schneier and B. Schneier, *Beyond Fear: Thinking Sensibly About Security in an Uncertain World* (Springer, 2003).

70. Costanza-Chock, *Design Justice.*

71. Costanza-Chock, *Design Justice*, 5.

72. Costanza-Chock, *Design Justice*, 4.

73. Foad Hamidi et al., "Gender Recognition or Gender Reductionism? The Social Implications of Embedded Gender Recognition Systems," in *Proceedings of the CHI Conference on Human Factors in Computing Systems* (ACM, 2018), 1–13.

74. Sharon M. Lee, "Racial Classifications in the US Census: 1890–1990," *Ethnic and Racial Studies* 16, no. 1 (1993): 75–94.

75. Lee, "Racial Classifications in the US Census."

76. Joe R. Feagin, *Racist America: Roots, Current Realities, and Future Reparations* (Routledge, 2000).

77. Eduardo Bonilla-Silva, "From Bi-Racial to Tri-Racial: Towards a New System of Racial Stratification in the USA," *Ethnic and Racial Studies* 27, no. 6 (2004): 931–950.

78. Bonilla-Silva, "From Bi-Racial to Tri-Racial."

79. Noel Ignatiev, *How the Irish Became White* (Routledge, 2012).

80. Bashi Treitler, *The Ethnic Project*, Ch. 4.

81. Karen Farquharson, "Racial Categories in Three Nations: Australia, South Africa and the United States," in *Proceedings of 'Public Sociologies: Lessons and Trans-Tasman Comparisons', the Annual Conference of the Australian Sociological Association (TASA)* (TASA, 2007), 1–6; and George M. Fredrickson, *White Supremacy: A Comparative Study of American and South African History* (Oxford University Press, 1981), as cited in Bashi Treitler, *The Ethnic Project*, 2020.

82. Bashi Treitler, *The Ethnic Project*. See Ch. 2, fn. 27 in Treitler for a Discussion of the Complexities of Racial Category Enumeration in Brazil, wherein the Brazilian

Government Uses a Set of Five Categories in its Census, but Brazilian Citizens Themselves Identify with More Than a Hundred Distinct Categorizations.

83. Godfried Asante et al., "Becoming 'Black': Exploring the Racialized Experiences of African Immigrants in the United States," *Howard Journal of Communications* 27, no. 4 (2016): 367–384; Isabel Wilkerson, *Caste: The origins of our discontents* (Random House, 2020); and Olaocha Nwadiuto Nwabara, "Becoming Black and African: Nigerian Diasporic Transformations of Racial and Ethnic Identities in the United States," *Africology: The Journal of Pan African Studies* 12, no. 1 (2018): 70–96.

84. Amanuel Isak Tewolde, "'Becoming Black': Racial Formation of Eritrean Migrants in Israel," *African Diaspora* 13, no. 1–2 (2021): 183–203.

85. Michelle Christian, "A Global Critical Race and Racism Framework: Racial Entanglements and Deep and Malleable Whiteness," *Sociology of Race and Ethnicity* 5, no. 2 (2019): 169–185, building on terminology introduced by Vrushali Patil and Bandana Purkayastha, "The Transnational Assemblage of Indian Rape Culture," *Ethnic and Racial Studies* 41, no. 11 (2018): 1952–1970.

86. Simone Browne, *Dark Matters* (Duke University Press, 2015).

87. Fanon, *Black Skin, White Masks.*

88. Simone Browne, "Everybody's Got a Little Light Under the Sun: Black Luminosity and the Visual Culture of Surveillance," *Cultural Studies* 26, no. 4 (2012): 542–564.

89. Morgan Klaus Scheuerman et al., "Auto-Essentialization: Gender in Automated Facial Analysis as Extended Colonial Project," *Big Data & Society* 8, no. 2 (2021), 1–15.

90. Thiems Nanavati, *Biometrics* (John Wiley, 2002), 36–37, as cited in Simone Browne, "Digital Epidermalization: Race, Identity and Biometrics," *Critical Sociology* 36, no. 1 (2010): 135.

91. Nanavati, *Biometrics*, 37, as cited in Browne, "Digital Epidermalization," 135. See also the Extensive Discussion of These Histories in *Dark Matters.*

92. Bernd Simon and David L. Hamilton, "Self-Stereotyping and Social Context: The Effects of Relative In-Group Size and In-Group Status," *Journal of Personality and Social Psychology* 66, no. 4 (1994): 699–711, as cited in Orla T. Muldoon et al., "Personal and Political: Post-Traumatic Stress Through the Lens of Social Identity, Power, and Politics," *Political Psychology* 42, no. 3 (2021): 501–533.

93. Tom Williams et al., "Voice in the Machine: Ethical Considerations for Language-Capable Robots," *Communications of the ACM (CACM)* 66, no. 8 (2023): 20–23.

94. Browne, "Digital Epidermalization."

95. Morgan Klaus Scheuerman et al., "Do Datasets Have Politics? Disciplinary Values in Computer Vision Dataset Development," in *Proceedings of the ACM on Human–Computer Interaction*, 5(CSCW2) (ACM, 2021), 1–37.

96. Ben Hutchinson et al., "Towards Accountability for Machine Learning Datasets: Practices from Software Engineering and Infrastructure," in *Proceedings of the ACM Conference on Fairness, Accountability, and Transparency (FAccT)* (ACM, 2021), 560–575, as quoted in Scheuerman et al., "Do Datasets Have Politics?"

97. Morgan Klaus Scheuerman et al., "How We've Taught Algorithms to See Identity: Constructing Race and Gender in Image Databases for Facial Analysis," in *Proceedings of the ACM on Human–Computer Interaction*, 4(CSCW1) (ACM, 2020), 1–35.

98. Morgan Klaus Scheuerman et al., "How Computers See Gender: An Evaluation of Gender Classification in Commercial Facial Analysis Services," in *Proceedings of the ACM on Human–Computer Interaction*, 3(CSCW) (ACM, 2019), 1–33.

99. Hamidi et al., "Gender Recognition or Gender Reductionism?"

100. Cynthia L. Bennett et al., "'It's Complicated': Negotiating Accessibility and (Mis) representation in Image Descriptions of Race, Gender, and Disability," in *Proceedings of the CHI Conference on Human Factors in Computing Systems* (ACM, 2021), 1–19.

101. Council of European Union, "Regulation of the European Parliament and of the Council Laying Down Harmonised Rules on Artificial Intelligence and Amending Regulations (EC) no 300/2008, (EU) no 167/2013, (EU) no 168/2013, (EU) 2018/858, (EU) 2018/1139 and (EU) 2019/2144 and directives 2014/90/eu, (EU) 2016/797 and (EU) 2020/1828," 2024, https://www.europarl.europa.eu/doceo/document/TA-9-2024-0138-FNL-COR01_EN.pdf. 2024.

102. Scheuerman et al., "How Computers See Gender."

103. Bonilla-Silva, *Racism Without Racists.*

104. See also Os Keyes et al., "Human–Computer Insurrection: Notes on an Anarchist HCI," in *Proceedings of the CHI Conference on Human Factors in Computing Systems* (ACM, 2019), 1–13.

105. Maria Teresa Parreira et al., "How Did We Miss This? A Case Study on Unintended Biases in Robot Social Behavior," in *Proceedings of the ACM/IEEE International Conference on Human–Robot Interaction* (ACM, 2023), 11–20.

106. Kathleen Villaluz, "The World's First Robot Police Officer Begins Patrolling in Dubai," *Interesting Engineering*, May 30, 2017.

107. Human Rights Watch, "UAE: Stop Policing Gender Expression," *Human Rights Watch*, September 7, 2017.

108. Browne, *Dark Matters.*

Chapter 6

1. A brief note for non-roboticists: the phrase "end effector" in this chapter's title refers to a gripper or other device affixed to the end of a robot's arm so that it can

interact with its environment. The phrase is also used here as a reference to the ways that the police have been using robots to violent ends.

2. Craig D. Uchida, "The Development of the American Police: An Historical Overview," *Critical Issues in Policing: Contemporary Readings* 3 (1993): 19–35.

3. Hadden, *Slave Patrols.*

4. Hadden, *Slave Patrols*, 16.

5. Hadden, *Slave Patrols*, 19.

6. Hadden, *Slave Patrols*; and Richard C. Wade, *Slavery in the Cities: The South 1820–1860* (Oxford University Press, 1967).

7. Ritchie, *Invisible No More.*

8. Hadden, *Slave Patrols*, 38.

9. Hadden, *Slave Patrols*, 55.

10. Hadden, *Slave Patrols*, 56.

11. Hadden, *Slave Patrols*, 82.

12. Matthew Guariglia, *Police and the Empire City: Race and the Origins of Modern Policing in New York* (Duke University Press, 2023), 5.

13. Paddy Hillyard, "Policing Divided Societies: Trends and Prospects in Northern Ireland and Britain," in *Policing Futures: The Police, Law Enforcement and the Twenty-First Century*, ed. Peter Francis, Pamela Davies, and Victor Jupp (Springer, 1997), 166.

14. Alex S. Vitale, *The End of Policing* (Verso Books, 2017), 35.

15. Galen Broeker, "Robert Peel and the Peace Preservation Force," *The Journal of Modern History* 33, no. 4 (1961): 363–373.

16. Uchida, "The Development of the American police."

17. Uchida, "The Development of the American police."

18. Judith Rowbotham et al., "The Beginnings of Crime Intelligence: 1800–1860," in *Crime news in Modern Britain: Press Reporting and Responsibility, 1820–2010* (Springer, 2013), 16–17.

19. Vitale, *The End of Policing*, 35.

20. Kristian Williams, *Our Enemies in Blue: Police and Power in America* (AK Press, 2015), 54.

21. Roger Lane, *Policing the City: Boston, 1822–1885* (Harvard University Press, 2013).

22. Raymond Blaine Fosdick, *Crime in America and the Police* (Century Company, 1920).

23. Roger G. Dunham et al., *Critical Issues in Policing: Contemporary Readings* (Waveland Press, 2020).

24. James F. Richardson, *Urban Police in the United States* (Kennikat Press, 1974), as cited in Williams, *Our Enemies in Blue*, 114.

25. Ben Jones and Eduardo Mendieta, "Introduction: Police Ethics after Ferguson," in *The ethics of policing* (New York University Press, 2021), 26.

26. Philip L. Reichel, "The Misplaced Emphasis on Urbanization in Police Development," *Policing and Society: An International Journal* 3, no. 1 (1992): 1–12.

27. Williams, *Our Enemies in Blue*, Ch. 3.

28. Hadden, *Slave Patrols*, 177.

29. Blackmon, *Slavery by Another Name*.

30. Ritchie, *Invisible no More*, 28–29.

31. Hadden, *Slave Patrols*, 198.

32. Hadden, *Slave Patrols*, 104.

33. Hadden, *Slave Patrols*, 199–200.

34. Hadden, *Slave Patrols*, 219.

35. Hadden, *Slave Patrols*, 202.

36. Louis Bernard Cei, *Law Enforcement in Richmond: A History of Police–Community Relations, 1737–1974* (Florida State University, 1975), as cited in Hadden, *Slave Patrols*.

37. Hadden, *Slave Patrols*, 216–220.

38. See also Williams, *Our Enemies in Blue*, Ch. 4.

39. Williams, *Our Enemies in Blue*, Ch. 4.

40. Hadden, *Slave Patrols*, 216.

41. Hadden, *Slave Patrols*, 219.

42. Hadden, *Slave Patrols*, 215.

43. Vida B. Johnson, "KKK in the PD: White Supremacist Police and What to do About It," *Lewis and Clark Law Review* 23, no. 1 (2019): 205–261; and Alice Speri, "The FBI Has Quietly Investigated White Supremacist Infiltration of Law Enforcement," *The Intercept*, January 31, 2017.

44. Michael Walzer, "Soldiers and Police," in *The Ethics of Policing: New Perspectives on Law enforcement*, ed. Ben Jones and Eduardo Mendieta (New York University Press, 2021), 98.

45. Michael German, *Hidden in Plain Sight: Racism, White Supremacy, and Far-Right Militancy in Law Enforcement* (Brennan Center for Justice, 2020).

46. Emily Hoerner and Rick Tulsky, "Cops Around the Country Are Posting Racist and Violent Comments on Facebook," *Injustice Watch* (2019).

47. Vitale, *The End of Policing*, 43.

48. Mike Cox, *The Texas Rangers: Wearing the Cinco Peso, 1821–1900* (Forge Books, 2008).

49. Julian Samora et al., *Gunpowder Justice: A Reassessment of the Texas Rangers* (University of Notre Dame Press, 1979).

50. Philip V. McHarris, "Disrupting Order: Race, Class, and the Roots of Policing," in *Violent order: Essays on the Nature of Police*, ed. David Correia and Tyler Wall (Haymarket Books, 2021), 31–54.

51. Michael Sierra-Arévalo, "Reward and 'Real' Police Work," in *The Ethics of Policing: New Perspectives on Law Enforcement*, ed. Ben Jones and Eduardo Mendieta (New York University Press, 2021).

52. Elizabeth Hinton, *From the War on Poverty to the War on Crime: The Making of Mass Incarceration in America* (Harvard University Press, 2016), as cited in McHarris, "Disrupting Order."

53. McHarris, "Disrupting Order."

54. Sierra-Arévalo, "Reward and 'Real' Police Work"; and Walzer, "Soldiers and Police."

55. Lenora Lapidus et al., "Caught in the Net: The Impact of Drug Policies on Women and Families," Technical report (ACLU, 2005), as cited in Ritchie, *Invisible No More*, 46.

56. Lapidus et al., "Caught in the Net," as cited in Ritchie, *Invisible No More*, 46.

57. Katherine Tate et al., *Something's in the Air: Race, Crime, and the Legalization of Marijuana* (Routledge, 2013), as cited in Ritchie, *Invisible No More*, 46.

58. Dan Baum, "Legalize It All." *Harper's Magazine* 24 (2016): 21–32.

59. Alexander, *The New Jim Crow*.

60. Alexander, *The New Jim Crow*.

61. Bruce L. Benson et al., "Police Bureaucracies, Their Incentives, and the War on Drugs," *Public Choice* 83, no. 1–2 (1995): 21–45.

62. Dick M. Carpenter et al., *Policing for Profit: The Abuse of Civil Asset Forfeiture*. 2nd ed. (Institute for Justice, 2015), 20, as cited in Adam Crepelle, "Probable Cause to Plunder: Civil Asset Forfeiture and the Problems It Creates," *Wake Forest Journal of Law and Policy* 7 (2017): 315–316.

63. Christopher Ingraham, "New Report: In Tough Times, Police Start Seizing a Lot More Stuff from People," *Washington Post*, November 10, 2015, as cited in Crepelle, "Probable Cause to Plunder."

64. Lisa Knepper et al., *Policing for Profit: The Abuse of Civil Asset Forfeiture*. 3rd ed. (Institute for Justice, 2020).

65. John L. Worrall, "Addicted to the Drug War: The Role of Civil Asset Forfeiture as a Budgetary Necessity in Contemporary Law Enforcement," *Journal of Criminal Justice* 29, no. 3 (2001): 171–187.

66. United States v. Brignoni-Ponce, 422 U.S. 873 (1975).

67. Max Felker-Kantor, *Policing Los Angeles: Race, Resistance, and the Rise of the LAPD* (University of North Carolina Press Books, 2018), 2.

68. Felker-Kantor, *Policing Los Angeles*, 5.

69. Jake Monaghan, "Legitimate Policing and Professional Norms," in *The Ethics of Policing: New Perspectives on Law Enforcement*, ed. Ben Jones and Eduardo Mendieta (New York University Press, 2021), 46.

70. Alexander, *The New Jim Crow*.

71. John Balzar, "The System: Deals, Deadlines, Few Trials," *Los Angeles Times*, September 4, 2006.

72. Marc Mauer, *Race to Incarcerate* (New Press, 2006).

73. Matthew D. Lassiter, "Impossible Criminals: The Suburban Imperatives of America's War on Drugs," *Journal of American History* 102, no. 1 (2015): 126–140.

74. Christopher Schmitt, "Plea Bargaining Favors Whites, as Blacks, Hispanics Pay Price," *San Jose Mercury News*, December 8, 1991.

75. Becky Pettit et al., "Technical Report on Revised Population Estimates and NLSY 79 Analysis Tables for the Pew Public Safety and Mobility Project," Technical report (Harvard University, 2009), as cited in Bruce Western and Becky Pettit, "Incarceration and Social Inequality," *Daedalus* 139, no. 3 (2010): 8–19.

76. Alexander, *The New Jim Crow*.

77. This creates a further incentive for White American conservatives to invest in policing, as doing so ultimately leads to their political opponents being excluded from voter rolls, reinforcing their own structural power.

78. Vesla Weaver, "Policing Narratives in the Black Counterpublic," in *The Ethics of Policing: New Perspectives on Law Enforcement*, ed. Ben Jones and Eduardo Mendieta (New York University Press, 2021), 170.

79. Foucault, *Discipline and Punish*.

80. Edwin Black, *IBM and the Holocaust: The Strategic Alliance Between Nazi Germany and America's Most Powerful Corporation-Expanded Edition* (Dialog Press, 2012).

81. Black, *IBM and the Holocaust.*

82. Black, *IBM and the Holocaust.*

83. Charlton D. McIlwain, *Black Software: The Internet and Racial Justice, from the AfroNet to Black Lives Matter* (Oxford University Press, 2019), 198.

84. Sarah Brayne, *Predict and Surveil: Data, Discretion, and the Future of Policing* (Oxford University Press, 2020), 29.

85. Brayne, *Predict and Surveil*, 54.

86. Brayne, *Predict and Surveil*, 30.

87. Parks Associates, "Parks Associates: 25% of U.S. Broadband Households Plan to Buy a Smart Video Doorbell in 2019," *PR Newswire*, January 3, 2019, as cited in Grace Egger, "Ring, Amazon Calling: The State Action Doctrine and the Fourth Amendment," *Washington Law Review Online*, 95 (2020), 245–275.

88. Martine Powers and Elahe Izadi, "Security or Surveillance? How Smart Doorbell Company Ring Partners with Police," *Washington Post*, August 28, 2019, as cited in Egger, "Ring, Amazon Calling."

89. Caroline Haskins, "US Cities Are Helping People Buy Amazon Surveillance Cameras Using Taxpayer Money." *VICE*, August 2, 2019, as cited in Egger, "Ring, Amazon Calling."

90. Mike Murphy, "Amazon's Ring May Be Branching Out Beyond Outdoor Cameras," *Quartz*, 2019, and Cyrus Farivar, "Cute Videos, But Little Evidence: Police Say Amazon Ring Isn't Much of a Crime Fighter," *NBC News*, February 15, 2020, as Cited in Egger, "Ring, Amazon Calling."

91. Caroline Haskins, "Amazon Is Coaching Cops on How to Obtain Surveillance Footage Without a Warrant," *VICE*, August 5, 2019, as cited in Egger, "Ring, Amazon Calling."

92. Haskins, "Amazon is Coaching Cops," as cited in Egger, "Ring, Amazon Calling."

93. Haskins, "Amazon is Coaching Cops," as cited in Egger, "Ring, Amazon Calling."

94. Rachel Cericola, "Ring Neighbors is the Best and Worst Neighborhood Watch App," *New York Times*, February 14, 2020, as cited in Egger, "Ring, Amazon Calling."

95. Ben Gilbert, "Amazon Sells Facial Recognition Software to Police All Over the US, but Has No Idea How Many Departments Are Using It," *Business Insider*, February 21, 2020.

96. Fanna Gamal, "The Racial Politics of Protection: A Critical Race Examination of Police Militarization," *California Law Review* 104 (2016), 979–1008.

97. At the time of this writing, Palantir remains a member of my department's industry partnership program.

98. Luke Munn, "Seeing with Software: Palantir and the Regulation of Life," *Studies in Control Societies 2, no. 1* (2017): 1–17.

99. N'dea Yancey-Bragg, "Aurora Police Apologize After Drawing Guns on Black Family, Mistaking Vehicle as Stolen," *USA Today*, August 4, 2020.

100. Colleen Slevin and Thomas Peipert, "Prosecutors Investigating Police Who Handcuffed Black Girls," *Washington Post*, August 7, 2020.

101. Slevin and Peipert, "Prosecutors Investigating Police Who Handcuffed Black Girls."

102. Michael Levenson, "Officers Who Handcuffed Black Family Won't Be Charged, Prosecutors Say," *New York Times*, January 8, 2021.

103. Brayne, *Predict and Surveil.*

104. Patrick McGreevy, "LAPD Automatic License Plate Readers Pose a Massive Privacy Risk, Audit Says," *Los Angeles Times*, February 13, 2020.

105. McGreevy, "LAPD Automatic License Plate Readers Pose a Massive Privacy Risk."

106. Alex Najibi, "Racial Discrimination in Face Recognition Technology," *Harvard University Science in the News*, October 24, 2020.

107. Caroline Haskins, "Scars, Tattoos, and License Plates: This Is What Palantir and the LAPD Know About You," *BuzzFeed News*, September 29, 2020.

108. Linda S. Beres and Thomas D. Griffith, "Demonizing Youth," *Loyola of Los Angeles Law Review* 34 (2000), 747–766.

109. Stefanie Dazio, "Police Can't Access LAPD Gang Database Records After Scandal," *Washington Post*, July 14, 2020.

110. Dana Goodyear, "The L.A. County Sheriff's Deputy-Gang Crisis," *New Yorker*, May 30, 2022.

111. Lisa Guenther, "Police, Drones, and the Politics of Perception," in *The Ethics of Policing: New Perspectives on Law Enforcement*, ed. Ben Jones and Eduardo Mendieta (New York University Press, 2021), 256.

112. Guenther, "Police, Drones, and the Politics of Perception," 257.

113. Brayne, *Predict and Surveil*, 109.

114. Brayne, *Predict and Surveil*, 73.

115. Vesla Weaver, "Policing Narratives in the Black Counterpublic," in *The Ethics of Policing: New Perspectives on Law Enforcement*, ed. Ben Jones and Eduardo Mendieta (New York University Press, 2021), 168.

116. Brayne, *Predict and Surveil*, 101.

117. Jeff Asher and Ben Horwitz, "How Do the Police Actually Spend Their Time?" *New York Times*, June 19, 2020.

118. Brayne, *Predict and Surveil*, 73.

119. Virginia Eubanks, *Automating Inequality: How High-Tech Tools Profile, Police, and Punish the Poor* (St. Martin's Press, 2018), 215.

120. Eubanks, *Automating Inequality*, Ch. 4.

121. Yuval Abraham, "'Lavender': The AI Machine Directing Israel's Bombing Spree in Gaza," *+972 Magazine*, April 3, 2024.

122. McQuillan, *Resisting AI*, 63.

123. Hannah Arendt and Jens Kroh, *Eichmann in Jerusalem* (Viking Press, 1964).

124. McHarris, "Disrupting Order," 50.

125. David Lyon, *Surveillance as Social Sorting: Privacy, Risk, and Digital Discrimination* (Psychology Press, 2003), as cited in Brayne, *Predict and Surveil*, 107.

126. Guenther, "Police, Drones, and the Politics of Perception," 254.

127. David Correia, "On the Nature of Police," in *Violent Order: Essays on the Nature of Police*, ed. David Correia and Tyler Wall (Haymarket Books, 2021), 3. See also Walzer, "Soldiers and Police," 98.

128. Ben Jones and Eduardo Mendieta, eds., "Introduction: Police Ethics After Ferguson," in *The Ethics of Policing* (New York University Press, 2021), 1–22.

129. Patrick Ball, "Violence in Blue: Police Homicides in the United States," *Granta Magazine*, March 4, 2016.

130. Edward Lawson Jr., "Trends: Police Militarization and the Use of Lethal Force," *Political Research Quarterly* 72, no. 1 (2019): 177–189. See also Delehanty et al.'s analysis, which shows that the 1033 program has led to an increase in the number of civilians—*and dogs*—murdered by the police; Casey Delehanty et al., "Militarization and Police Violence: The Case of the 1033 Program," *Research & Politics* 4, no. 2 (2017), 1–7.

131. Campaign Zero, "Police Scorecard," 2016, https://policescorecard.org/ 2016.

132. Weaver, "Policing Narratives in the Black Counterpublic," 154.

133. Ritchie, *Invisible No More*, 113.

134. Ritchie, *Invisible No More*, 109.

135. Ritchie, *Invisible No More*, 169.

136. Dara E. Purvis and Melissa Blanco, "Police Sexual Violence," *California Law Review* 108, no. 5 (2020): 1487–1529.

137. Haven Orecchio-Egresitz, "A Defense Attorney Started Twitter Thread Documenting Police Using Force on Protesters and Media. There are more than 400 posts," *Business Insider*, June 8, 2020.

138. Joe Soss and Vesla Weaver, "Police Are Our Government: Politics, Political Science, and the Policing of Race–Class Subjugated Communities," *Annual Review of Political Science* 20 (2017): 565–591.

139. Browne, *Dark Matters*.

140. Chris Boyette and Zoe Sottile, "San Francisco Supervisors Vote to Allow Police to Use Robots to Kill," *CNN*, December 1, 2022.

141. Nicholas Reimann, "San Francisco Says Cops Can't Use Killer Robots After Public Outcry—For Now," *Forbes*, December 6, 2022.

142. Aiden H. While et al., "Urban Robotic Experimentation: San Francisco, Tokyo and Dubai," *Urban Studies* 58, no. 4 (2021): 769–786.

143. Kate Conger et al., "San Francisco Bans Facial Recognition Technology," *New York Times*, May 14, 2019.

144. Neil Smith, *The New Urban Frontier: Gentrification and the Revanchist City* (Psychology Press, 1996).

145. Manissa M. Maharawal, "Black Lives Matter, Gentrification and the Security State in the San Francisco Bay Area," *Anthropological Theory* 17, no. 3 (2017): 338–364.

146. Peter B. Kraska, "Militarization and Policing—Its Relevance to 21st Century Police," *Policing: A Journal of Policy and Practice* 1, no. 4 (2007): 501–513.

147. Scott Tighe and William Brown, "The Militarization of Law Enforcement: Bypassing the Posse Comitatus Act," *Justice Policy Journal* 12, no. 2 (2015): 1–39.

148. Gamal, "The Racial Politics of Protection."

149. Federal Bureau of Investigations, "Report on Prevention and Control of Major Disturbances," Technical report (Federal Bureau of Investigations, 1967), as reprinted in Michal R. Belknap, *Urban race riots* (Routledge, 1991), as cited in Gamal, "The Racial Politics of Protection."

150. Gamal, "The Racial Politics of Protection."

151. Radley Balko, "Overkill: The Rise of Paramilitary Police Raids in America," Technical report (The Cato Institute, 2006), as cited in Stacey Cotton et al., "Law Enforcement, Community, and Military Tactics: What's the Conflict?" in *Working in the Public Interest Conference* (2014).

152. Radley Balko, *Rise of the Warrior Cop: The Militarization of America's Police Forces* (PublicAffairs, 2013), as cited in Cotton et al., "Law Enforcement, Community, and Military Tactics."

153. Delehanty et al., "Militarization and Police Violence."

154. Ross Gianfortune, *10 Insane Things the Pentagon Gave to Local Law Enforcement* (Government Executive, 2014).

155. Gianfortune, *10 Insane Things.*

156. Gamal, "The Racial Politics of Protection."

157. Laura Withers, "How Bearcats Became Toys: The 1033 Program and its Effect on the Right to Protest," *George Washington Law Review* 84, no. 3 (2016): 812–841.

158. Nicky Woolf, "Did Police's Militarised Response to a Small Town Pumpkin Patch Riot Just Make It Worse?" *The Guardian*, October 22, 2014.

159. Brayne, *Predict and Surveil*, 26.

160. Center for the Study of the Drone at Bard College, "Law Enforcement Robots Datasheet," 2016, https://dronecenter.bard.edu/law-enforcement-robots-datasheet/2016.

161. Jonathan Mummolo, "Militarization Fails to Enhance Police Safety or Reduce Crime but May Harm Police Reputation," *Proceedings of the National Academy of Sciences*, 115, no. 37 (2018): 9181–9186.

162. Balko, *Rise of the Warrior Cop.*

163. John Bennett, "How Command Presence Affects Your Survival," 2010, https://www.police1.com/officer-safety/articles/how-command-presence-affects-your-survival-6cPA8v9EjGJZuWuH/, as cited in Seth Stoughton, "Law Enforcement's 'Warrior' Problem," *Harvard Law Review Forum* 128, no. 6 (2014): 225–234.

164. Sierra-Arévalo, "Reward and 'Real' Police Work," 68.

165. Sierra-Arévalo, "Reward and 'Real' Police Work," 72.

166. Michael Sierra-Arévalo, "American Policing and the Danger Imperative," *Law & Society Review* 55, no. 1 (2021): 70–103; and Dan Wang, *Data on police officers killed since 1961* (2015). https://danwang.co/statistics-on-police-fatalities/

167. Asher and Horwitz, "How do the Police Actually Spend Their Time?"

168. Sierra-Arévalo, "Reward and 'Real' Police Work," 70, although note that the police largely get to define and shape crime rates for themselves.

169. Kent A. Sepkowitz and Leon Eisenberg, "Occupational Deaths Among Healthcare Workers," *Emerging Infectious Diseases* 11, no. 7 (2005): 1003.

170. Mireille Kingma, "Workplace Violence in the Health Sector: A Problem of Epidemic Proportion," *International Nursing Review* 48, no. 3 (2001), 129–130, as cited in Brophy et al. 2018.

171. Alan Wright, *Policing: An Introduction to Concepts and Practice* (Routledge, 2002), although cp. Vincenzo Bove and Evelina Gavrilova, "Police Officer on the Frontline or a Soldier? The Effect of Police Militarization on Crime," *American Economic Journal: Economic Policy* 9, no. 3 (2017), 1–18.

172. Graeme Blair et al., "Community Policing Does Not Build Citizen Trust in Police or Reduce Crime in the Global South," *Science* 374, no. 6571 (2021), 1–14.

173. David H. Bayley, "Police Reform: Who Done It?" in *Police Reform from the Bottom Up*, ed. Monique Marks and David Sklansky (Routledge, 2014), 16–26.

174. Walzer, "Soldiers and Police," 95.

175. Center for the Study of the Drone, "Law Enforcement Robots Datasheet."

176. Dan Gettinger, *Public Safety Drones*, 3rd ed. (Center for the Study of the Drone at Bard College, 2020), as cited in Nancy K. Carr, "Programmed to Protect and Serve: The Dawn of Drones and Robots in Law Enforcement," *Journal of Air Law and Commerce* 86, no. 2 (2021), 183–218.

177. Boone Ashworth, "The NYPD Brings Robot Dogs Back," *Wired*, April 15, 2023.

178. Ron Amadeo, "The NYPD Retires 'Digidog' Robot After Public Backlash," *Ars Technica*, April 30, 2021.

179. Martin Cooney et al., "A Broad View on Robot Self-Defense: Rapid Scoping Review and Cultural Comparison," *Robotics* 12, no. 2 (2023), 1–37.

180. Jaclyn Jeffrey-Wilensky and David Freeman, "This Police Robot Could Make Traffic Stops Safer," *NBCUniversal News Group*, May 9, 2019, as cited in Cooney et al., "A Broad View on Robot Self-Defense," 2023.

181. Marie Donlon, "Automation at the Airport," *Electronics 360*, August 30, 2022, as cited in Cooney et al., "A Broad View on Robot Self-Defense," 2023.

182. Daniel W. Carruth and Cindy L. Bethel, "Challenges with the Integration of Robotics into Tactical Team Operations," in *Proceedings of the 15th IEEE International Symposium on Applied Machine Intelligence and Informatics (SAMI)* (IEEE, 2017) 27–32; and Nate Phillips et al., "A Robotic Augmented Reality Virtual Window for Law Enforcement Operations," in *Proceedings of the 12th International Conference on Virtual, Augmented and Mixed Reality. Held as Part of the 22nd HCI International Conference* (Springer, 2020), 591–610.

183. Matyos Kidane, "The LAPD Wants Robot Dogs. How Did We Get Here?" *VICE*, February 2, 2023.

184. Maria Cramer and Christine Hauser, "Digidog, a Robotic Dog Used by the Police, Stirs Privacy Concerns," *New York Times*, February 27, 2021.

185. Amy Shiner, "Police Helicopter Surveillance," *Arizona Law Review* 15, no. 1 (1973): 145–171.

186. Mike Davis, *City of Quartz: Excavating the Future in Los Angeles* (Verso Books, 2006), as cited in Hille Koskela, "'Cam Era"—The Contemporary Urban Panopticon," *Surveillance & Society* 1, no. 3 (2003): 292–313.

187. Brayne, *Predict and Surveil*, 72.

188. Rebecca Heilweil, "Members of Congress Want to Know More About Law Enforcement's Surveillance of Protesters," *Vox*, June 10, 2020; Ryan Songalia, "The Activist Who Streamed a Bizarre Standoff with the NYPD Will Keep Fighting," *VICE*, August 11, 2020; and Sam Biddle, "U.S. Marshals Used Drones to spy on Black Lives Matter Protests in Washington, DC," *The Intercept*, April 22, 2021.

189. Dhruv Mehrotra and Jesse Marx, "The Age of the Drone Police is Here," *Wired*, June 5, 2024.

190. Mehrotra and Marx, "The Age of the Drone Police is Here."

191. Peter Finn, "Domestic Use of Aerial Drones by Law Enforcement Likely to Prompt Privacy Debate," *Washington Post*, January 22, 2011.

192. James Randerson, "Eye in the Sky: Police Use Drone to Spy on V Festival," *The Guardian*, August 21, 2007.

193. Todd Feathers, "Honolulu Police Used a Robot Dog to Patrol a Homeless Encampment," *VICE*, June 14, 2021. See also Sarah Buhr, "Security Robots Are Being Used to Ward Off San Francisco's Homeless Population," *TechCrunch*, December 13, 2017.

194. Tyler Wall, "Unmanning the Police Manhunt: Vertical Security as Pacification," *Socialist Studies/Études Socialistes* 9, no. 2 (2013): 47. This Alabaman surveillance of drug transactions is likely racialized, given that Black people in Alabama are four times more likely to be arrested for marijuana possession (Leah Nelson et al., "Alabama's War on Marijuana," Technical report [Alabama Appleseed Center for Law & Justice and the Southern Poverty Law Center, 2018]) and forty-five times more likely to be jailed for marijuana use in counties such as Dekalb County (ACLU [American Civil Liberties Union] et al., *A Tale of Two Countries: Racially Targeted Arrests in the Era of Marijuana Reform* [American Civil Liberties Union, 2020]; despite equal or lesser use of marijuana by Black people compared to White people). Moreover, police use of drones in particular is acutely informed by the racialized surveillance contexts in which military surveillance drones were originally deployed; Atanasoski and Vora, *Surrogate Humanity*, 176.

195. Police departments are already expressing excitement at the ability for aerial robots to target and track individuals; Monte Reel, "Secret Cameras Record

Baltimore's Every Move from above," *Bloomberg Businessweek*, August 23, 2016, as cited in Jake Laperruque, "Preventing an Air Panopticon: A Proposal for Reasonable Legal Restrictions on Aerial Surveillance," *University of Richmond Law Review* 51, no. 3 (2016): 705–726.

196. Xin Ye and Lionel Robert, "Human Security Robot Interaction and Anthropomorphism: An Examination of Pepper, RAMSEE, and Knightscope Robots," in *Proceedings of the IEEE International Conference on Robot & Human Interactive Communication (RO-MAN)* (IEEE, 2023), 982–987.

197. Tom Ward, "Dubai Wants Robots to Make Up 25% of Its Police Force by 2030," *Futurism*, May 23, 2017.

198. Mangai Prabakar and Jong-Hoon Kim, "Telebot: Design Concept of Telepresence Robot for Law Enforcement," in *Proceedings of the World Congress on Advances in Nano, Biomechanics, Robotics, and Energy Research (ANBRE 2013), Seoul, Korea* (IASEM, 2013), 34–42.

199. Prabakar and Kim, "Telebot."

200. Brian Heater, "Amazon Cites First Amendment Protection for Alexa in Arkansas Murder Case," *TechCrunch*, February 23, 2017.

201. Ritchie, *Invisible No More*; Johanna Wald and Daniel J. Losen, "Defining and Redirecting a School-to-Prison Pipeline," *New Directions for Youth Development* 2003, no. 99 (2003): 9–15; and Alex Lu Jiahong et al., "Coding Bias in the Use of Behavior Management Technologies: Uncovering Socio-Technical Consequences of Data-Driven Surveillance in Classrooms," in *Designing Interactive Systems Conference 2021* (ACM, 2021), 508–522.

202. David Axe, "Drug Raid Turns Ugly as SWAT Guns Down Marine Vet," *Wired*, May 30, 2011.

203. Richard J. Berry and Gorden E. Eden Jr, "Albuquerque Police Department Monthly Report November 2014," Technical report (Albuquerque Police Department, 2014).

204. Andy Beale, "A New Weapon for SWAT Teams: Bomb-Squad Robots," *VICE*, May 24, 2015.

205. Carl Lundberg and Henrik I. Christensen, "Assessment of Man-Portable Robots for Law Enforcement Agencies," in *Proceedings of the Workshop on Performance Metrics for Intelligent Systems* (ACM, 2007), 76–83.

206. KUSA Staff, "One Year Later: Signs of PTSD, Mental Illness; Search for Treatments," *9News*, July 7, 2017.

207. Brian Thevenot and Erwin Seba, "Dallas Sniper Plotted Bigger Assault, Taunted Officers: Police chief," *Reuters*, July 11, 2016.

208. Associated Press, "Dallas Cops Told Sniper Explosive-Strapped Robot Had a Phone," *Associated Press*, February 2, 2018.

209. Sara Sidner and Mallory Simon, "How Robot, Explosives Took Out Dallas Sniper in Unprecedented Way," *CNN*, July 12, 2016.

210. Caitlin Rogers, "The Aftermath of Police Blowing Up a Maine Man's Home with a Bomb Robot," *Bangor Daily News*, August 28, 2018.

211. Associated Press, "Documents Detail Use of Bomb Robot in Maine Standoff," *Associated Press*, September 1, 2018.

212. Associated Press, "Documents Detail Use of Bomb Robot in Maine Standoff."

213. Boyette and Sottile, "San Francisco Supervisors Vote."

214. Eduardo Kochenborger Duarte et al., "Robot Self-Defense: Robots Can Use Force on Human Attackers to Defend Victims," in *Proceedings of the IEEE International Conference on Robot and Human Interactive Communication (RO-MAN)* (IEEE, 2022), 1606–1613.

215. Franklin Zimring, "When Police Do Not Need to Kill," in *The ethics of policing: New perspectives on law enforcement*, ed. Ben Jones and Eduardo Mendieta (New York University Press, 2021), 110–111.

216. Walzer, "Soldiers and Police," 100.

217. Zimring, "When Police Do Not Need to Kill," 108.

218. Deborah Young, "Unnecessary Evil: Police Lying in Interrogations," *Connecticut Law Review* 28 (1995):425; and Carl B. Klockars, "Blue Lies and Police Placebos: The Moralities of Police Lying," *American Behavioral Scientist* 27, no. 4 (1984): 529–544.

219. Jennifer Hunt and Peter K. Manning, "The Social Context of Police Lying," *Symbolic Interaction* 14, no. 1 (1991): 51–70.

220. Hunt and Manning, "The Social Context of Police Lying"; Christopher Slobogin, "Testilying: Police Perjury and What to Do About It," *University of Colorado Law Review*, 67 (1996):1037–1060; and Gabriel J. Chin and Scott C. Wells, "The Blue Wall of Silence as Evidence of Bias and Motive to Lie: A New Approach to Police Perjury," *University of Pittsburgh Law Review* 59 (1997), 233–299.

221. Hunt and Manning, "The Social Context of Police Lying."

222. Steve Martinot, "On the Epidemic of Police Killings," *Social Justice* 39, no. 4 (2012): 52–75.

223. Kate Knibbs, "The Future of Police Robots Began Last Year—Where Is It Now?" *The Ringer*, July 28, 2017.

224. Guenther, "Police, Drones, and the Politics of Perception," 259.

225. Mary L. Cummings, "Automation and Accountability in Decision Support System Interface Design," *Journal of Technology Studies 32, no. 1* (2006): 23–31.

226. Guenther, "Police, Drones, and the Politics of Perception," 250.

227. Travis Linnemann, "Proof of Death: Police Power and the Visual Economies of Seizure, Accumulation and Trophy," *Theoretical Criminology* 21, no. 1 (2017): 57–77.

228. Grégoire Chamayou, *Manhunts: A Philosophical History* (Princeton University Press, 2012).

229. Gina Martinez, "'Come on Guys, He's Right There!': Moment Bumbling LAPD Cops Are Seen in News Helicopter Footage Searching for Suspect Hiding Right in Front of Them as Bemused News Anchor Watches in Disbelief," *Daily Mail*, March 11, 2022.

230. Guenther, "Police, Drones, and the Politics of Perception," 250.

231. Wall, "Unmanning the Police Manhunt."

232. Peter-Paul Verbeek, *Moralizing Technology: Understanding and Designing the Morality of Things* (University of Chicago Press, 2011); and Peter-Paul Verbeek, "Materializing Morality: Design Ethics and Technological Mediation," *Science, Technology, & Human Values* 31, no. 3 (2006): 361–380.

233. Wall, "Unmanning the Police Manhunt," 48.

234. Alexander, *The New Jim Crow*.

235. Guenther, "Police, Drones, and the Politics of Perception," 257.

236. See also Elizabeth E. Joh, "Policing Police Robots," *UCLA Law Review Discourse* 64 (2016): 516–543.

237. Williams and Haring, "No Justice, No Robots."

238. Brayne, *Predict and Surveil*, 5–6.

239. Logan D. A. Williams, "Data Racism, Machine Learning and Contextualizing Design," in *Because Technology Discriminates: Anti-Racist Counter-Expertise* (Springer, 2023), 127–156.

240. McIlwain, *Black Software*, 222.

241. Kate Stevens, "Police, Pringle Robotics Collaborate for Red White & Boom Safety," *Central Illinois Proud*, June 30, 2023.

242. Campaign Zero, "Police Scorecard: Peoria, il," 2023, https://policescorecard.org/il/police-department/peoria.

243. Correia, "On the Nature of Police," 5.

244. Ben Wagner, "Ethics as an Escape from Regulation. From 'ethics-washing' to ethics-shopping?" in *Being Profiled: Cogitas ergo sum: 10 Years of Profiling the European*

Citizen, ed. Emre Bayamlioglu, Irina Baraliuc, Liisa Albertha Wilhelmina Janssens, and Mireille Hildebrandt (Amsterdam University Press, 2018), 84–89.

245. Ronan Farrow, "How an Élite University Research Center Concealed Its Relationship with Jeffrey Epstein," *New Yorker*, September 6, 2019.

246. Joi Ito, "My Apology Regarding Jeffrey Epstein," 2019, https://www.media.mit.edu/posts/myapology-regarding-jeffrey-epstein/.

247. Russell Brandom, "AI Pioneer Accused of Having Sex with Trafficking Victim on Jeffrey Epstein's Island," *The Verge*, August 9, 2019.

248. Ethan Zuckerman, "On Me, and the Media Lab," 2019, https://ethanzuckerman.com/2019/08/20/on-me-and-the-media-lab/.

249. Alyssa Vaughn, "After Saying Sex with Minors is not Always Sexual Assault, MIT Scientist Resigns," *Boston Magazine*, September 18, 2019.

250. Only around 20 percent of US robotics-related undergraduate degrees go to Black, Latinx, or Indigenous students (Christopher Pannier et al., "Diversity and Inclusion in Mechatronics and Robotics Engineering Education," in *ASEE Annual Conference Exposition Proceedings* [ASEE, 2020]), reflecting trends seen across science and engineering fields, in which Black students receive only around 7 percent of all undergraduate degrees (Fry et al. 2021), and less than 2 percent of doctorates (National Science Foundation 2018).

251. Guariglia, *Police and the Empire City*, building on concepts introduced by Micol Seigel, *Violence Work: State Power and the Limits of Police* (Duke University Press, 2018).

252. Tom Williams, "No Justice, No Robots," 2020, https://nojusticenorobots.github.io/

253. Tanja Heuer et al., "Privacy-Centered Design for Social Robots," *Interaction Studies* 20, no. 3 (2019): 509–529.

254. Tad Hirsch and John Henry, "Txtmob: Text Messaging for Protest Swarms," in *Extended Abstracts of the CHI Conference on Human Factors in Computing Systems* (ACM, 2005), 1455–1458.

255. Steve Mann et al., "Sousveillance: Inventing and Using Wearable Computing Devices for Data Collection in Surveillance Environments," *Surveillance & Society* 1, no. 3 (2003): 331–355.

256. Kathleen Bartzen Culver, "From Battlefield to Newsroom: Ethical Implications of Drone Technology in Journalism," *Journal of Mass Media Ethics* 29, no. 1 (2014): 52–64, as cited in Dennis Zuev and Gary Bratchford, "The Citizen Drone: Protest, Sousveillance and Droneviewing," *Visual Studies* 35, no. 5 (2020): 442–456.

257. Hagit Keysar, "A Spatial Testimony: The Politics of Do-It-Yourself Aerial Photography in East Jerusalem," *Environment and Planning D: Society and Space* 37, no. 3 (2019): 523–541, as described in Zuev and Bratchford, "The Citizen Drone."

258. Zuev and Bratchford, "The Citizen Drone."

259. Matt Franchi et al., "Detecting Disparities in Police Deployments Using Dashcam Data," in *Proceedings of the ACM Conference on Fairness, Accountability, and Transparency (FAccT)* (ACM, 2023), 534–544.

260. Bayley, "Police Reform."

261. Bernard E. Harcourt, *Illusion of Order: The False Promise of Broken Windows Policing* (Harvard University Press, 2005); Elizabeth Brown and George Barganier, *Race and crime: Geographies of Injustice* (University of California Press, 2018); Tammy Rinehart Kochel, "Constructing Hot Spots Policing: Unexamined Consequences for Disadvantaged Populations and for Police Legitimacy," *Criminal Justice Policy Review* 22, no. 3 (2011): 350–374.

262. Vitale, *The End of Policing.*

263. Rebecca Tiger, *Judging Addicts: Drug Courts and Coercion in the Justice System* (New York University Press, 2013).

264. Johann Hari, *Chasing the Scream: The First and Last Days of the War on Drugs* (Bloomsbury, 2015).

265. Drug Policy Alliance, "Approaches to Decriminalizing Drug Use and Possession" (2015). https://www.unodc.org/documents/ungass2016/Contributions/Civil/DrugPolicyAlliance/DPA_Fact_Sheet_Approaches_to_Decriminalization_Feb2015_1.pdf

266. H. G. Levine and D. Small, *Marijuana Arrest Crusade Racial Bias and Police Policy in New York City, 1997–2007* (New York Civil Liberties Union, 2008).

267. Lindsay Graef et al., "Systemic Failure to Appear in Court," *University of Pennsylvania Law Review* 172, no. 1 (2023): 1–60.

268. Resources for learning about abolition efforts can be found at http://criticalresistance.org, an organization founded by Angela Davis, Ruth Wilson Gilmore, and Rose Braz that works toward abolition of the prison-industrial complex. See also CR10 Publications Collective, *Abolition now!: Ten years of strategy and struggle against the Prison Industrial Complex* (AK Press, 2008).

269. Angela Y. Davis, *Are Prisons Obsolete?* (Seven Stories Press, 2011).

270. Ruth Wilson Gilmore and Chenjerai Kumanyika, "Ruth Wilson Gilmore Makes the Case for Abolition," *The Intercept*, Intercepted podcast by Jeremy Scahill, June 10, 2020.

271. Mariame Kaba, *We Do This 'til We Free Us: Abolitionist Organizing and Transforming Justice* (Haymarket Books, 2021); Mariame Kaba, *Yes, We Mean Literally Abolish the Police* (Abolition Library Commons, 2020); and Mariame Kaba and Andrea J. Ritchie, *No More Police: A Case for Abolition* (New Press, 2022).

272. Angela Y. Davis, *Abolition Democracy: Beyond Empire, Prisons, and Torture* (Seven Stories Press, 2011); and Allegra M. McLeod, "Prison Abolition and Grounded Justice," *UCLA Law Review* 62 (2015): 1156.

273. Lisa Lowe, "The Dialectics of Abolition," *American Quarterly* 75, no. 2 (2023): 371–376.

274. Ruth Wilson Gilmore, *Abolition Geography: Essays Towards Liberation* (Verso Books, 2022).

275. Lowe, "The Dialectics of Abolition."

276. Tracey L. Meares, "Clashing Narratives of Policing?" in *The ethics of policing: New perspectives on Law Enforcement*, ed. Ben Jones and Eduardo Mendieta (New York University Press, 2021), 33.

277. Leydens and Lucena, *Engineering Justice*.

278. Solomon Gustavo, "Four takeaways from the Minneapolis DFL's mayoral endorsement process," *MinnPost*, June 18, 2021, as cited in Jack Post Gramlich, "Are Police Obsolete in Minneapolis? Racial Capitalism and Police Abolition," *Social Science Quarterly* 102, no. 7 (2021): 3149–3157.

279. Haodan Tan et al., "Raven: A Street Robot to Address Homelessness," in *Proceedings of the Companion of the 2017 ACM/IEEE International Conference on Human–Robot Interaction* (ACM, 2017), 397–398; and Joshi and Sabanovic 2018, 2019.

280. Eduard Fosch-Villaronga and Michiel Heldeweg, "'Regulation, I Presume?' Said the Robot—Towards an Iterative Regulatory Process for Robot Governance," *Computer Law & Security Review* 34, no. 6 (2018): 1258–1277; and Stefan Larsson et al., "Towards a Sociolegal Robotics: A Theoretical Framework on Norms and Adaptive Technologies," *International Journal of Social Robotics* 15 (2023): 1755–1768.

281. Tiera Tanksley, "Employing an Abolitionist, Critical Race Pedagogy in CS: Centering the Voices, Experiences and Technological Innovations of Black youth," *Journal of Computer Science Integration* 6, no. 1 (2023): 1–16.

282. Williams and Haring, "No Justice, No Robots."

283. Khalifa Rashid Al Shaali and Neil Kibble, "Policing and Police Accountability in the UAE: The Case for Reform," *Arab Law Quarterly* 15, no. 3 (2000): 272–303.

284. Al Shaali and Kibble, "Policing and Police Accountability in the UAE."

285. Harsha Walia, *Border and Rule: Global Migration, Capitalism, and the Rise of Racist Nationalism* (Haymarket Books, 2021), 152. See also David Batty, "Conditions for Abu Dhabi's Migrant Workers 'Shame the West,'" *The Guardian*, 21, December 22, 2013.

286. Williams, "No Justice, No Robots."

287. Heuer et al., "Privacy-Centered Design for Social Robots."

288. Mann et al., "Sousveillance."

Chapter 7

1. Tom Williams, "Understanding Roboticists' Power Through Matrix Guided Technology Power Analysis," in *Companion of the 2024 ACM/IEEE International Conference on Human–Robot Interaction (alt.HRI)* (ACM, 2024), 46–56.

2. I apologize to my students Terran Mott, Mark Higger, and Alexa Bejarano, who already used this phrase as the title of their robotics education tool and tabletop role-playing game of the same name (Terran Mott et al., "Degrees of Freedom: A Storytelling Game That Supports Technology Literacy About Social Robots," in *Proceedings of the IEEE International Conference on Robot and Human Interactive Communication [RO-MAN]* [IEEE, 2024], 2095–2102), but the metaphor is too apt for it to not be reused here.

3. We should inherently be wary of such arguments, which are a key rhetorical strategy used to support today's system of color-blind racism; Bonilla-Silva, *Racism Without Racists*.

4. Zhu et al., "Robots for Social Justice (R4SJ)."

5. Williams, "Understanding Roboticists' Power."

6. Katie Winkle et al., "Feminist Human–Robot Interaction: Disentangling Power, Principles and Practice for Better, More Ethical HRI," in *Proceedings of the ACM/IEEE International Conference on Human–Robot Interaction* (ACM, 2023), 72–82.

7. Zhu et al., "Robots for Social Justice (R4SJ)."

8. Leydens and Lucena, *Engineering Justice*.

9. Costanza-Chock, *Design Justice*.

10. Zhu et al., "Robots for Social Justice (R4SJ)."

11. David Touretzky et al., "Special Session: AI for K-12 Guidelines Initiative," in *Proceedings of the 50th ACM Technical Symposium on Computer Science Education* (ACM, 2019), 492–493; Ismaila Temitayo Sanusi and Sunday Olaleye, "An Insight into Cultural Competence and Ethics in K-12 Artificial Intelligence Education," in *2022 IEEE Global Engineering Education Conference (EDUCON)* (IEEE, 2022), 790–794; Randi Williams, "How to Train Your Robot: Project-Based AI and Ethics Education for Middle School Classrooms, in *Proceedings of the 52nd ACM Technical Symposium on Computer Science Education* (ACM, 2021), 1382–1382; Jasmine DeHart, "Building Cultural Competency for Computing: Identity Computing Lab, Duke University," *XRDS: Crossroads, The ACM Magazine for Students* 28, no. 4 (2022): 38–39; and Amy J. Ko et al.,

"Critically Conscious Computing: Methods for Secondary Education," 2022, https://criticallyconsciouscomputing.org 2022.

12. William F. McComas, "Benchmarks for Science Literacy," in *The Language of Science Education: An Expanded Glossary of Key Terms and Concepts in Science Teaching and Learning* (Springer, 2014), 12–12; Terran Mott and Tom Williams, "Community Futures with Morally Capable Robotic Technology," in *HRI Workshop on Perspectives on Moral Agency in Human–Robot Interaction* (2023); and Mott et al., "Degrees of Freedom."

13 Inioluwa Deborah Raji et al., "Closing the AI Accountability Gap: Defining an End-to-End framework for Internal Algorithmic Auditing," in *Proceedings of the ACM conference on Fairness, Accountability, and Transparency (FAccT)* (ACM, 2020), 33–44; Danaë Metaxa et al., "Auditing Algorithms: Understanding Algorithmic Systems from the Outside In," *Foundations and Trends in Human–Computer Interaction* 14, no. 4 (2021): 272–344; Ellen P. Goodman and Julia Trehu, "AI Audit Washing and Accountability," available at SSRN 4227350, 2022; Abeba Birhane et al., "AI Auditing: The Broken Bus on the Road to AI Accountability," in *Proceedings of the IEEE Conference on Secure and Trustworthy Machine Learning (SaTML)* (IEEE, 2024), 612–664; Sasha Costanza-Chock et al., "Who Audits the Auditors? Recommendations from a Field Scan of the Algorithmic Auditing Ecosystem," in *Proceedings of the ACM Conference on Fairness, Accountability, and Transparency* (ACM, 2022), 1571–1583.

14. Fosch-Villaronga and Heldeweg, "'Regulation, I presume?' Said the Robot"; Ronald Leenes et al., "Regulatory Challenges of Robotics: Some Guidelines for Addressing Legal and Ethical Issues," *Law, Innovation and Technology* 9, no. 1 (2017): 1–44; Ryan Calo et al., *Robot law* (Edward Elgar, 2016); Ryan Calo, "Robotics and the Lessons of Cyberlaw," *California Law Review* 103 (2015): 513; and Ryan Calo, "Artificial Intelligence Policy: A Primer and Roadmap," *UCDL Review* 51 (2017): 399.

15. Council of European Union, "Regulation of the European Parliament and of the Council"; and Colorado General Assembly, "Consumer Protections for Artificial Intelligence, bill sb24–205," 2024, https://leg.colorado.gov/bills/sb24–205. See also https://fpf.org/wp-content/uploads/2024/05/FPF-FINAL-CO-SB-205-Two-Pager-.pdf.

Bibliography

Abercrombie, Gavin, Amanda Cercas Curry, Mugdha Pandya, and Verena Rieser. "Alexa, Google, Siri: What Are Your Pronouns? Gender and Anthropomorphism in the Design and Perception of Conversational Assistants." In *Proceedings of the 3rd Workshop on Gender Bias in Natural Language Processing*, 24–33, ACL, 2021.

Abid, Abubakar, Maheen Farooqi, and James Zou. "Persistent Anti-Muslim Bias in Large Language Models." In *Proceedings of the AAAI/ACM Conference on AI, Ethics, and Society*, 298–306, ACM, 2021.

Abnet, Dustin A. *The American Robot: A Cultural History*. University of Chicago Press, 2020.

Abraham, Yuval. "'Lavender': The AI Machine Directing Israel's Bombing Spree in Gaza." *+972 Magazine*, April 3, 2024.

ACLU (American Civil Liberties Union). "The School-to-Prison Pipeline in Black and White." Technical report. American Civil Liberties Union of Rhode Island, 2015.

ACLU (American Civil Liberties Union), Ezekiel Edwards, Emily Greytak, Brooke Madubuonwu, Thania Sanchez, Sophie Beiers, Charlotte Resing, Paige Fernandez, and Sagiv Galai. *A Tale of Two Countries: Racially Targeted Arrests in the Era of Marijuana Reform*. American Civil Liberties Union, 2020.

Addison, Arifah, Christoph Bartneck, and Kumar Yogeeswaran. "Robots Can Be More Than Black and White: Examining Racial Bias Towards Robots." In *Proceedings of the AAAI/ACM Conference on AI, Ethics, and Society*, 493–498, ACM, 2019.

Ajaykumar, Gopika, Maureen Steele, and Chien-Ming Huang. "A Survey on End-User Robot Programming." *ACM Computing Surveys (CSUR)*, 54(8), 1–36. 2021.

Al Moubayed, Samer, Jonas Beskow, Gabriel Skantze, and Björn Granström. "Furhat: A Back-Projected Human-Like Robot Head for Multiparty Human-Machine Interaction." In *Cognitive Behavioural Systems: COST 2102 International Training School*, edited by Anna Esposito, Antoinetta M. Esposito, Alessandro Vinciarelli, Rüdiger Hoffmann, and Vincent C. Müller, 114–130. Springer, 2012.

Al-Qadri, Syed Iylia Hariz, and Hafzi Hasani Mohamed. "Meet ADAM, Malaysia's Own Version of Sophia, the Humanoid Robot." 2019. http://energy.bernama.com/news.php?id=1748097.

Al Shaali, Khalifa Rashid, and Neil Kibble. "Policing and Police Accountability in the UAE: The Case for Reform." *Arab Law Quarterly*, 15(3): 272–303, 2000.

Aldebaran Robotics. "Nao." 2008. https://www.aldebaran.com/en/nao.

Aldebaran Robotics. "ALFaceCharacteristics API." 2014. http://doc.aldebaran.com/2-4/naoqi/peopleperception/alfacecharacteristics-api.html.

Alesich, Simone, and Michael Rigby. "Gendered Robots: Implications for Our Humanoid Future." *IEEE Technology and Society Magazine*, 36(2): 50–59, 2017.

Alexander, Michelle. *The New Jim Crow: Mass Incarceration in the Age of Colorblindness*. New Press, 2020.

Allyn, Bobby. "'The Computer Got It Wrong': How Facial Recognition Led to False Arrest of Black Man." *NPR*, 2020. https://www.npr.org/2020/06/24/882683463/the-computer-got-it-wrong-how-facial-recognition-led-to-a-false-arrest-in-michig

Amadeo, Ron. "The NYPD Retires 'Digidog' Robot After Public Backlash." *Ars Technica*, April 30, 2021.

Ames, Roger T., and Henry Rosemont Jr. *The Analects of Confucius: A Philosophical Translation*. Ballantine Books, 2010.

Amini, Alexander, Ava P Soleimany, Wilko Schwarting, Sangeeta N. Bhatia, and Daniela Rus. "Uncovering and Mitigating Algorithmic Bias Through Learned Latent Structure." In *Proceedings of the 2019 AAAI/ACM Conference on AI, Ethics, and Society*, 289–295, ACM, 2019.

Aminzade, Ronald. "Reinterpreting Capitalist Industrialization: A Study of Nineteenth-Century France: Proletarianization, Small-Scale Industry and Capitalist Industrialization." *Social History*, 9(3): 329–350, 1984.

Ammari, Tawfiq, Jofish Kaye, Janice Y. Tsai, and Frank Bentley. "Music, Search, and IoT: How People (Really) Use Voice Assistants." *ACM Transactions on Computer–Human Interaction (TOCHI)*, 26(3): 1–28, 2019.

Anderson, Kevin. "Marcuse's and Fromm's Correspondence with the Socialist Feminist Raya Dunayevskaya: A New Window on Critical Theory." *Logos: A Journal of Modern Society and Culture*, 11(2), 2012.

Angwin, Julia, Jeff Larson, Surya Mattu, and Lauren Kirchner. "Machine Bias." In *Ethics of Data and Analytics: Concepts and Cases*, edited by Kirsten Martin, 254–264. Auerbach, 2022.

Ansara, Y. Gavriel. "Cisgenderism in Medical Settings: How Collaborative Partnerships Can Challenge Structural Violence." In *Out of the Ordinary: Representations of LGBT Lives*, edited by Ian Rivers and Richard Ward, 102–122. Cambridge Scholars, 2012.

Ansari, Aziz, and Alan Yang. "Master of None: Season 1." 2015.

Anwar, Saira, Nicholas Alexander Bascou, Muhsin Menekse, and Asefeh Kardgar. "A Systematic Review of Studies on Educational Robotics." *Journal of Pre-College Engineering Education Research (J-PEER)*, 9(2): 2, 2019.

Arendt, Hannah, and Jens Kroh. *Eichmann in Jerusalem*. Viking Press, 1964.

Asante, Godfried, Sachi Sekimoto, and Christopher Brown. "Becoming 'Black': Exploring the Racialized Experiences of African Immigrants in the United States." *Howard Journal of Communications*, 27(4): 367–384, 2016.

Ascham, Roger. *The Scholemaster*. 1570.

Aschenbrenner, Karl. "Moral Judgment." In *The Concepts of Value: Foundations of Value Theory*, 95–110. Springer, 1971.

Asher, Jeff, and Ben Horwitz. "How Do the Police Actually Spend Their Time?" *New York Times*, June 19, 2020.

Ashworth, Boone. "The NYPD Brings Robot Dogs Back." *Wired*, April 15, 2023.

Asimov, Isaac. Liar! *Astounding Science Fiction*, 28(3): 43–87, 1941

Asimov, Isaac. *I, Robot*. Gnome Press, 1950.

Asimov, Isaac. *In Memory Yet Green: The Autobiography of Isaac Asimov, 1920–1954*. Doubleday, 1979.

Asimov, Isaac. *The Rest of the Robots*. Doubleday, 1964.

Asimov, Isaac. *Robot Visions*. Penguin, 1991.

Asimov, Isaac. "Runaround." *Astounding Science Fiction*, 29(1): 94–103, 1942.

Associated Press. "Dallas Cops Told Sniper Explosive-Strapped Robot Had a Phone." *Associated Press*, February 2, 2018.

Associated Press. "Documents Detail Use of Bomb Robot in Maine Standoff." *Associated Press*, September 1, 2018.

Associated Press. "LAPD Used Facial Recognition Nearly 30,000 Times." *Associated Press*, September 22, 2020.

Association for Advancing Automation. "Unimate // The First Industrial Robot. A Tribute to Joseph Engelberger." 2015. https://www.automate.org/a3-content/joseph-engelberger-unimate.

Atanasoski, Neda, and Kalindi Vora. *Surrogate Humanity: Race, Robots, and the Politics of Technological Futures*. Duke University Press, 2019.

Axe, David. "Drug Raid Turns Ugly as SWAT Guns Down Marine Vet." *Wired*, May 30, 2011.

Baig, Edward C. "Kids Were Being Rude to Alexa, so Amazon Updated It." *USA Today*, April 25, 2018.

Bainbridge, Wilma, Phillip Isola, Idan Blank, and Aude Oliva. "Establishing a Database for Studying Human Face Photograph Memory." In *Proceedings of the Annual Meeting of the Cognitive Science Society*, 1302–1307, The Cognitive Science Society, 2012.

Bainbridge, Wilma A., Phillip Isola, and Aude Oliva. "The Intrinsic Memorability of Face Photographs." *Journal of Experimental Psychology: General*, 142(4): 1323–1334, 2013.

Baker, David. "Systemic White Racism and the Brutalization of Executed Black Women in the United States." *It's a Crime: Women and Justice,* edited by Roslyn Muraskin. Pearson Prentice Hall, 2007.

Balko, Radley. "Overkill: The Rise of Paramilitary Police Raids in America." Technical report. The Cato Institute, 2006.

Balko, Radley. *Rise of the Warrior Cop: The Militarization of America's Police Forces*. PublicAffairs, 2013.

Ball, Patrick. "Violence in Blue: Police Homicides in the United States." *Granta Magazine*, March 4, 2016.

Balzar, John. "The System: Deals, Deadlines, Few Trials." *Los Angeles Times*, September 4, 2006.

Banks, Jaime, and Kevin Koban. "A Kind Apart: The Limited Application of Human Race and Sex Stereotypes to a Humanoid Social Robot." *International Journal of Social Robotics*, 15(11): 1949–1961, 2022.

Bannon, Liam J. "The Politics of Design: Representing Work." *Communications of the ACM*, 38(9): 66–68, 1995.

Barfield, Jessica. "Designing Social Robots to Accommodate Diversity, Equity, and Inclusion in Human–Robot Interaction." In *Proceedings of the ACM SIGIR Conference on Human Information Interaction and Retrieval*, 463–466, ACM, 2023.

Barocas, Solon, Moritz Hardt, and Arvind Narayanan. "Fairness and Machine Learning." 2019. https://fairmlbook.org/.

Baroni, Ilaria, Marco Nalin, Mattia Coti Zelati, Elettra Oleari, and Alberto Sanna. "Designing Motivational Robot: How Robots Might Motivate Children to Eat Fruits and Vegetables." In *Proceedings of the 23rd IEEE International Conference on Robot and Human Interactive Communication*, 796–801, IEEE, 2014.

Barreto, Manuela, and Naomi Ellemers. The Burden of Benevolent Sexism: How it Contributes to the Maintenance of Gender Inequalities." *European Journal of Social Psychology*, 35(5): 633–642, 2005.

Bartneck, Christoph, Kumar Yogeeswaran, Qi Min Ser, Graeme Woodward, Robert Sparrow, Siheng Wang, and Friederike Eyssel. "Robots and Racism." In *Proceedings of the ACM/IEEE International Conference on Human–Robot Interaction*, 196–204, ACM, 2018.

Bashi Treitler, Vilna. "Racialization: Paradigmatic Frames from British Colonization to Today, and Beyond." Working Paper Series. 2015. desiguALdades.net.

Bashi Treitler, Vilna. *The Ethnic Project*. Stanford University Press, 2020.

Batty, David. "Conditions for Abu Dhabi's Migrant Workers 'Shame the West.'" *The Guardian*, 21, December 22, 2013.

Bauer, Natalee Kehaulani. *(En)Gendering Whiteness: A Historical Analysis of White Womanhood, Colonial Anxieties, and "Tender Violence" in US Schools*. University of California, Berkeley, 2017.

Baum, Dan. "Legalize It All." *Harper's Magazine*, 24: 21–32, 2016.

Bayley, David H. "Police Reform: Who done it?" In *Police Reform from the Bottom Up*, edited by Monique Marks and David Sklansky, 16–26. Routledge, 2014.

BBC. "Amazon Alexa to Reward Kids Who Say: 'Please.'" 2018. https://www.bbc.com/news/technology-43897516.

Beal, Carole R., Andrew Garrod, Kate Ruben, Terri L. Stewart, and Dawn J. Dekle. "Children's Moral Orientation: Does the Gender of Dilemma Character Make a Difference?" *Journal of Moral Education*, 26(1): 45–58, 1997.

Beale, Andy. "A New Weapon for SWAT Teams: Bomb-Squad Robots." *VICE*, May 24, 2015.

Bean, Annemarie, James V. Hatch, and Brooks McNamara. *Inside the Minstrel Mask: Readings in Nineteenth-Century Blackface Minstrelsy*. Wesleyan University Press, 1996.

Beauchamp, Gorman. "The Frankenstein Complex and Asimov's Robots." *Mosaic: A Journal for the Interdisciplinary Study of Literature*, 13(3–4): 83–94, 1980.

Belkhir, Jean. "Race, Sex, Class and 'Intelligence': Scientific Racism, Sexism, and Classism." *Race, Gender, & Class*, 1(2): 53–83, 1994.

Belknap, Michal R. *Urban Race Riots*. Routledge, 1991.

Bell, Derrick. *Faces at the Bottom of the Well: The Permanence of Racism*. Basic Books, 1992.

Bello, Paul. "Attention, Self-Control, and (Virtuous) Machine Ethics." In *Proceedings of the 52nd Annual Cal State Fullerton Philosophy Symposium*, 2024.

Benjamin, L. T. "The Birth of American Intelligence Testing." *Monitor on Psychology*, 40(1): 20–21, 2009.

Benjamin, Ruha. *Race After Technology: Abolitionist Tools for the New Jim Code*. Social Forces, 2019.

Bennett, Cynthia, and Os Keyes. "What is the Point of Fairness?" *Interactions*, 27(3): 35–39, 2020.

Bennett, Cynthia L., Cole Gleason, Morgan Klaus Scheuerman, Jeffrey P. Bigham, Anhong Guo, and Alexandra To. "'It's Complicated': Negotiating Accessibility and (Mis)representation in Image Descriptions of Race, Gender, and Disability." In *Proceedings of the CHI Conference on Human Factors in Computing Systems*, 1–19, ACM, 2021.

Bennett, John. "How Command Presence Affects Your Survival." 2010. https://www.police1.com/officer-safety/articles/how-command-presence-affects-your-survival-6cPA8v9EjGJZuWuH/.

Benson, Bruce L., David W. Rasmussen, and David L. Sollars. "Police Bureaucracies, Their Incentives, and the War on Drugs." *Public Choice*, 83(1–2): 21–45, 1995.

Beres, Linda S., and Thomas D. Griffith. "Demonizing Youth." *Loyola of Los Angeles Law Review*, 34: 747–766, 2000.

Bergen, Hilary. "I'd Blush if I Could": Digital Assistants, Disembodied Cyborgs and the Problem of Gender." *Word and Text, a Journal of Literary Studies and Linguistics*, 6(1): 95–113, 2016.

Berk, Richard, Hoda Heidari, Shahin Jabbari, Michael Kearns, and Aaron Roth. "Fairness in Criminal Justice Risk Assessments: The State of the Art." *Sociological Methods & Research*, 50(1): 3–44, 2021.

Bernotat, Jasmin, Friederike Eyssel, and Janik Sachse. "The (Fe)Male Robot: How Robot Body Shape Impacts First Impressions and Trust Towards Robots." *International Journal of Social Robotics*, 13: 477–489, 2021.

Berry, Richard J., and Gorden E. Eden Jr. "Albuquerque Police Department Monthly Report November 2014." Technical report. Albuquerque Police Department, 2014.

Biddle, Sam. "U.S. Marshals Used Drones to Spy on Black Lives Matter Protests in Washington, DC." *The Intercept*, April 22, 2021.

Binder, Earl, and Otto Binder. "I, Robot." In *Amazing Stories*, edited by B. G. Davis, Experimenter Publishing, 13(1): 8–21, 1939.

Binder, Otto O. "You'll Own Slaves by 1965." *Mechanix Illustrated*, 62–65, January 1957.

Binet, Alfred, and Theophile Simon. "The Development of the Binet–Simon Scale, 1905–1908." In *Readings in the History of Psychology*, edited by Wayne Dennis, Appleton-Century-Crofts, 412–424, 1948.

Birhane, Abeba. "Algorithmic Injustice: A Relational Ethics Approach." *Patterns*, 2(2): 1–9, 2021.

Birhane, Abeba, Ryan Steed, Victor Ojewale, Briana Vecchione, and Inioluwa Deborah Raji. "AI Auditing: The Broken Bus on the Road to AI Accountability." In *Proceedings of the IEEE Conference on Secure and Trustworthy Machine Learning (SaTML)*, 612–643. IEEE, 2024.

Black, Edwin. *IBM and the Holocaust: The Strategic Alliance Between Nazi Germany and America's Most Powerful Corporation-Expanded Edition*. Dialog Press, 2012.

Blackmon, Douglas A. *Slavery by Another Name: The Re-Enslavement of Black Americans from the Civil War to World War II*. Anchor, 2009.

Blair, Graeme, Jeremy M. Weinstein, Fotini Christia, Eric Arias, Emile Badran, Robert A. Blair, Ali Cheema, Ahsan Farooqui, Thiemo Fetzer, Guy Grossman, et al. "Community Policing Does Not Build Citizen Trust in Police or Reduce Crime in the Global South." *Science*, 374(6571): 1–14, 2021.

Bleiler, Everett F. "From the Newark Steam Man to Tom Swift." *Extrapolation*, 30(2): 101–116, 1989.

Bolton, Christopher, Istvan Csicsery-Ronay Jr., and Takayuki Tatsumi. *Robot Ghosts and Wired Dreams: Japanese Science Fiction from Origins to Anime*. University of Minnesota Press, 2007.

Bolukbasi, Tolga, Kai-Wei Chang, James Y. Zou, Venkatesh Saligrama, and Adam T. Kalai. "Man Is to Computer Programmer as Woman Is to Homemaker? Debiasing Word Embeddings." *Advances in Neural Information Processing Systems (NeurIPS)*, 29: 4349–4357, 2016.

Bonilla-Silva, Eduardo. "From Bi-Racial to Tri-Racial: Towards a New System of Racial Stratification in the USA." *Ethnic and Racial Studies*, 27(6): 931–950, 2004.

Bonilla-Silva, Eduardo. *Racism Without Racists: Color-Blind Racism and the Persistence of Racial Inequality in the United States*. Rowman & Littlefield, 2006.

Bonilla-Silva, Eduardo. "Rethinking Racism: Toward a Structural Interpretation." *American Sociological Review*, 62(3): 465–480, 1997.

Borenstein, Jason, and Ronald C. Arkin. "Nudging for Good: Robots and the ethical Appropriateness of Nurturing Empathy and Charitable Behavior." *AI & Society*, 32: 499–507, 2017.

Boryczka, Jocelyn M. "The Virtues of Vice: The Lowell Mill Girl Debate and Contemporary Feminist Ethics." *Feminist Theory*, 7(1): 49–67, 2006.

Bosco, Carla. "Harvard University and the Fugitive Slave Act." *The New England Quarterly*, 79(2): 227–247, 2006.

Bove, Vincenzo, and Evelina Gavrilova. "Police Officer on the Frontline or a Soldier? The Effect of Police Militarization on Crime." *American Economic Journal: Economic Policy*, 9(3): 1–18, 2017.

Boyette, Chris, and Zoe Sottile. "San Francisco Supervisors Vote to Allow Police to Use Robots to Kill." *CNN*, December 1, 2022.

Brandom, Russell. "AI Pioneer Accused of Having Sex with Trafficking Victim on Jeffrey Epstein's Island." *The Verge*, August 9, 2019.

Brayne, Sarah. *Predict and Surveil: Data, Discretion, and the Future of Policing*. Oxford University Press, 2020.

Briggs, Gordon, and Matthias Scheutz. "How Robots Can Affect Human Behavior: Investigating the Effects of Robotic Displays of Protest and Distress." *International Journal of Social Robotics*, 6: 343–355, 2014.

Briggs, Gordon, Tom Williams, Ryan Blake Jackson, and Matthias Scheutz. "Why and How Robots Should Say 'No.'" *International Journal of Social Robotics*, 14(2): 323–339, 2022.

Brigham, Carl Campbell. *A Study of American Intelligence*. Princeton University Press, 1922.

Brighenti, Andrea. "Visibility: A Category for the Social Sciences." *Current Sociology*, 55(3): 323–342, 2007.

Brignull, Harry. "Dark Patterns." https://darkpatterns.org/. Accessed 2023.

Broeker, Galen. "Robert Peel and the Peace Preservation Force." *The Journal of Modern History*, 33(4): 363–373, 1961.

Brophy, James T., Margaret M. Keith, and Michael Hurley. "Assaulted and Unheard: Violence Against Healthcare Staff." *New Solutions: A Journal of Environmental and Occupational Health Policy*, 27(4): 581–606, 2018.

Brown, Elizabeth, and George Barganier. *Race and Crime: Geographies of Injustice*. University of California Press, 2018.

Brown, Landon, Jared Hamilton, Zhao Han, Albert Phan, Thao Phung, Eric Hansen, Nhan Tran, and Tom Williams. "Best of Both Worlds? Combining Different Forms of Mixed Reality Deictic Gestures." *ACM Transactions on Human–Robot Interaction (T-HRI)*, 12(1): 1–23, 2022.

Browne, Simone. *Dark Matters*. Duke University Press, 2015.

Browne, Simone. "Digital Epidermalization: Race, Identity and Biometrics." *Critical Sociology*, 36(1): 131–150, 2010.

Browne, Simone. "Everybody's Got a Little Light Under the Sun: Black Luminosity and the Visual Culture of Surveillance." *Cultural Studies*, 26(4): 542–564, 2012.

Browne, Simone. "Thinking AI Through Rastus Robot, the Westinghouse Mechanical Slave. AI Now Experts Workshop, 2017." https://www.youtube.com/watch?v=HvC6kvrwsN4.

Bruxelle, Marc. "The Problem with the "Rainbow-Washing" of LGBTQ+ Pride." *Wired*, June 21, 2018.

Bryson, Joanna J. "Robots Should Be Slaves." In *Close Engagements with Artificial Companions: Key Social, Psychological, Ethical and Design Issues*, edited by Yorick Wilks, Natural Language Processing 8, 63–74. John Benjamins, 2010.

Buhr, Sarah. "Security Robots Are Being Used to Ward Off San Francisco's Homeless Population." *TechCrunch*, December 13, 2017.

Buolamwini, Joy. "When the Robot Doesn't See Dark Skin." *New York Times*, June 21, 2018.

Buolamwini, Joy, and Timnit Gebru. Gender Shades: Intersectional Accuracy Disparities in Commercial Gender Classification." In *Proceedings of the ACM Conference on Fairness, Accountability and Transparency (FAccT)*, 77–91, ACM, 2018.

Burris, Heather H., Molly Passarella, Sara C. Handley, Sindhu K. Srinivas, and Scott A. Lorch. "Black–White Disparities in Maternal In-Hospital Mortality According to Teaching and Black-Serving Hospital Status." *American Journal of Obstetrics and Gynecology*, 225(1): 83e1–83e9, 2021.

Butler, Samuel. *Darwin Among the Machines*. The Press, 1863.

Butler, Samuel. *Erewhon: Or, Over the Range*. Trübner & Co, 1872.

C. deBaca, Luis, and Griffin Thomas Black. "The Development of US Anti-Slavery Law: A Historical Review." In *The Historical Roots of Human Trafficking: Informing Primary Prevention of Commercialized Violence*, edited by Makini Chisolm-Straker and Katherine Chon, 127–168. Springer, 2021.

Calo, Ryan, A. Michael Froomkin, and Ian Kerr. *Robot Law*. Edward Elgar, 2016.

Calo, Ryan. "Artificial Intelligence Policy: A Primer and Roadmap." *UCDL Review*, 51: 399, 2017.

Calo, Ryan. "Robotics and the Lessons of Cyberlaw." *California Law Review*, 103: 513, 2015.

Campaign Zero. "Police Scorecard." 2016. https://policescorecard.org/.

Campaign Zero. "Police Scorecard: Peoria, il." 2023. https://policescorecard.org/il/police-department/peoria.

Čapek, Karel. *R.U.R. (Rossum's Universal Robots)*. Aventinum, 1920.

Cappuccio, Massimiliano L., Eduardo B. Sandoval, Omar Mubin, Mohammad Obaid, and Mari Velonaki. "Can Robots Make Us Better Humans? Virtuous Robotics and the Good Life with Artificial Agents." *International Journal of Social Robotics*, 13: 7–22, 2021.

Carpenter, Dick M., Lisa Knepper, Angela C. Erickson, and Jennifer McDonald. *Policing for Profit: The Abuse of Civil Asset Forfeiture*. 2nd ed. Institute for Justice, 2015.

Carpenter, Julie. *Culture and Human–Robot Interaction in Militarized Spaces: A War Story*. Routledge, 2016.

Carpenter, Julie. "Why Project Q is More Than the World's First Nonbinary Voice for Technology." *Interactions*, 26(6): 56–59, 2019.

Carr, Nanci K. "Programmed to Protect and Serve: The Dawn of Drones and Robots in Law Enforcement." *Journal of Air Law and Commerce*, 86(2): 183–218, 2021.

Carruth, Daniel W., and Cindy L. Bethel. "Challenges with the Integration of Robotics into Tactical Team Operations." In *Proceedings of the 15th IEEE International Symposium on Applied Machine Intelligence and Informatics (SAMI)*, 27–32, IEEE, 2017.

Castañeda, Claudia. "Robotic Skin: The Future of Touch?" In *Thinking Through the Skin*, edited by Sara Ahmed and Jackie Stacey, 237–250. Routledge, 2003.

Castoriadis, Cornelius. *The Imaginary Institution of Society*. MIT Press, 1997.

Cave, Stephen, and Kanta Dihal. "The Whiteness of AI." *Philosophy & Technology*, 33(4): 685–703, 2020.

Cei, Louis Bernard. *Law Enforcement in Richmond: A History of Police–Community Relations, 1737–1974*. Florida State University, 1975.

Center for the Study of the Drone at Bard College. "Law Enforcement Robots Datasheet." 2016. https://dronecenter.bard.edu/law-enforcement-robots-datasheet/.

Cericola, Rachel. "Ring Neighbors Is the Best and Worst Neighborhood Watch App." *New York Times*, February 14, 2020.

Chamayou, Grégoire. *Manhunts: A Philosophical History*. Princeton University Press, 2012.

Chapuis, Alfred, Edmond Droz, and Alec Reid. *Automata: A Historical and Technological Study*. Neuchatel / Editions Du Griffon. Central Book Company, 1958.

Cheshire, Jenny, Paul Kerswill, Sue Fox, and Eivind Torgersen. "Contact, the Feature Pool and the Speech Community: The Emergence of Multicultural London English." *Journal of Sociolinguistics*, 15(2): 151–196, 2011.

Chidambaram, Vijay, Yueh-Hsuan Chiang, and Bilge Mutlu. "Designing Persuasive Robots: How Robots Might Persuade People Using Vocal and Nonverbal Cues." In *Proceedings of the ACM/IEEE International Conference on Human–Robot Interaction*, 293–300, ACM, 2012.

Chin, Gabriel J., and Scott C. Wells. "The Blue Wall of Silence as Evidence of Bias and Motive to Lie: A New Approach to Police Perjury." *University of Pittsburgh Law Review*, 59: 233–299, 1997.

Chomanski, Bartek. "What's Wrong with Designing People to Serve?" *Ethical Theory and Moral Practice*, 22(4): 993–1015, 2019.

Christian, Michelle. "A Global Critical Race and Racism Framework: Racial Entanglements and Deep and Malleable Whiteness." *Sociology of Race and Ethnicity*, 5(2): 169–185, 2019.

Chude-Sokei, Louis. "Race and Robotics." In *Cyborg Futures: Cross-Disciplinary Perspectives on Artificial Intelligence and Robotics*, edited by Teresa Heffernan, 159–171. Palgrave Macmillan, 2019.

Chude-Sokei, Louis. *The Sound of Culture: Diaspora and Black Technopoetics*. Wesleyan University Press, 2015.

Chun, Wendy Hui Kyong. "Race and/as Technology, or How to Do Things to Race." In *Race After the Internet*, edited by Lisa Nakamura and Peter A. Chow-White, 44–66. Routledge, 2013.

Coeckelbergh, Mark. "The Ubuntu Robot: Towards a Relational Conceptual Framework for Intercultural Robotics." *Science and Engineering Ethics*, 28(2): 16, 2022.

Coffee, Kevin. "Industrial Lowell and the Dawn of the Anthropocene." *Industrial Archaeology Review*, 43(1): 20–33, 2021.

Coleman, Beth. "Race as Technology." *Camera Obscura: Feminism, Culture, and Media Studies*, 24(1): 177–207, 2009.

Collins, Patricia Hill. "Black Feminist Thought in the Matrix of Domination." In *Black Feminist Thought: Knowledge, Consciousness, and the Politics of Empowerment*, 221–238. Routledge, 1990.

Collins, Patricia Hill. *Black Feminist Thought: Knowledge, Consciousness, and the Politics of Empowerment*. Routledge, 2022.

Collins, Patricia Hill. "It's All in the Family: Intersections of Gender, Race, and Nation." *Hypatia*, 13(3): 62–82, 1998.

Collins, Stephanie. *The Core of Care Ethics*. Springer, 2015.

Colorado General Assembly. "Consumer Protections for Artificial Intelligence, Bill sb24–205." 2024. https://leg.colorado.gov/bills/sb24-205. See also https://fpf.org/wp-content/uploads/2024/05/FPF-FINAL-CO-SB-205-Two-Pager-.pdf.

Common Crawl. "Common Crawl." 2023. https://commoncrawl.org/.

Conger, Kate, Richard Fausset, and Serge F. Kovaleski. "San Francisco Bans Facial Recognition Technology." *New York Times*, May 14, 2019.

Connolly, Randy. "Why Computing Belongs Within the Social Sciences." *Communications of the ACM*, 63(8): 54–59, 2020.

Cooley, Amanda Harmony. "Quiet Encroachments on School Prayer Jurisprudence." *West Virginia Law Review*, 126(2): 395–486, 2024.

Cooney, Martin, Masahiro Shiomi, Eduardo Kochenborger Duarte, and Alexey Vinel. "A Broad View on Robot Self-Defense: Rapid Scoping Review and Cultural Comparison." *Robotics*, 12(2):1–37, 2023.

Correia, David. "On the Nature of Police." In *Violent Order: Essays on the Nature of Police*, edited by David Correia and Tyler Wall, 1–12. Haymarket Books, 2021.

Correll, Joshua, Bernadette Park, Charles M. Judd, and Bernd Wittenbrink. "The Police Officer's Dilemma: Using Ethnicity to Disambiguate Potentially Threatening Individuals." *Journal of Personality and Social Psychology*, 83(6): 1314–1329, 2002.

Costanza-Chock, Sasha. *Design Justice: Community-Led Practices to Build the Worlds We Need*. MIT Press, 2020.

Costanza-Chock, Sasha, Inioluwa Deborah Raji, and Joy Buolamwini. "Who Audits the Auditors? Recommendations from a Field Scan of the Algorithmic Auditing Ecosystem." In *Proceedings of the ACM Conference on Fairness, Accountability, and Transparency*, 1571–1583, ACM, 2022.

Cotton, Stacey, Peter Kraska, and James A. Pikl. "Law Enforcement, Community, and Military Tactics: What's the Conflict?" In *Working in the Public Interest Conference*, 2014.

Couldry, Nick, and Ulises A. Mejias. *The Costs of Connection: How Data Are Colonizing Human Life and Appropriating It for Capitalism*. Oxford University Press, 2020.

Council of European Union. Regulation of the European parliament and of the council laying down harmonised rules on artificial intelligence and amending regulations (EC) no 300/2008, (EU) no 167/2013, (EU) no 168/2013, (EU) 2018/858, (EU) 2018/1139 and (EU) 2019/2144 and directives 2014/90/eu, (EU) 2016/797 and (EU) 2020/1828, 2024. https://www.europarl.europa.eu/doceo/document/TA-9-2024-0138-FNL-COR01_EN.pdf.

Cowley, Charles. *Illustrated History of Lowell*. Lee & Shepard, 1868.

Cox, Mike. *The Texas Rangers: Wearing the Cinco Peso, 1821–1900*. Forge Books, 2008.

Cox-George, Chantal, and Susan Bewley. "I, Sex Robot: The Health Implications of the Sex Robot Industry." *BMJ Sexual & Reproductive Health*, 44(3): 161–164, 2018.

CR10 Publications Collective. *Abolition Now!: Ten Years of Strategy and Struggle Against the Prison Industrial Complex*. AK Press, 2008.

Crain, Marion, Winifred Poster, and Miriam Cherry. *Invisible Labor: Hidden Work in the Contemporary World*. University of California Press, 2016.

Cramer, Maria, and Christine Hauser. "Digidog, a Robotic Dog Used by the Police, Stirs Privacy Concerns." *New York Times*, February 27, 2021.

Creager, Elliot, David Madras, Jörn-Henrik Jacobsen, Marissa Weis, Kevin Swersky, Toniann Pitassi, and Richard Zemel. "Flexibly Fair Representation Learning by Disentanglement." In *Proceedings of the International Conference on Machine Learning*, 1436–1445, PMLR, 2019.

Crenshaw, Kimberlé. *Demarginalizing the Intersection of Race and Sex: A Black Feminist Critique of Antidiscrimination Doctrine, Feminist Theory and Antiracist Politics*. University of Chicago Legal Forum, 1989.

Crepelle, Adam. "Probable Cause to Plunder: Civil Asset Forfeiture and the Problems It Creates." *Wake Forest Journal of Law and Policy*, 7: 315–316, 2017.

Croizet, Jean-Claude. "The Racism of Intelligence: How Mental Testing Practices Have Constituted an Institutionalized Form of Group Domination." In *The Oxford Handbook of African American Citizenship, 1865–Present*, edited by Lawrence D. Bobo, Lisa Crooms-Robinson, Linda Darling-Hammond, Michael C. Dawson, Henry Louis Gates, Gerald Jaynes, and Claude Steele, 769–816. Oxford Academic, 2012.

Csikszentmihalyi, Mark. "Confucius." In *The Stanford Encyclopedia of Philosophy*, edited by Edward N. Zalta. Metaphysics Research Lab, Stanford University, Summer 2020. https://plato.stanford.edu/archives/sum2020/entries/confucius/

Culver, Kathleen Bartzen. "From Battlefield to Newsroom: Ethical Implications of Drone Technology in Journalism." *Journal of Mass Media Ethics*, 29(1): 52–64, 2014.

Cummings, Mary L. "Automation and Accountability in Decision Support System Interface Design." *Journal of Technology Studies*, 32: 1(23–31), 2006.

Cunningham, Chris. "All is Full of Love." *Leonardo*, 35(5): 558–558, 2002.

Dain, Bruce R. *A Hideous Monster of the Mind*. Harvard University Press, 2021.

Danaher, John. "Robotic Rape and Robotic Child Sexual Abuse: Should They Be Criminalised?" *Criminal Law and Philosophy*, 11(1): 71–95, 2017.

Danaher, John, and Neil McArthur, eds. *Robot Sex: Social and Ethical Implications*. MIT Press, 2017.

Danaher, John, Brian D. Earp, and Anders Sandberg. "Should We Campaign Against Sex Robots?" In *Robot Sex: Social and Ethical Implications*, edited by John Danaher and Neil McArthur, 47–72. MIT Press, 2017.

Dancy, T. Elon. (Un)doing Hegemony in Education: Disrupting School-to-Prison Pipelines for Black Males." *Equity & Excellence in Education*, 47(4): 476–493, 2014.

Danescu-Niculescu-Mizil, Cristian, Moritz Sudhof, Dan Jurafsky, Jure Leskovec, and Christopher Potts. "A Computational Approach to Politeness with Application to

Social Factors." In *Proceedings of the 51st Annual Meeting of the Association for Computational Linguistics*, 250–259, ACL, 2013.

Darling, Kate. *The New Breed: What Our History with Animals Reveals About Our Future with Robots*. Henry Holt, 2021.

Darling, Kate. "'Who's Johnny?' Anthropomorphic Framing in Human–Robot Interaction, Integration, and Policy." In *Robot ethics 2.0,* edited by Patrick Lin, Ryan Jenkins, and Keith Abney, 173–192. Oxford University Press, 2015.

Davis, Angela Y. *Abolition Democracy: Beyond Empire, Prisons, and Torture*. Seven Stories Press, 2011.

Davis, Angela Y. *Are Prisons Obsolete?* Seven Stories Press, 2011.

Davis, Mike. *City of Quartz: Excavating the Future in Los Angeles*. Verso Books, 2006.

Dazio, Stefanie. "Police Can't Access LAPD Gang Database Records after Scandal." *Washington Post,* July 14, 2020.

de l'Isle-Adam, Auguste Villiers. *L'Ève Future*. Paris, 1886.

DeHart, Jasmine. "Building Cultural Competency for Computing: Identity Computing Lab, Duke University." *XRDS: Crossroads, The ACM Magazine for Students*, 28(4): 38–39, 2022.

Delaunay, Frédéric, Joachim De Greeff, and Tony Belpaeme. "Towards Retro-Projected Robot Faces: An Alternative to Mechatronic and Android Faces." In *Proceedings of the IEEE International Symposium on Robot and Human Interactive Communication (RO-MAN)*, 306–311, IEEE, 2009.

Delehanty, Casey, Jack Mewhirter, Ryan Welch, and Jason Wilks. "Militarization and Police Violence: The Case of the 1033 Program." *Research & Politics*, 4(2): 1–7, 2017.

DeLoache, Judy S., Deborah J. Cassidy, and C. Jan Carpenter. "The Three Bears Are All Boys: Mothers' Gender Labeling of Neutral Picture Book Characters." *Sex Roles,* 17(3–4): 163–178, 1987.

D'Ignazio, Catherine, and Lauren F. Klein. *Data Feminism*. MIT Press, 2020.

Doane, Woody., and Eduardo Bonilla-Silva. "Rethinking Whiteness Studies." In *White Out: The Continuing Significance of Racism,* edited by Ashley W. Doane and Eduardo Bonilla-Silva, 11–26. Routledge, 2013.

Dole, Lorin. "The Influence of a Robot's Mere Presence on Human Communication." PhD thesis, Stanford University, 2017.

Donlon, Marie. "Automation at the Airport." *Electronics 360*, August 30, 2022.

Dooley, Samuel, George Z. Wei, Tom Goldstein, and John Dickerson. "Robustness Disparities in Face Detection." *Advances in Neural Information Processing Systems*, 35: 38245–38259, 2022.

Döring, Nicola. "Vom internetsex zum robotersex: Forschungsstand und herausforderungen für die sexualwissenschaft." *Zeitschrift für Sexualforschung*, 30(01): 35–57, 2017.

Döring, Nicola, and Sandra Pöschl. "Sex Toys, Sex Dolls, Sex Robots: Our Under-Researched Bed-Fellows." *Sexologies*, 27(3): 51–55, 2018.

Drug Policy Alliance. "Approaches to Decriminalizing Drug Use and Possession," 2015. https://www.unodc.org/documents/ungass2016/Contributions/Civil/DrugPolicyAlliance/DPA_Fact_Sheet_Approaches_to_Decriminalization_Feb2015_1.pdf.

Du Bois, William Edward Burghardt. *The Souls of Black Folk*. Routledge, 2015.

Duarte, Eduardo Kochenborger, Masahiro Shiomi, Alexey Vinel, and Martin Cooney. "Robot Self-Defense: Robots Can Use Force on Human Attackers to Defend Victims." In *Proceedings of the IEEE International Conference on Robot and Human Interactive Communication (RO-MAN)*, 1606–1613. IEEE, 2022.

Dublin, Thomas. *Lowell: The Story of an Industrial City: A Guide to Lowell National Historical Park and Lowell Heritage State Park, Lowell, Massachusetts*. Vol. 140. Government Printing Office, 1992.

Dublin, Thomas. "Women, Work, and Protest in the Early Lowell Mills: 'The Oppressing Hand of Avarice Would Enslave US.'" *Labor History*, 16(1): 99–116, 1975.

Duckworth, Dexter, Zachary Henkel, Stephanie Wuisan, Brendan Cogley, Christopher Collins, and Cindy L. Bethel. "Therabot: The Initial Design of a Robotic Therapy Support System." In *Companion of the ACM/IEEE International Conference on Human–Robot Interaction (HRI Extended Abstracts)*, 13–14, ACM, 2015.

Dugger, Karen. Social Location and Gender-Role Attitudes: A Comparison of Black and White Women." *Gender & Society*, 2(4): 425–448, 1988.

Dunham, Roger G., Geoffrey P. Alpert, and Kyle D. McLean. *Critical Issues in Policing: Contemporary Readings*. Waveland Press, 2020.

Dunn, Anthony and Fiona Raby. *Speculative Everything: Design, Fiction, and Social Dreaming*. MIT Press, 2024.

Dunstan, Belinda J., and Guy Hoffman. "Social Robot Morphology: Cultural Histories of Robot Design." In *Cultural Robotics: Social Robots and Their Emergent Cultural Ecologies*, edited by Belinda J. Dunstan, Jeffrey T. K. V. Koh, Deborah Turnbull Tillman, and Scott Andrew Brown, 13–34. Springer, 2023.

Edwards, Chad, Autumn Edwards, Patric R. Spence, and David Westerman. "Initial Interaction Expectations with Robots: Testing the Human-to-Human Interaction Script." *Communication Studies*, 67(2): 227–238, 2016.

Edwards, Jared F. "Color-Blind Racial Attitudes: Microaggressions in the Context of Racism and White Privilege." *Administrative Issues Journal*, 7(1): 5–18, 2017.

Egger, Grace. "Ring, Amazon Calling: The State Action Doctrine and the Fourth Amendment." *Washington Law Review Online*, 95: 245–275, 2020.

Eglash, Ron, Audrey Bennett, Michael Lachney, and William Babbitt. "Race-Positive Design: A Generative Approach to Decolonizing Computing." In *Proceedings of the CHI Workshop on Race in HCI*, 2020.

Elbeleidy, Saad, Elizabeth Reddy, and Tom Williams. "The Invisible Labor of Authoring Dialogue for Teleoperated Socially Assistive Robots." In *Proceedings of the IEEE International Conference on Robot and Human Interactive Communication (RO-MAN)*, 2462–2469, IEEE, 2023.

Elbeleidy, Saad, Terran Mott, and Tom Williams. "Practical, Ethical, and Overlooked: Tele-operated Socially Assistive Robots in the Quest for Autonomy." In *Companion of the 17th ACM/IEEE International Conference on Human–Robot Interaction (alt.HRI)*, 577–587, ACM, 2022.

Elbeleidy, Saad, Terran Mott, Dan Liu, Ellen Yi-Luen Do, Elizabeth Reddy, and Tom Williams. "Beyond the Session: Centering Teleoperators in Robot-Assisted Therapy Reveals the Bigger Picture." In *Proceedings of the ACM on Human–Computer Interaction*, (CSCW), 1–33, ACM, 2023.

Elliot, Charles Harrington. *The Republican Refuted: In a Series of Biographical, Critical and Political Strictures on Thomas Paine's Rights of Man*. London,1791.

Ellis, Edward S. *The Steam Man of the Prairies*. Beadle and Adams, 1868.

Elzway, Salem. "Arms of the State: A History of the Industrial Robot in Postwar America." PhD thesis, University of Michigan, 2024.

Elzway, Salem. "Technoliberal Machines: Robotic Work(ers) from Science Fiction to the Assembly Line." *Work: The Labors of Language, Culture, and History in North America*, edited by J. Jesse Ramírez and Sixta Quassdorf, 159–186, 2021.

Elzway, Salem, and Jason Resnikoff. "Whence Automation? The History (and Possible Futures) of a Concept." *Labor*, 21(1): 27–41, 2024.

Emerson, Ralph Waldo. *Journals of Ralph Waldo Emerson with Annotations*. Vol. 7. Houghton Mifflin, 1913.

Engelberger, J. F. "Economic and Sociological Impact of Industrial Robots." *Transactions of the American Nuclear Society*, 14(2): 873 1971.

Engelberger, Joseph F. "Sociological Impact of Robots." In *Robotics in Practice: Management and Applications of Industrial Robots*, edited by Dennis Lock and Kenneth Willis, 111–116. Springer, 1980.

Epictetus. *Discourses*. 108.

Eubanks, Virginia. *Automating Inequality: How High-Tech Tools Profile, Police, and Punish the Poor*. St. Martin's Press, 2018.

Evans, Taylor Scott. "The Race of Machines: A Prehistory of the Posthuman." PhD thesis, University of California, Riverside, 2018.

Eyssel, Friederike, and Dieta Kuchenbrandt. "Social Categorization of Social Robots: Anthropomorphism as a Function of Robot Group Membership." *British Journal of Social Psychology*, 51(4): 724–731, 2012.

Eyssel, Friederike, and Steve Loughnan. "It Don't Matter If You're Black or White?" In *Proceedings of the International Conference on Social Robotics*, 422–431, Springer, 2013.

Fairlie, Robert W. "Have We Finally Bridged the Digital Divide? Smart Phone and Internet Use Patterns by Race and Ethnicity." *First Monday*, September 1, 2017.

Fanon, Frantz. *Black Skin, White Masks*. Grove Press, 2008.

Farina, Mirko, Petr Zhdanov, Artur Karimov, and Andrea Lavazza. "AI and Society: A Virtue Ethics Approach." *AI & Society*, 39: 1127–1140, 2024.

Farivar, Cyrus. "Cute videos, but Little Evidence: Police Say Amazon Ring Isn't Much of a Crime Fighter." *NBC News*, February 15, 2020.

Farquharson, Karen. "Racial Categories in Three Nations: Australia, South Africa and the United States." In *Proceedings of 'Public Sociologies: Lessons and Trans-Tasman Comparisons', the Annual Conference of the Australian Sociological Association*, TASA, 1–6, 2007.

Farrow, Ronan. "How an Élite University Research Center Concealed Its Relationship with Jeffrey Epstein." *New Yorker*, September 6, 2019.

Feagin, Joe. *Systemic Racism: A Theory of Oppression*. Routledge, 2013.

Feagin, Joe R. *Racist America: Roots, Current Realities, and Future Reparations*. Routledge, 2000.

Feagin, Joe R. *The White Racial Frame: Centuries of Racial Framing and Counter-Framing*. Routledge, 2020.

Feathers, Todd. "Honolulu Police Used a Robot Dog to Patrol a Homeless Encampment." *VICE*, June 14, 2021.

Federal Bureau of Investigations. "Report on Prevention and Control of Major Disturbances." Technical report. Federal Bureau of Investigations, 1967.

Feine, Jasper, Ulrich Gnewuch, Stefan Morana, and Alexander Maedche. "Gender Bias in Chatbot Design." In *Proceedings of the Third International Workshop on Chatbot Research and Design*, 79–93, Springer, 2020.

Felker-Kantor, Max. *Policing Los Angeles: Race, Resistance, and the Rise of the LAPD*. University of North Carolina Press Books, 2018.

Fet, Yakov. "Norbert Wiener in Moscow." In *Proceedings of the 3rd International Conference on Computer Technology in Russia and in the Former Soviet Union (SoRuCom)*, 194–195. IEEE Computer Society, 2014.

Finn, Peter. "Domestic Use of Aerial Drones by Law Enforcement Likely to Prompt Privacy Debate." *Washington Post*, January 22, 2011.

Floridi, Luciano, and J. W. Sanders. "On the Morality of Artificial Agents." *Minds and Machines*, 14: 349–379, 2004.

Foggia, Pasquale, Antonio Greco, Gennaro Percannella, Mario Vento, and Vincenzo Vigilante. "A system for Gender Recognition on Mobile Robots." In *Proceedings of the 2nd International Conference on Applications of Intelligent Systems*, 1–6, ACM, 2019.

Fong, Terrence, and Illah Nourbakhsh. "Interaction Challenges in Human–Robot Space Exploration." *Interactions*, 12(2): 42–45, 2005.

Forbes, Bryan. *The Stepford Wives*. 1975.

Forrest, Derek William. *Francis Galton: The Life and Work of a Victorian Genius*. Taplinger, 1974.

Fortunati, Leopoldina, Autumn Edwards, Chad Edwards, Anna Maria Manganelli, and Federico de Luca. "Is Alexa Female, Male, or Neutral? A Cross-National and Cross-Gender Comparison of Perceptions of Alexa's Gender and Status as a Communicator." *Computers in Human Behavior*, 137: 1–9, 2022.

Fosch-Villaronga, Eduard, and Michiel Heldeweg. "Regulation, I Presume?" said the Robot—Towards an Iterative Regulatory Process for Robot Governance." *Computer Law & Security Review*, 34(6): 1258–1277, 2018.

Fosdick, Raymond Blaine. *Crime in America and the Police*. Century Company, 1920.

Foucault, Michel. *Discipline and Punish: The Birth of the Prison*. Vintage, 1979.

Franchi, Matt, J. D. Zamfirescu-Pereira, Wendy Ju, and Emma Pierson. "Detecting Disparities in Police Deployments Using Dashcam Data." In *Proceedings of the ACM Conference on Fairness, Accountability, and Transparency (FAccT)*, 534–544, ACM, 2023.

Frankfurt, Harry G. *On Bullshit*. Princeton University Press, 2009.

Fredrickson, George M. *White Supremacy: A Comparative Study of American and South African History*. Oxford University Press, 1981.

Friedman, Lawrence M. *Crime and Punishment in American History*. Basic Books, 1994.

Frumer, Yulia. "The Short, Strange Life of the First Friendly Robot: Japan's Gakutensoku Was a Giant Pneumatic Automaton That Toured Through Asia—Until it Mysteriously Disappeared." *IEEE Spectrum*, 57(6): 42–48, 2020.

Fry, Richard, Brian Kennedy, and Cary Funk. "STEM Jobs See Uneven Progress in Increasing Gender, Racial and Ethnic Diversity." *Pew Research Center*, April 1, 2021.

Galatolo, Alessio, Gaspar I. Melsión, Iolanda Leite, and Katie Winkle. "The Right (Wo) Man for the Job? Exploring the Role of Gender When Challenging Gender Stereotypes with a Social Robot." *International Journal of Social Robotics*, 15: 1933–1947, 2022.

Gamal, Fanna. "The Racial Politics of Protection: A Critical Race Examination of Police Militarization." *California Law Review*, 104, 979–1008, 2016.

Garg, Radhika, Hua Cui, and Yash Kapadia. "Learn, Use, and (Intermittently) Abandon": Exploring the Practices of Early Smart Speaker Adopters in Urban India." In *Proceedings of the ACM on Human–Computer Interaction*, 5(CSCW2): 1–28, ACM, 2021.

Garland, Alex. *Ex Machina*. 2015.

Garrison, William Lloyd, and James Brown Yerrinton. "Convention in Lowell." *The Liberator*, 14(9): 35, 1844.

Gee, Harvey. "Surveillance State: Fourth Amendment Law, Big Data Policing, and Facial Recognition Technology." *Berkeley Journal of African-American Law & Policy*, 21(2): 43–83, 2020.

German, Michael. *Hidden in Plain Sight: Racism, White Supremacy, and Far-Right Militancy in Law Enforcement*. Brennan Center for Justice, 2020.

Gettinger, Dan. *Public Safety Drones*, 3rd ed. Center for the Study of the Drone at Bard College, 2020.

Giaccardi, Elisa, Pieter Desmet, and Olya Kudina. "The Repertoire of Meaningful Voice Interactions. How to Design Good Smart Speakers." *The State of Responsible IoT 2020*, ThingsCon e.V., 53–57, 2020, https://thingscon.org/wp-content/uploads/2020/12/RIOT20_iVersion.pdf

Gianfortune, Ross. *10 Insane Things the Pentagon Gave to Local Law Enforcement*. Government Executive, 2014.

Gibert, Martin. "The Case for Virtuous Robots." *AI and Ethics*, 3(1): 135–144, 2023.

Giger, Jean-Christophe, Nuno Piçarra, Patrícia Alves-Oliveira, Raquel Oliveira, and Patrícia Arriaga. "Humanization of Robots: Is It Really Such a Good Idea?" *Human Behavior and Emerging Technologies*, 1(2): 111–123, 2019.

Gilbert, Ben. "Amazon Sells Facial Recognition Software to Police All Over the US, but Has No Idea How Many Departments Are Using It." *Business Insider*, February 21, 2020.

Gillet, Sarah, Ronald Cumbal, André Pereira, José Lopes, Olov Engwall, and Iolanda Leite. "Robot Gaze Can Mediate Participation Imbalance in Groups with Different Skill Levels." In *Proceedings of the ACM/IEEE International Conference on Human–Robot Interaction*, 303–311, ACM, 2021.

Gillet, Sarah, Wouter van den Bos, and Iolanda Leite. "A Social Robot Mediator to Foster Collaboration and Inclusion Among Children." In *Proceedings of Robotics: Science and Systems*, RSS Foundation, 2020.

Gilmore, Ruth Wilson. *Abolition Geography: Essays Towards Liberation*. Verso Books, 2022.

Gilmore, Ruth Wilson. *Golden Gulag: Prisons, Surplus, Crisis, and Opposition in Globalizing California*. Vol. 21. University of California Press, 2007.

Gilmore, Ruth Wilson, and Chenjerai Kumanyika. "Ruth Wilson Gilmore Makes the Case for Abolition." *The Intercept*, Intercepted podcast by Jeremy Scahill, June 10, 2020.

Glenn, Evelyn Nakano. "From Servitude to Service Work: Historical Continuities in the Racial Division of Paid Reproductive Labor." *Signs: Journal of Women in Culture and Society*, 18(1):1–44, 1992.

Glickman, Deanna J. "Fashioning Children: Gender Restrictive Dress Codes as an Entry Point for the Trans School to Prison Pipeline." *American University Journal of Gender, Social Policy, and the Law*, 24: 263, 2015.

Göckeritz, Susanne, Marco F. H. Schmidt, and Michael Tomasello. "Young Children's Creation and Transmission of Social Norms." *Cognitive Development*, 30: 81–95, 2014.

Goel, Naman, Mohammad Yaghini, and Boi Faltings. "Non-Discriminatory Machine Learning Through Convex Fairness Criteria." In *Proceedings of the 2018 AAAI/ACM Conference on AI, Ethics, and Society (AIES)*, 116–116, AAAI, 2018.

Goo, Hideki Garcia, Katie Winkle, Tom Williams, and Megan Strait. "Victims and Observers: How Identity, Experience, and Biases Shape Perceptions of Robot Abuse." In *Proceedings of the IEEE International Conference on Robot and Human Interactive Communication (RO-MAN)*, 2439–2446 IEEE, 2023.

Goodman, Kylie L., and Christopher B. Mayhorn. "It's Not What You Say but How You Say It: Examining the Influence of Perceived Voice Assistant Gender and Pitch on Trust and Reliance." *Applied Ergonomics*, 106: 1–8, 2023.

Goodwin, Michele. "The Thirteenth Amendment: Modern Slavery, Capitalism, and Mass Incarceration." *Cornell Law Review*, 104: 899–990, 2018.

Goodyear, Dana. "The L.A. County Sheriff's Deputy-Gang Crisis." *New Yorker*, May 30, 2022.

Gordon, Ken. "Alexa and the Age of Casual Rudeness." *The Atlantic*, April 23, 2018.

Graef, Lindsay, Sandra G. Mayson, Aurelie Ouss, and Megan T. Stevenson. "Systemic Failure to Appear in Court." *University of Pennsylvania Law Review*, 172(1): 1–60, 2023.

Graham, Jennifer. "Are Alexa and Siri Making Your Kids Rude?" *Deseret News*, June 24, 2019.

Graham, Mark. "Inequitable Distributions in Internet Geographies: The Global South is Gaining Access, but Lags in Local Content." *Innovations: Technology, Governance, Globalization*, 9(3–4): 3–19, 2014.

Gramlich, Jack Post. "Are Police Obsolete in Minneapolis? Racial Capitalism and Police Abolition." *Social Science Quarterly*, 102(7): 3149–3157, 2021.

Gras, Louis Joseph. *Histoire de la rubanerie.* Société de l'imprimerie Théolier, 1906.

Green, Alan W. C. "'Jim Crow,' 'Zip Coon': The Northern Origins of Negro Minstrelsy." *The Massachusetts Review*, 11(2): 385–397, 1970.

Gregory, Judith. "Scandinavian Approaches to Participatory Design." *International Journal of Engineering Education*, 19(1): 62–74, 2003.

Grim, Valerie. "The Impact of Mechanized Farming on Black Farm Families in the Rural South: A Study of Farm Life in the Brooks Farm Community, 1940–1970." *Agricultural History*, 68(2): 169–184, 1994.

Gross, Laurence F. *The Course of Industrial Decline: The Boott Cotton Mills of Lowell, Massachusetts, 1835–1955*. Johns Hopkins University Press, 1993.

Grother, Patrick, Mei Ngan, and Kayee Hanaoka. "Face Recognition Vendor Test (FVRT): Part 3, Demographic Effects." Technical report. National Institute of Standards and Technology Gaithersburg, 2019.

Grudin, Jonathan, and Richard Jacques. "Chatbots, Humbots, and the Quest for Artificial General Intelligence." In *Proceedings of the CHI Conference on Human Factors in Computing Systems*, 1–11, ACM, 2019.

Guariglia, Matthew. *Police and the Empire City: Race and the Origins of Modern Policing in New York.* Duke University Press, 2023.

Guenther, Lisa. "Police, Drones, and the Politics of Perception." In *The Ethics of Policing: New Perspectives on Law Enforcement*, edited by Ben Jones and Eduardo Mendieta, 248–267. New York University Press, 2021.

Guevarra, Anna Romina. "Techno-Modeling Care: Racial Branding, Dis/embodied Labor, and 'CybraceroS' in South Korea." *Frontiers: A Journal of Women Studies*, 36(3): 139–159, 2015.

Guizzo, Erico. "Cynthia Breazeal Unveils Jibo, a Social Robot for the Home." *IEEE Spectrum*, July 16, 2014.

Guo, Yandong, Lei Zhang, Yuxiao Hu, Xiaodong He, and Jianfeng Gao. "MS-CELEB-1M: A Dataset and Benchmark for Large-Scale Face Recognition." In *Proceedings of the European Conference on Computer Vision (ECCV)*, 87–102, Springer, 2016.

Gustavo, Solomon. "Four Takeaways from the Minneapolis DFL's Mayoral Endorsement Process." *MinnPost*, June 18, 2021.

Ha, Inyong, Yusuke Tamura, Hajime Asama, Jeakweon Han, and Dennis W Hong. "Development of Open Humanoid Platform DARwIn-OP." In *Proceedings of the SICE Annual Conference*, 2178–2181. IEEE, 2011.

Hadden, Sally E. *Slave Patrols: Law and Violence in Virginia and the Carolinas*. Harvard University Press, 2003.

Halpern, Bence Mark, Rob van Son, Michiel van den Brekel, and Odette Scharenborg. "Detecting and Analysing Spontaneous Oral Cancer Speech in the Wild." In *Proceedings of Interspeech*, 4826–4830, Curran Associates, 2020.

Ham, Jaap. "Influencing Robot Influence: Personalization of Persuasive Robots." *Interaction Studies*, 22(3): 464–487, 2021.

Ham, Jaap, René Bokhorst, Raymond Cuijpers, David Van der Pol, and John-John Cabibihan. "Making Robots Persuasive: The Influence of Combining Persuasive Strategies (Gazing and Gestures) by a Storytelling Robot on Its Persuasive Power." In *Proceedings of the International Conference on Social Robotics*, 71–83, Springer, 2011.

Hamidi, Foad, Morgan Klaus Scheuerman, and Stacy M. Branham. "Gender Recognition or Gender Reductionism? The Social Implications of Embedded Gender Recognition Systems." In *Proceedings of the CHI Conference on Human Factors in Computing Systems*, 1–13, ACM, 2018.

Hamilton, Jared, Thao Phung, Nhan Tran, and Tom Williams. "What's the Point? Tradeoffs Between Effectiveness and Social Perception When Using Mixed Reality to Enhance Gesturally Limited Robots." In *Proceedings of the 16th ACM/IEEE International Conference on Human–Robot Interaction (HRI)*, 177–186, ACM, 2021.

Han, Zhao, Yifei Zhu, Albert Phan, Fernando Sandoval Garza, Amia Castro, and Tom Williams. "Crossing Reality: Comparing Physical and Virtual Robot Deixis." In *Proceedings of the ACM/IEEE International Conference on Human–Robot Interaction*, 152–161, ACM, 2023.

Hancock, Peter A., Deborah R. Billings, Kristin E. Schaefer, Jessie Y. C. Chen, Ewart J. De Visser, and Raja Parasuraman. "A Meta-Analysis of Factors Affecting Trust in Human–Robot Interaction. *Human Factors*, 53(5): 517–527, 2011.

Hanson, Alyssa, Nichole Starr, Cloe Emnett, Ruchen Wen, Bertram Malle, and Tom Williams. "The Power of Advice: Differential Blame for Human and Robot Advisors and Deciders in a Moral Advising Context. In *Proceedings of the ACM/IEEE International Conference on Human–Robot Interaction*, 240–249, ACM, 2024.

Hanson Robotics. Bina48. 2010. https://www.hansonrobotics.com/bina48-9/.

Harcourt, Bernard E. *Illusion of Order: The False Promise of Broken Windows Policing*. Harvard University Press, 2005.

Hari, Johann. *Chasing the Scream: The First and Last Days of the War on Drugs*. Bloomsbury, 2015.

Harrington, Christina N., Radhika Garg, Amanda Woodward, and Dimitri Williams. "'It's Kind of Like Code-Switching': Black Older Adults' Experiences with a Voice Assistant for Health Information Seeking." In *Proceedings of the CHI Conference on Human Factors in Computing Systems*, 1–15, ACM, 2022.

Hartman, Saidiya V. *Scenes of Subjection: Terror, Slavery, and Self-Making in Nineteenth-Century America*. Oxford University Press, 1997.

Haskins, Caroline. "Amazon is Coaching Cops on How to Obtain Surveillance Footage Without a Warrant." *VICE*, August 5, 2019.

Haskins, Caroline. "Scars, Tattoos, and License Plates: This is What Palantir and the LAPD Know About You." *BuzzFeed News*, September 29, 2020.

Haskins, Caroline. "US Cities Are Helping People Buy Amazon Surveillance Cameras Using Taxpayer Money." *VICE*, August 2, 2019.

Hayes, K. E., and M. Kaba. *Let This Radicalize You: Organizing and the Revolution of Reciprocal Care*. Haymarket Books, 2023.

Heater, Brian. "Amazon Cites First Amendment Protection for Alexa in Arkansas Murder Case." *TechCrunch*, February 23, 2017.

Heffernan, Teresa. "Fiction Meets Science: Ex Machina, Artificial Intelligence, and the Robotics Industry." In *Cyborg Futures: Cross-Disciplinary Perspectives on Artificial Intelligence and Robotics*, edited by Teresa Heffernan, 127–140, Springer, 2019.

Heilweil, Rebecca. "Members of Congress Want to Know More About Law Enforcement's Surveillance of Protesters." *Vox*, June 10, 2020.

Heitzeg, Nancy A. *The School-to-Prison Pipeline: Education, Discipline, and Racialized Double Standards*. Bloomsbury, 2016.

Held, Virginia. *The Ethics of Care: Personal, Political, and Global*. Oxford University Press, 2006.

Heuer, Tanja, Ina Schiering, and Reinhard Gerndt. "Privacy-Centered Design for Social Robots." *Interaction Studies*, 20(3): 509–529, 2019.

Higbie, Tobias. "Why do Robots Rebel? The Labor History of a Cultural Icon." *Labor: Studies in Working-Class History of the Americas*, 10(1): 99–121, 2013.

Higgins, David M., and Matthew Iung. "Comic Books 1." In *The Routledge Companion to Cyberpunk Culture*, edited by Anna McFarlane, Graham J. Murphy, and Lars Schmeink, 91–100. Routledge, 2019.

Hillyard, Paddy. "Policing Divided Societies: Trends and Prospects in Northern Ireland and Britain." In *Policing Futures: The Police, Law Enforcement and the Twenty-First Century*, edited Peter Francis, Pamela Davies, and Victor Jupp, 163–185. Springer, 1997.

Hinson, Waymon R., and Edward Robinson. "'We Didn't Get Nothing': The Plight of Black Farmers." *Journal of African American Studies*, 12: 283–302, 2008.

Hinton, Elizabeth. *From the War on Poverty to the War on Crime: The Making of Mass Incarceration in America*. Harvard University Press, 2016.

Hipólito, Inês, Katie Winkle, and Merete Lie. "Enactive Artificial Intelligence: Subverting Gender Norms in Robot and Human Interaction." *Frontiers in Neurorobotics*, 17: 1–11, 2023.

Hirsch, Tad, and John Henry. "Txtmob: Text Messaging for Protest Swarms." In *Extended Abstracts of the CHI Conference on Human Factors in Computing Systems*, 1455–1458, ACM, 2005.

Hitron, Tom, Noa Morag Yaar, and Hadas Erel. "Implications of AI Bias in HRI: Risks (and Opportunities) When Interacting with a Biased Robot." In *Proceedings of the ACM/IEEE International Conference on Human–Robot Interaction*, 83–92, ACM, 2023.

Hoerner, Emily, and Rick Tulsky. "Cops Around the Country Are Posting Racist and Violent Comments on Facebook." *Injustice Watch*, 2019.

Hoffman, Kelly M., Sophie Trawalter, Jordan R. Axt, and M. Norman Oliver. "Racial Bias in Pain Assessment and Treatment Recommendations, and False Beliefs About Biological Differences Between Blacks and Whites." *Proceedings of the National Academy of Sciences*, 113(16): 4296–4301, 2016.

Hoffmann, Anna Lauren. "Terms of Inclusion: Data, Discourse, Violence." *New Media & Society*, 23(12): 3539–3556, 2021.

Hollister, Sean. "Google Contractors Reportedly Targeted Homeless People for Pixel 4 Facial Recognition." *The Verge*, October 3, 2019.

hooks, bell. *Feminist Theory: From Margin to Center Boston: South End*. South End Press, 1984.

hooks, bell. *Representing Whiteness in the Black Imagination*. Routledge, 1992.

hooks, bell. *Talking Back: Thinking Feminist, Thinking Black*. South End Press, 1989.

Hou, Yoyo Tsung-Yu, EunJeong Cheon, and Malte F. Jung. "Power in Human–Robot Interaction." In *Proceedings of the ACM/IEEE International Conference on Human–Robot Interaction*, 269–282, ACM, 2024.

Human Rights Watch. "UAE: Stop Policing Gender Expression." *Human Rights Watch*, September 7, 2017.

Hunt, Jennifer, and Peter K. Manning. "The Social Context of Police Lying." *Symbolic Interaction*, 14(1): 51–70, 1991.

Husband, Julie. "The White Slave of the North": Lowell Mill Women and the Reproduction of Free Labor." *Legacy*, 16(1): 11–21, 1999.

Hutchinson, Ben, Andrew Smart, Alex Hanna, Emily Denton, Christina Greer, Oddur Kjartansson, Parker Barnes, and Margaret Mitchell. "Towards Accountability for Machine Learning Datasets: Practices from Software Engineering and Infrastructure." In *Proceedings of the ACM Conference on Fairness, Accountability, and Transparency (FAccT)*, 560–575, ACM, 2021.

Ignatiev, Noel. *How the Irish Became White*. Routledge, 2012.

Ingraham, Christopher. "New Report: In Tough Times, Police Start Seizing a Lot More Stuff from People." *Washington Post*, November 10, 2015.

Ings, Simon. "Clones Are Not Us." *New Scientist*, 244(3254): 30, 2019.

Ito, Joi. "My Apology Regarding Jeffrey Epstein." 2019. https://www.media.mit.edu/posts/myapology-regarding-jeffrey-epstein/.

Jackson, Ryan Blake, Ruchen Wen, and Tom Williams. "Tact in Noncompliance: The Need for Pragmatically Apt Responses to Unethical Commands." In *Proceedings of the AAAI/ACM Conference on AI, Ethics, and Society*, 499–505, ACM, 2019.

Jackson, Ryan Blake, Sihui Li, Santosh Balajee Banisetty, Sriram Siva, Hao Zhang, Neil Dantam, and Tom Williams. "An Integrated Approach to Context-Sensitive Moral Cognition in Robot Cognitive Architectures." In *Proceedings of the IEEE/RSJ International Conference on Intelligent Robots and Systems (IROS)*, 1911–1918, IEEE, 2021.

Jackson, Ryan Blake, and Tom Williams. "A Theory of Social Agency for Human–Robot Interaction." *Frontiers in Robotics and AI*, 8: 1–15, 2021.

Jackson, Ryan Blake, and Tom Williams. "Enabling Morally Sensitive Robotic Clarification Requests." *ACM Transactions on Human–Robot Interaction (T-HRI)*, 11(2): 1–18, 2022.

Jackson, Ryan Blake, and Tom Williams. "Language-Capable Robots May Inadvertently Weaken Human Moral Norms." In *Proceedings of the 14th ACM/IEEE International Conference on Human–Robot Interaction (HRI)*, 401–410, IEEE, 2019.

Jackson, Ryan Blake, Tom Williams, and Nicole Smith. "Exploring the Role of Gender in Perceptions of Robotic Noncompliance." In *Proceedings of the ACM/IEEE International Conference on Human–Robot Interaction*, 559–567, ACM, 2020.

Janssens, Ruben, Eva Verhelst, Giulio Antonio Abbo, Qiaoqiao Ren, Maria Jose Pinto Bernal, and Tony Belpaeme. "Child Speech Recognition in Human–Robot Interaction: Problem Solved?" In *Proceedings of the International Symposium on Technological Advances in Human–Robot Interaction (TAHRI)*, 2024.

Jany, Libor. "LAPD Doesn't Fully Track Its Use of Facial Recognition, Report Finds." *Los Angeles Times*, December 14, 2022.

Jeffrey-Wilensky, Jaclyn, and David Freeman. "This Police Robot Could Make Traffic Stops Safer." *NBCUniversal News Group*, May 9, 2019.

Jennings, Thelma. "'Us Colored Women Had to Go Through a Plenty': Sexual Exploitation of African-American Slave Women." *Journal of Women's History*, 1(3): 45–74, 1990.

Joh, Elizabeth E. "Policing Police Robots." *UCLA Law Review Discourse*, 64: 516–543, 2016.

Johnson, Bob. "Energy Slaves: Carbon Technologies, Climate Change, and the Stratified History of the Fossil Economy." *American Quarterly*, 68(4): 955–979, 2016.

Johnson, Brian E. *The Role Ethics of Epictetus: Stoicism in Ordinary Life*. Lexington Books, 2013.

Johnson, Khari. "ExamSoft's Remote Bar Exam Sparks Privacy and Facial Recognition Concerns." *VentureBeat*, September 29, 2020.

Johnson, Vida B. "KKK in the PD: White Supremacist Police and What to Do About It." *Lewis and Clark Law Review*, 23(1): 205–261, 2019.

Johnston, Derek. "Genre, Taste and the BBC: The Origins of British Television Science Fiction." PhD thesis, University of East Anglia, 2009.

Jois, Himavath, and Alan R. Wagner. "What Happens When Robots Punish? Evaluating Human Task Performance During Robot-Initiated Punishment." *ACM Transactions on Human–Robot Interaction (T-HRI)*, 10(4): 1–18, 2021.

Jones, Ben, and Eduardo Mendieta, eds. "Introduction: Police Ethics After Ferguson." In *The Ethics of Policing*, 1–22. New York University Press, 2021.

Jonze, Spike. "Her." 2014.

Joshi, Swapna, and Selma Šabanović. "A Tactical Urbanist Approach to Facilitate Exploratory HRI Research in Public Spaces." In *Proceedings of the 17th European Conference on Computer-Supported Cooperative Work-Demos and Posters*. European Society for Socially Embedded Technologies (EUSSET), 2019.

Joshi, Swapna, and Selma Šabanović. "Pop-up Robotics: Facilitating HRI in Public Spaces." In *Adjunct Proceedings of the 31st Annual ACM Symposium on User Interface Software and Technology*, 131–133, ACM, 2018.

Kaba, Mariame. *We Do This 'til We Free Us: Abolitionist Organizing and Transforming Justice*. Haymarket Books, 2021.

Kaba, Mariame. *Yes, We Mean Literally Abolish the Police*. Abolition Library Commons, 2020.

Kaba, Mariame, and Andrea J. Ritchie. *No More Police: A Case for Abolition*. New Press, 2022.

Kahn, Peter H. Jr., Takayuki Kanda, Hiroshi Ishiguro, Brian T. Gill, Solace Shen, Heather E. Gary, and Jolina H. Ruckert. "Will People Keep the Secret of a Humanoid Robot? Psychological Intimacy in HRI." In *Proceedings of the Tenth Annual ACM/IEEE International Conference on Human–Robot Interaction*, 173–180, ACM, 2015.

Kamishima, Toshihiro, Shotaro Akaho, Hideki Asoh, and Jun Sakuma. "Fairness-Aware Classifier with Prejudice Remover Regularizer." In *Proceedings of the European Conference on Machine Learning and Knowledge Discovery in Databases*, 35–50. Springer, 2012.

Kant, Immanuel. *Observations on the Feeling of the Beautiful and Sublime.* Translated by John T. Goldthwait, University of California Press, 1960 [1764].

Kant, Immanuel. *Of the Different Human Races, 1777.* Translated by Jon Mark Mikkelsen, Albany: SUNY Press, 2014.

Kantayya, Shalini, and Joy Buolamwini. "Coded Bias." 2020.

Kanter, Rosabeth. *Men and Women of the Corporation.* Basic Books, 1977.

Kasirzadeh, Atoosa. "Algorithmic Fairness and Structural Injustice: Insights from Feminist Political Philosophy." In *Proceedings of the 2022 AAAI/ACM Conference on AI, Ethics, and Society*, 349–356, ACM, 2022.

Kasson, John F. *Civilizing the Machine: Technology and Republican Values in America, 1776–1900.* Macmillan, 1999.

Katz, Yarden. *Artificial Whiteness: Politics and Ideology in Artificial Intelligence.* Columbia University Press, 2020.

Keay, Andrea. "The Naming of Robots: Biomorphism, Gender and Identity." PhD thesis, Master thesis in digital cultures, University of Sydney, 2012.

Kendall, Linus, Bidisha Chaudhuri, and Apoorva Bhalla. "Understanding Technology as Situated Practice: Everyday Use of Voice User Interfaces Among Diverse Groups of Users in Urban India." *Information Systems Frontiers*, 22: 585–605, 2020.

Kensing, Finn. *The Trade Unions Influence on Technological Change.* North Holland, 1983.

Keyes, Os. "The Misgendering Machines: Trans/HCI Implications of Automatic Gender Recognition." In *Proceedings of the ACM on Human–Computer Interaction*, 2(CSCW), 1–22, ACM, 2018.

Keyes, Os, Josephine Hoy, and Margaret Drouhard. "Human–Computer Insurrection: Notes on an Anarchist HCI." In *Proceedings of the CHI Conference on Human Factors in Computing Systems*, 1–13, ACM, 2019.

Keysar, Hagit. "A Spatial Testimony: The Politics of Do-it-Yourself Aerial Photography in East Jerusalem." *Environment and Planning D: Society and Space*, 37(3): 523–541, 2019.

Kidane, Matyos. "The LAPD Wants Robot Dogs. How Did We Get Here?" *VICE*, February 2, 2023.

Kilgore, Ed. "Tax Dollars Are Now Funding Christian-Nationalist Schools." *Intelligencer*, June 2024.

Kim, Boyoung, Ruchen Wen, Qin Zhu, Tom Williams, and Beth Phillips. "Can Robot Advisers Encourage Honesty? Considering the Impact of Rule, Identity, and Role-Based Moral Advice." *International Journal of Human–Computer Studies*, 184: 1–16, 2024.

Kim, Boyoung, Ruchen Wen, Qin Zhu, Tom Williams, and Elizabeth Phillips. "The Impact of Different Ethical Frameworks Underlying a Robot's Advice on Charitable Donations." In *Proceedings of the IEEE International Conference on Robot and Human Interactive Communication (RO-MAN)*, 2433–2438, IEEE, 2023.

Kim, Eugenia, De'Aira Bryant, Deepak Srikanth, and Ayanna Howard. "Age Bias in Emotion Detection: An Analysis of Facial Emotion Recognition Performance on Young, Middle-Aged, and Older Adults." In *Proceedings of the AAAI/ACM Conference on AI, Ethics, and Society*, 638–644, ACM, 2021.

Kim, Yanghee, Sherry Marx, Hung Viet Pham, and Tung Nguyen. "Designing for Robot-Mediated Interaction Among Culturally and Linguistically Diverse Children." *Educational Technology Research and Development*, 69: 3233–3254, 2021.

King, Martin Luther Jr. "Illinois Wesleyan University Student Senate Convocation." 1966. https://www.iwu.edu/mlk/page-3.html

Kingma, Mireille. "Workplace Violence in the Health Sector: A Problem of Epidemic Proportion." *International Nursing Review*, 48(3): 129–130, 2001.

Klare, Brendan F., Mark J. Burge, Joshua C. Klontz, Richard W. Vorder Bruegge, and Anil K. Jain. "Face Recognition Performance: Role of Demographic Information." *IEEE Transactions on Information Forensics and Security*, 7(6): 1789–1801, 2012.

Kline, Ronald R. *The Cybernetics Moment: Or Why We Call Our Age the Information Age*. Johns Hopkins University Press, 2015.

Klockars, Carl B. "Blue Lies and Police Placebos: The Moralities of Police Lying." *American Behavioral Scientist*, 27(4): 529–544, 1984.

Knepper, Lisa, Jennifer McDonald, Kathy Sanchez, and Elyse Smith Pohl. *Policing for Profit: The Abuse of Civil Asset Forfeiture*, 3rd ed. Institute for Justice, 2020.

Knibbs, Kate. "The Future of Police Robots Began Last Year—Where Is It Now?" *The Ringer*, July 28, 2017.

Ko, Amy J., Anne Beitlers, Brett Wortzman, Matt Davidson, Alannah Oleson, Mara Kirdani-Ryan, Stefania Druga, and Jayne Everson. "Critically Conscious Computing: Methods for Secondary Education." 2022. https://criticallyconsciouscomputing.org.

Koenecke, Allison, Andrew Nam, Emily Lake, Joe Nudell, Minnie Quartey, Zion Mengesha, Connor Toups, John R. Rickford, Dan Jurafsky, and Sharad Goel. "Racial Disparities in Automated Speech Recognition." *Proceedings of the National Academy of Sciences*, 117(14): 7684–7689, 2020.

Korsgaard, Christine M. "The Reasons We Can Share: An Attack on the Distinction Between Agent-Relative and Agent-Neutral Values." *Social Philosophy and Policy*, 10(1): 24–51, 1993.

Koskela, Hille. "'Cam era'—The Contemporary Urban Panopticon." *Surveillance & Society*, 1(3): 292–313, 2003.

Kovacs, Anja. "Reading Surveillance Through a Gendered Lens: Some Theory." *Gendering Surveillance*, February 2017.

Krasanakis, Emmanouil, Eleftherios Spyromitros-Xioufis, Symeon Papadopoulos, and Yiannis Kompatsiaris. "Adaptive Sensitive Reweighting to Mitigate Bias in Fairness-Aware Classification." In *Proceedings of the 2018 World Wide Web Conference (WWW)*, 853–862, ACM, 2018.

Kraska, Peter B. "Militarization and Policing—Its Relevance to 21st Century Police." *Policing: A Journal of Policy and Practice*, 1(4): 501–513, 2007.

Kratz, Robert, Maximilian Stelzer, Martin Friedmann, and Oskar von Stryk. "Control Approach for a Novel High Power-to-Weight Robotic Actuator Scalable in Force and Length." In *2007 IEEE/ASME International Conference on Advanced Intelligent Mechatronics*. IEEE, 2007.

KUSA Staff. "One Year Later: Signs of PTSD, Mental Illness; Search for Treatments." *9News*, July 7, 2017.

Kyriakou, Kyriakos, Styliani Kleanthous, Jahna Otterbacher, and George A. Papadopoulos. "Emotion-Based Stereotypes in Image Analysis Services." In *Adjunct Publication of the 28th ACM Conference on User Modeling, Adaptation and Personalization*, 252–259, ACM, 2020.

Lacey, Cherie, and Catherine Caudwell. "Cuteness as a 'dark pattern' in Home Robots." In *Proceedings of the 14th ACM/IEEE International Conference on Human–Robot Interaction (HRI)*, 374–381, ACM, 2019.

Lakoff, Robin Tolmach. *Language and Woman's Place: Text and Commentaries*. Vol. 3. Oxford University Press, 2004.

Lam, Khang Nhut, Trong Thanh La, Khang Duy Nguyen, Man Minh Le, Vy Trieu Truong, and Andrew Ware. "Monitoring Attendance and Checking School Uniforms Using Yolov8." In *International Conference on Intelligent Systems and Data Science*, 200–207. Springer, 2023.

Lambdin, Jennifer R., Kristen M. Greer, Kari Selby Jibotian, Kelly Rice Wood, and Mykol C. Hamilton. "The Animal = Male Hypothesis: Children's and Adults' Beliefs About the Sex of Non-Sex-Specific Stuffed Animals." *Sex Roles*, 48: 471–482, 2003.

Lane, Roger. *Policing the City: Boston, 1822–1885*. Harvard University Press, 2013.

Lang, Fritz. "Metropolis." 1973.

Laperruque, Jake. "Preventing an Air Panopticon: A Proposal for Reasonable Legal Restrictions on Aerial Surveillance. *University of Richmond Law Review*, 51(3): 705–726, 2016.

Lapidus, Lenora, Namita Luthra, Anjuli Verma, Deborah Small, Patricia Allard, and Kirsten Levingston. "Caught in the Net: The Impact of Drug Policies on Women and Families." Technical report. ACLU, 2005.

Larson, Jeff, Surya Mattu, Lauren Kirchner, and Julia Angwin. "How We Analyzed the COMPAS Recidivism Algorithm." *ProPublica*, May 23, 2016.

Larsson, Stefan, Mia Liinason, Laetitia Tanqueray, and Ginevra Castellano. "Towards a Sociolegal Robotics: A Theoretical Framework on Norms and Adaptive Technologies. *International Journal of Social Robotics*, 15: 1755–1768, 2023.

Lassiter, Matthew D. "Impossible Criminals: The Suburban Imperatives of America's War on Drugs." *Journal of American History*, 102(1): 126–140, 2015.

Lau, Dim Cheuk. *Mencius*. Chinese University Press, 2003.

Lawson, Edward Jr. "Trends: Police Militarization and the Use of Lethal Force." *Political Research Quarterly*, 72(1): 177–189, 2019.

Le Bui, Matthew, and Safiya Umoja Noble. "We're Missing a Moral Framework of Justice in Artificial Intelligence." *The Oxford Handbook of Ethics of AI*, edited by Markus D. Dubber, Frank Pasquale, and Sunit Das, 537–553. Oxford Academic, 2020.

Le Guin, Ursula K. *The Ones Who Walk Away from Omelas*. Harper Perennial, 2017.

Lee, Hak-Chong. "The Lordstown Plant of General Motors." In *Folder: Histories—1974–1975—Lordstown historical documents, box 3, Lordstown collection*. General Motors Heritage Center, 1974.

Lee, Min Kyung, Sara Kiesler, Jodi Forlizzi, and Paul Rybski. "Ripple Effects of an Embedded Social Agent: A Field Study of a Social Robot in the Workplace." In *Proceedings of the SIGCHI Conference on Human Factors in Computing Systems*, 695–704, ACM, 2012.

Lee, Sharon M. "Racial Classifications in the US Census: 1890–1990." *Ethnic and Racial Studies*, 16(1): 75–94, 1993.

Leenes, Ronald, Erica Palmerini, Bert-Jaap Koops, Andrea Bertolini, Pericle Salvini, and Federica Lucivero. "Regulatory Challenges of Robotics: Some Guidelines for Addressing Legal and Ethical Issues." *Law, Innovation and Technology*, 9(1): 1–44, 2017.

Lemburg, Johannes, José de Gea Fernández, Markus Eich, Dennis Mronga, Peter Kampmann, Andreas Vogt, Achint Aggarwal, Yuping Shi, and Frank Kirchner. "AILA-Design of an Autonomous Mobile Dual-Arm Robot." In *Proceedings of the IEEE International Conference on Robotics and Automation (ICRA)*, 5147–5153. IEEE, 2011.

Lemons, J. Stanley. "Black Stereotypes as Reflected in Popular Culture, 1880–1920." *American Quarterly*, 29(1): 102–116, 1977.

Leonardo, Zeus. "The Color of Supremacy: Beyond the Discourse of 'White Privilege.'" *Educational Philosophy and Theory*, 36(2): 137–152, 2004.

Levenson, Michael. "Officers Who Handcuffed Black Family Won't Be Charged, Prosecutors Say." *New York Times*, January 8, 2021.

Levine, H. G., and D. Small. *Marijuana Arrest Crusade Racial Bias and Police Policy in New York City, 1997–2007*. New York Civil Liberties Union, 2008.

Lewis, Patricia, and Ruth Simpson. "Introduction: Theoretical Insights into the Practices of Revealing and Concealing Gender Within Organizations." In *Revealing and Concealing Gender: Issues of Visibility in Organizations*, edited by Patricia Lewis and Ruth Simpson, 1–22. Springer, 2010.

Lewis, Patricia, and Ruth Simpson. "Kanter Revisited: Gender, Power and (In)visibility." *International Journal of Management Reviews*, 14(2): 141–158, 2012.

Ley, Madelaine. "Care Ethics and the Future of Work: A Different Voice." *Philosophy & Technology*, 36(1): 7, 2023.

Leydens, Jon A., and Juan C. Lucena. *Engineering Justice: Transforming Engineering Education and Practice*. John Wiley, 2017.

Liao, Yuting, and Jiangen He. "Racial Mirroring Effects on Human–Agent Interaction in Psychotherapeutic Conversations." In *Proceedings of the 25th International Conference on Intelligent User Interfaces*, 430–442, ACM, 2020.

Libbey, Peter. "John W. Campbell award is Renamed." *New York Times*, August 28, 2019.

Lind, Roger. "The Robots Are Coming, the Robots Are Coming." *Oui* 3(2): 40, 1974.

Linder, Timm, Sven Wehner, and Kai O. Arras. "Real-Time Full-Body Human Gender Recognition in (RGB)-D Data." In *Proceedings of the IEEE International Conference on Robotics and Automation (ICRA)*, 3039–3045, IEEE, 2015.

Linnemann, Travis. "Proof of Death: Police Power and the Visual Economies of Seizure, Accumulation and Trophy." *Theoretical Criminology*, 21(1): 57–77, 2017.

Lipman, Francine J. "Tax Audits, Economics, and Racism." In *Oxford Research Encyclopedia of Economics and Finance*. 2022. https://doi.org/10.1093/acrefore/9780190625979.013.759

Liu, JeeLoo. "Confucian Robotic Ethics." In *Proceedings of the International Conference on the Relevance of the Classics under the Conditions of Modernity: Humanity and Science*, 2017.

Louizos, Christos, Kevin Swersky, Yujia Li, Max Welling, and Richard Zemel. "The Variational Fair Autoencoder." In *Proceedings of the International Conference on Learning Representations*, Curran Associates, 2016.

Lowe, Lisa. "The Dialectics of Abolition." *American Quarterly*, 75(2): 371–376, 2023.

Lowe, Roy. "The Educational Impact of the Eugenics Movement." *International Journal of Educational Research*, 27(8): 647–660, 1998.

Lu Jiahong, Alex, Gabriela Marcu, Mark S. Ackerman, and Tawanna R. Dillahunt. "Coding Bias in the Use of Behavior Management Technologies: Uncovering Socio-Technical Consequences of Data-Driven Surveillance in Classrooms." In *Designing Interactive Systems Conference 2021*, 508–522, ACM, 2021.

Lucy, Li, and David Bamman. "Gender and Representation Bias in GPT-3 Generated Stories." In *Proceedings of the Third Workshop on Narrative Understanding*, 48–55, ACL, 2021.

Lum, Kristian, Yunfeng Zhang, and Amanda Bower. "De-Biasing 'Bias' Measurement." In *Proceedings of the ACM Conference on Fairness, Accountability, and Transparency (FAccT)*, 379–389, ACM, 2022.

Lundberg, Carl, and Henrik I. Christensen. "Assessment of Man-Portable Robots for Law Enforcement Agencies." In *Proceedings of the Workshop on Performance Metrics for Intelligent Systems*, 76–83, ACM, 2007.

Lynch, Casey R. "Glitch Epistemology and the Question of (Artificial) Intelligence: Perceptions, Encounters, Subjectivities." *Dialogues in Human Geography*, 12(3): 379–383, 2022.

Lyon, David. *Surveillance as Social Sorting: Privacy, Risk, and Digital Discrimination*. Psychology Press, 2003.

Maass, Dave, and Matthew Guariglia. *San Francisco Police Accessed Business District Camera Network to Spy on Protestors*. Electronic Frontier Foundation, 2020.

Maharawal, Manissa M. "Black Lives Matter, Gentrification and the Security State in the San Francisco Bay Area." *Anthropological Theory*, 17(3): 338–364, 2017.

Maldonado-Torres, Nelson. "Race, Religion, and Ethics in the Modern/Colonial World." *Journal of Religious Ethics*, 42(4): 691–711, 2014.

Malle, Bertram F., and Matthias Scheutz. "Moral Competence in Social Robots." In *Machine Ethics and Robot Ethics*, edited by Wendell Wallach and Peter Asaro, 225–230. Routledge, 2020.

Malle, Bertram F., Matthias Scheutz, and Joseph L. Austerweil. "Networks of Social and Moral Norms in Human and Robot Agents." In *Proceedings of the International Conference on Robot Ethics*, 3–17, CLAWAR, 2017.

Mann, Steve, Jason Nolan, and Barry Wellman. "Sousveillance: Inventing and Using Wearable Computing Devices for Data Collection in Surveillance Environments." *Surveillance & Society*, 1(3): 331–355, 2003.

Marcus, Gary, Francesca Rossi, and Manuela Veloso. "Beyond the Turing test." *AI Magazine*, 37(1): 3–4, 2016.

Marino, Mark C. "The Racial Formation of Chatbots." *CLCWeb: Comparative Literature and Culture*, 16(5):1481–4374, 2014.

Martin, Joshua L., and Kevin Tang. "Understanding Racial Disparities in Automatic Speech Recognition: The Case of Habitual 'Be.'" In *Proceedings of Interspeech*, 626–630, Curran Associates, 2020.

Martinez, Gina. "'Come on Guys, He's Right There!': Moment Bumbling LAPD Cops Are Seen in News Helicopter Footage Searching for Suspect Hiding Right in Front of Them as Bemused News Anchor Watches in Disbelief." *Daily Mail*, March 11, 2022.

Martinot, Steve. "On the Epidemic of Police Killings." *Social Justice*, 39(4): 52–75, 2012.

Mauer, Marc. *Race to Incarcerate*. New Press, 2006.

Mbiti, John S. *African Religions and Philosophy*. Heinemann, 1990.

McCluney, Courtney L., Kathrina Robotham, Serenity Lee, Richard Smith, and Myles Durkee. "The Costs of Code-Switching." *Harvard Business Review*, 15, 2019.

McComas, William F. "Benchmarks for Science Literacy." In *The Language of Science Education: An Expanded Glossary of Key Terms and Concepts in Science Teaching and Learning*, 12–12. Springer, 2014.

McCulloch, Warren S., and Walter Pitts. "A Logical Calculus of the Ideas Immanent in Nervous Activity." *The Bulletin of Mathematical Biophysics*, 5: 115–133, 1943.

McCurry, Justin. "Erica, the 'Most Beautiful and Intelligent' Android, Leads Japan's Robot Revolution." *The Guardian*, December 31, 2015.

McGlynn, Sean A., Denise Geiskkovitch, Tracy L. Mitzner, and Wendy A. Rogers. "Paro's Stress-Reduction Potential for Older Adults." In *Proceedings of the Human Factors and Ergonomics Society Annual Meeting*, 1799–1803, Sage, 2016.

McGreevy, Patrick. "LAPD Automatic License Plate Readers Pose a Massive Privacy Risk, Audit Says." *Los Angeles Times*, February 13, 2020.

McHarris, Philip V. "Disrupting Order: Race, Class, and the Roots of Policing." In *Violent Order: Essays on the Nature of Police*, edited by David Correia and Tyler Wall, 31–54 Haymarket Books, 2021.

McIlwain, Charlton D. *Black Software: The Internet and Racial Justice, from the AfroNet to Black Lives Matter*. Oxford University Press, 2019.

McLemore, Kevin A. "Experiences with Misgendering: Identity Misclassification of Transgender Spectrum Individuals." *Self and Identity*, 14(1): 51–74, 2015.

McLeod, Allegra M. "Prison Abolition and Grounded Justice." *UCLA Law Review*, 62: 1156, 2015.

McNeely, Clea A., Besufekad Alemu, Won Fy Lee, and Isis West. "Exploring an Unexamined Source of Racial Disparities in Juvenile Court Involvement: Unexcused Absenteeism Policies in US Schools." *AERA Open*, 7, 2021.

McQuillan, Dan. *Resisting AI: An Anti-Fascist Approach to Artificial Intelligence*. Bristol University Press, 2022.

McRae, Elizabeth Gillespie. *Mothers of Massive Resistance: White Women and the Politics of White Supremacy*. Oxford University Press, 2018.

Meares, Tracey L. "Clashing Narratives of Policing?" In *The Ethics of Policing: New Perspectives on Law Enforcement*, edited by Ben Jones and Eduardo Mendieta, 25–38. New York University Press, 2021.

Mehrabi, Ninareh, Fred Morstatter, Nripsuta Saxena, Kristina Lerman, and Aram Galstyan. "A Survey on Bias and Fairness in Machine Learning." *ACM Computing Surveys (CSUR)*, 54(6): 1–35, 2021.

Mehrotra, Dhruv, and Jesse Marx. "The Age of the Drone Police Is Here." *Wired*, June 5, 2024.

Mengesha, Zion, Courtney Heldreth, Michal Lahav, Juliana Sublewski, and Elyse Tuennerman. "'I Don't Think These Devices Are Very Culturally Sensitive.'—Impact of Automated Speech Recognition Errors on African Americans." *Frontiers in Artificial Intelligence*, 4:1–12, 2021.

Menon, Aditya Krishna, and Robert C. Williamson. "The Cost of Fairness in Binary Classification." In *Proceedings of the ACM Conference on Fairness, Accountability and Transparency (FAccT)*, 107–118, ACM, 2018.

Metaxa, Danaë, Joon Sung Park, Ronald E. Robertson, Karrie Karahalios, Christo Wilson, Jeff Hancock, Christian Sandvig, et al. "Auditing Algorithms: Understanding Algorithmic Systems from the Outside in." *Foundations and Trends in Human–Computer Interaction*, 14(4): 272–344, 2021.

Mian, Zahra N. "Black Identity Extremist" or Black Dissident?: How United States v. Daniels Illustrates FBI Criminalization of Black Dissent of Law Enforcement, from COINTELPRO to Black Lives Matter." *Rutgers Race and Law Review*, 21: 53, 2020.

Mills, Charles W. *Black Rights/White Wrongs: The Critique of Racial Liberalism*. Oxford University Press, 2017.

Mills, Charles W. Kant's untermenschen. In *Race and Racism in Modern Philosophy*, edited by Andrew Valls, 169–193. Cornell University Press, 2005.

Mills, Charles W. *The Racial Contract*. Cornell University Press, 2014.

Mills, Sara. Gender and impoliteness. *Journal of Politeness Research*, 1(2): 263–280, 2005.

Mills, Sara. *Gender and Politeness*. Cambridge University Press, 2003.

Minsky, Marvin, and Doug Riecken. "A Conversation with Marvin Minsky About Agents." *Communications of the ACM*, 37(7): 22–29, 1994.

Miranda, Lux, Ginevra Castellano, and Katie Winkle. "A Case for Diverse Social Robot Identity Performance in Education." In *Companion of the 2024 ACM/IEEE International Conference on Human–Robot Interaction*, 28–35, ACM, 2024.

Mitchell, Shira, Eric Potash, Solon Barocas, Alexander D'Amour, and Kristian Lum. "Algorithmic Fairness: Choices, Assumptions, and Definitions." *Annual Review of Statistics and Its Application*, 8: 141–163, 2021.

Moberg, David F. "Rattling the Golden Chains: Conflict and Consciousness of Auto Workers." PhD thesis, University of Chicago, 1978.

Monaghan, Jake. "Legitimate Policing and Professional Norms." In *The Ethics of Policing: New Perspectives on Law Enforcement*, edited by Ben Jones and Eduardo Mendieta, 39–65. New York University Press, 2021.

Moniuszko, Sara. "It Remains Purely Symbolic": Why Rainbow Capitalism Can Be Harmful to LGBTQ People." *USA Today*, June 7, 2021.

Moor, James H. "The Nature, Importance, and Difficulty of Machine Ethics." *IEEE Intelligent Systems*, 21(4): 18–21, 2006.

Moore, Ronald D. Battlestar Galactica, 2004.

Mori, Masahiro, Karl F. MacDorman, and Norri Kageki. "The Uncanny Valley [From the Field]." *IEEE Robotics & Automation Magazine*, 19(2): 98–100, 2012.

Morin, Roc. "Can Child Dolls Keep Pedophiles from Offending." *The Atlantic*, January 11, 2016.

Morris, Nyomi, and Williams, Tom, and Ben Jelen. "Trauma-Informed Insights from Co-Design of Self-Disclosure Robots with Domestic Abuse Survivors." In *Proceedings of the ACM/IEEE International Conference on Human–Robot Interaction (HRI)*, IEEE, 2025.

Mott, Terran, and Tom Williams. "Community Futures with Morally Capable Robotic Technology." In *HRI Workshop on Perspectives on Moral Agency in Human–Robot Interaction*, 2023.

Mott, Terran, and Tom Williams. "Confrontation and Cultivation: Understanding Perspectives on Robot Responses to Norm Violations. In *32nd IEEE International Conference on Robot and Human Interactive Communication (RO-MAN)*, 2336–2343, IEEE, 2023.

Mott, Terran, Mark Higger, Alexandra Bejarano, and Tom Williams. "Degrees of Freedom: A Storytelling Game That Supports Technology Literacy About Social Robots." In *Proceedings of the IEEE International Conference on Robot and Human Interactive Communication (RO-MAN)*, 2095–2102, IEEE, 2024.

Mowatt, Rasul A., Bryana H. French, and Dominique A. Malebranche. "Black/Female/Body Hypervisibility and Invisibility: A Black Feminist Augmentation of Feminist Leisure Research." *Journal of Leisure Research*, 45(5): 644–660, 2013.

Mubin, Omar, Catherine J. Stevens, Suleman Shahid, Abdullah Al Mahmud, and Jian-Jie Dong. A Review of the Applicability of Robots in Education." *Journal of Technology in Education and Learning*, 1: 1–7, 2013.

Muldoon, Orla T., Robert D. Lowe, Jolanda Jetten, Tegan Cruwys, and S. Alexander Haslam. "Personal and Political: Post-Traumatic Stress Through the Lens of Social Identity, Power, and Politics." *Political Psychology*, 42(3): 501–533, 2021.

Mummolo, Jonathan. "Militarization Fails to Enhance Police Safety or Reduce Crime But May Harm Police Reputation." *Proceedings of the National Academy of Sciences*, 115(37): 9181–9186, 2018.

Munn, Luke. "Seeing with Software: Palantir and the Regulation of Life." *Studies in Control Societies* 2(1): 1–17, 2017.

Murphy, Mike. "Amazon's Ring May Be Branching out Beyond Outdoor Cameras." *Quartz*, 2019.

Nagasawa, Tadashi. "The Reception of American Science Fiction in Japan." In *Oxford research encyclopedia of literature*. 2016. https://doi.org/10.1093/acrefore/9780190201098.013.166

Najibi, Alex. Racial Discrimination in Face Recognition Technology. *Harvard University Science in the News*, October 24, 2020.

Nakamura, Lisa. "Race in/for Cyberspace: Identity Tourism and Racial Passing on the Internet." *Works and Days*, 13(1–2): 181–193, 1995.

Nanavati, Thiems. *Biometrics*. John Wiley, 2002.

Natale, Simone. *Deceitful Media: Artificial Intelligence and Social Life After the Turing Test*. Oxford University Press, 2021.

National Science Foundation. "Doctorate Recipients from US Universities: 2017." Technical report. Special Report NSF 19–301, 2018.

Nelson, Leah, Frank Knaack, and Will Tucker. "Alabama's War on Marijuana." Technical report. Alabama Appleseed Center for Law & Justice and the Southern Poverty Law Center, 2018.

Neto, Isabel, Filipa Correia, Filipa Rocha, Patricia Piedade, Ana Paiva, and Hugo Nicolau. "The Robot Made Us Hear Each Other: Fostering Inclusive Conversations Among Mixed-Visual Ability Children." In *Proceedings of the ACM/IEEE International Conference on Human–Robot Interaction*, 13–23, ACM, 2023.

Newell, Allen, John C. Shaw, and Herbert A. Simon. Report on a General Problem Solving Program. In *Proceedings of the International Conference on Information Processing*, 256–264, RAND Corporation, 1959.

Newton, Veronica A. Hypervisibility and Invisibility: Black Women's Experiences with Gendered Racial Microaggressions on a White Campus. *Sociology of Race and Ethnicity*, 9(2): 164–178, 2023.

Ng, Nyx L., Dillon M. Luke, and Bertram Gawronski. "Thinking About Reasons for One's Choices Increases Sensitivity to Moral Norms in Moral-Dilemma Judgments." *Personality and Social Psychology Bulletin*, 51(1): 33–48, 2024.

Ngueajio, Mikel K., and Gloria Washington. "Hey ASR System! Why Aren't You More Inclusive? Automatic Speech Recognition Systems' Bias and Proposed Bias Mitigation Techniques. A Literature Review." In *Proceedings of the International Conference on Human–Computer Interaction*, 421–440, Springer, 2022.

Nias, Jaye, Lelia Hampton, Princess Sampson, and Margie Ruffin. "Decolonizing Technologies for Preserving Cultural and Societal Diversity." In *Proceedings of the CHI 2020 Workshop on Engaging in Race in HCI*, 2020.

Nickerson, Raymond S. "On Conversational Interaction with Computers." In *Proceedings of the ACM/SIGGRAPH Workshop on User-oriented Design of Interactive Graphics Systems*, 101–113, ACM, 1976.

Nilsson, Nils J. "Shakey the Robot." Technical report. SRI International Menlo Park, California, 1984.

Nilsson, Nils J. *The Quest for Artificial Intelligence*. Cambridge University Press, 2009.

Nishio, Shuichi, Hiroshi Ishiguro, and Norihiro Hagita. "Geminoid: Teleoperated Android of an Existing Person." *Humanoid Robots: New Developments*, 14(343–352): 10–1109, 2007.

Nolan, Jonathan, and Lisa Joy. "Westworld," 2016.

Noone, Kristin. "Rossum's Universal Robots: Karel Capek Meets Joss Whedon in the Dollhouse." In *Inside Joss' Dollhouse: From Alpha to Rossum*, edited by Jane Espenson, 21–32. Smart Pop, 2011.

Nwabara, Olaocha Nwadiuto. "Becoming Black and African: Nigerian Diasporic Transformations of Racial and Ethnic Identities in the United States." *Africology: The Journal of Pan African Studies*, 12(1): 70–96, 2018.

O'Brien, Fitz James. "The Wondersmith." *The Atlantic Monthly*. 1859

Ogunyale, Tobi, De'Aira Bryant, and Ayanna Howard. "Does Removing Stereotype Priming Remove Bias? A Pilot Human-Robot Interaction Study." In 5th Workshop on Fairness, Accountability, and Transparency in Machine Learning, 2018.

Oleksy, Tomasz, and Anna Wnuk. "Do Women Perceive Sex Robots as Threatening? The Role of Political Views and Presenting the Robot as a Female-vs Male-Friendly Product." *Computers in Human Behavior*, 117: 1–9, 2021.

Omi, Michael, and Howard Winant. "On the Theoretical Status of the Concept of Race." In *Race, Identity and Representation in Education*, edited by Cameron McCarthy and Warren Crichlow, 3–10. Routledge, 1993.

Omiye, Jesutofunmi A., Jenna C. Lester, Simon Spichak, Veronica Rotemberg, and Roxana Daneshjou. "Large Language Models Propagate Race-Based Medicine." *npj Digital Medicine*, 6 (195), 2023.

O'Neil, Cathy. *Weapons of Math Destruction: How Big Data Increases Inequality and Threatens Democracy*. Crown, 2017.

Orecchio-Egresitz, Haven. "A Defense Attorney Started Twitter Thread Documenting Police Using Force on Protesters and Media. There are More Than 400 Posts." *Business Insider*, June 8, 2020.

Ostrowski, Anastasia K., Raechel Walker, Madhurima Das, Maria Yang, Cynthia Breazea, Hae Won Park, and Aditi Verma. "Ethics, Equity, and Justice in Human–Robot Interaction: A Review and Future Directions." In *Proceedings of the 31st IEEE International Conference on Robot and Human Interactive Communication (RO-MAN)*, 969–976, IEEE, 2022.

Otis, Ginger Adams, and Nancy Dillon. "Google Using Dubious Tactics to Target People with 'Darker Skin' in Facial Recognition Project: Sources." *New York Daily News*, October 3, 2019.

Palanica, Adam, Anirudh Thommandram, Andrew Lee, Michael Li, and Yan Fossat. "Do You Understand the Words That Are Comin Outta My Mouth? Voice Assistant Comprehension of Medication Names." *NPJ Digital Medicine*, 2(1): 1–6, 2019.

Pandey, Amit Kumar, and Rodolphe Gelin. "A Mass-Produced Sociable Humanoid Robot: Pepper: The First Machine of Its Kind." *IEEE Robotics & Automation Magazine*, 25(3): 40–48, 2018.

Pannier, Christopher, C. Berry, Melissa Morris, and Xiaopeng Zhao. "Diversity and Inclusion in Mechatronics and Robotics Engineering Education." In *ASEE Annual Conference Exposition Proceedings*, ASEE, 2020.

Panzarino, Matthew. "Apple Adds Two Brand New Siri Voices and Will no Longer Default to a Female or Male Voice in IOS." *Tech Crunch*, March 31, 2021.

Parent, Anthony S. *Foul Means: The Formation of a Slave Society in Virginia, 1660–1740*. University of North Carolina Press, 2003.

Parks Associates. "Parks Associates: 25% of U.S. Broadband Households Plan to Buy a Smart Video Doorbell in 2019." *PR Newswire*, January 3, 2019.

Parreira, Maria Teresa, Sarah Gillet, Katie Winkle, and Iolanda Leite. "How Did We Miss This? A Case Study on Unintended Biases in Robot Social Behavior." In *Proceedings of the ACM/IEEE International Conference on Human–Robot Interaction*, 11–20, ACM, 2023.

Patil, Vrushali, and Bandana Purkayastha. "The Transnational Assemblage of Indian Rape Culture." *Ethnic and Racial Studies*, 41(11): 1952–1970, 2018.

Pearce, Jeremy. "George C. Devol, Inventor of Robot Arm, Dies at 99." *New York Times*, August 15, 2011.

Perrigo, Billy. "OpenAI Used Kenyan Workers on Less Than $2 Per Hour to Make ChatGPT Less Toxic." *Time*, January 18, 2023.

Perugia, Giulia, Stefano Guidi, Margherita Bicchi, and Oronzo Parlangeli. The Shape of Our Bias: Perceived Age and Gender in the Humanoid Robots of the ABOT Database." In *Proceedings of the ACM/IEEE International Conference on Human–Robot Interaction*, 110–119, IEEE, 2022.

Petersen, Steve. "Designing People to Serve." In *Robot Ethics: The Ethical and Social Implications of Robotics*, edited by Patrick Linn, Keith Abney, and George A. Bekey, 283–298. MIT Press, 2011.

Pettit, Becky, Bryan Sykes, and Bruce Western. "Technical Report on Revised Population Estimates and NLSY 79 Analysis Tables for the Pew Public Safety and Mobility Project." Technical report. Harvard University, 2009.

Phillips, Elizabeth, Xuan Zhao, Daniel Ullman, and Bertram F Malle. "What is Human-Like?: Decomposing Robots' Human-Like Appearance Using the Anthropomorphic Robot (ABOT) Database." In *Proceedings of the 13th ACM/IEEE International Conference on Human–Robot Interaction (HRI)*, 105–113, ACM, 2018.

Phillips, Nate, Brady Kruse, Farzana Alam Khan, J. Edward Swan II, and Cindy L. Bethel. "A Robotic Augmented Reality Virtual Window for Law Enforcement Operations." In *Proceedings of the 12th International Conference on Virtual, Augmented and Mixed Reality. Held as Part of the 22nd HCI International Conference*, 591–610, Springer, 2020.

Pierce, Edward Lillie. *Memoir and Letters of Charles Sumner*. Vol. 4. Roberts Brothers, 1893.

Pilný, Ondřej, Ruud van den Beuken, and Ian R. Walsh. *Cultural Convergence: The Dublin Gate Theatre, 1928–1960*. Springer Nature, 2021.

Pohflepp, Sascha. "The Golden Institute." *Thresholds*, 38: 70–75, 2010.

Pontin, Jason. "Artificial Intelligence, with Help from the Humans." *New York Times*, March 25, 2007.

Poulsen, Adam, Eduard Fosch-Villaronga, and Roger Andre Søraa. "Queering Machines." *Nature Machine Intelligence*, 2(3): 152–152, 2020.

Powell, Laura C. "The Good, the Bad, and the Ugly: Black Lives Matter Protests, the January 6th Insurrection, and Facial Recognition Technology as Admissible Evidence." *American University Law Review*, 72: 277, 2022.

Powers, Martine, and Elahe Izadi. "Security or Surveillance? How Smart Doorbell Company Ring Partners with Police." *Washington Post*, August 28, 2019.

Poynton, Bella. "From RUR to Westworld: Personal Revolt, Digital Technology, and the Making of a New Robot Ur-Text." *Comparative Drama*, 56(4): 363–388, 2022.

Prabakar, Mangai, and Jong-Hoon Kim. "Telebot: Design Concept of Telepresence Robot for Law Enforcement." In *Proceedings of the World Congress on Advances in Nano, Biomechanics, Robotics, and Energy Research (ANBRE 2013), Seoul, Korea*, 34–42, IASEM, 2013.

Prassl, Jeremias. *Humans as a Service: The Promise and Perils of Work in the Gig Economy*. Oxford University Press, 2018.

Pullum, Geoffrey K. "African American Vernacular English is not Standard English with Mistakes." In *The workings of Language: From Prescriptions to Perspectives,* edited by Rebecca S. Wheeler, Praeger, 59–66, 1999.

Purvis, Dara E., and Melissa Blanco. "Police Sexual Violence." *California Law Review*, 108(5): 1487–1529, 2020.

Qian, Yao, Keelan Evanini, Xinhao Wang, Chong Min Lee, and Matthew Mulholland. "Bidirectional LSTM-RNN for Improving Automated Assessment of Non-Native Children's Speech." In *Proceedings of Interspeech*, 1417–1421, Curran Associates, 2017.

Rabelo, Verónica Caridad, and Ramaswami Mahalingam. "'They Really Don't Want to See Us': How Cleaners Experience Invisible 'Dirty' Work." *Journal of Vocational Behavior*, 113: 103–114, 2019.

Raji, Inioluwa Deborah, Andrew Smart, Rebecca N. White, Margaret Mitchell, Timnit Gebru, Ben Hutchinson, Jamila Smith-Loud, Daniel Theron, and Parker Barnes. "Closing the AI Accountability Gap: Defining an End-To-End Framework for Internal Algorithmic Auditing." In *Proceedings of the ACM conference on Fairness, Accountability, and Transparency (FAccT)*, 33–44, ACM, 2020.

Ramachandran, Aditi, Chien-Ming Huang, Edward Gartland, and Brian Scassellati. "Thinking Aloud with a Tutoring Robot to Enhance Learning." In *Proceedings of the ACM/IEEE International Conference on Human–Robot Interaction*, 59–68, ACM, 2018.

Ramey, Arnaud, and Miguel A. Salichs. "Morphological Gender Recognition by a Social Robot and Privacy Concerns." In *Proceedings of the 9th ACM/IEEE International Conference on Human–Robot Interaction (HRI)*, 272–273, ACM, 2014.

Randall, Thomas E. "Justifying Partiality in Care Ethics." *Res Publica*, 26(1): 67–87, 2020.

Randerson, James. "Eye in the Sky: Police Use Drone to Spy on V Festival." *The Guardian*, August 21, 2007.

Rangarajan, Sinduja. "Hey Siri—Why Don't You Understand More People Like Me?" *Mother Jones*, February 23, 2021.

Reel, Monte. "Secret Cameras Record Baltimore's Every Move from Above." *Bloomberg Businessweek*, August 23, 2016.

Reich-Stiebert, Natalia, and Friederike Eyssel. (Ir)relevance of Gender? On the Influence of Gender Stereotypes on Learning with a Robot." In *Proceedings of the ACM/IEEE International Conference on Human–Robot Interaction*, 166–176, ACM, 2017.

Reichel, Philip L. "The Misplaced Emphasis on Urbanization in Police Development." *Policing and Society: An International Journal*, 3(1): 1–12, 1992.

Reilly, Kara. "From Automata to Automation: The Birth of the Robot in *RUR (Rossum's Universal Robots)*." In *Automata and Mimesis on the Stage of Theatre History*, 148–176. Springer, 2011.

Reimann, Nicholas. "San Francisco Says Cops Can't Use Killer Robots After Public Outcry—For Now." *Forbes*, December 6, 2022.

Remick, Carl. "Robots: New Faces on the Production Line." *Management Review*, 68(5): 24–39, 1979.

Resnikoff, Jason. *Labor's End: How the Promise of Automation Degraded Work*. University of Illinois Press, 2022.

Reynolds, Emily. "This Robot Could Be Your Next Receptionist." *Wired*, January 4, 2016.

Rhee, Jennifer. *The Robotic Imaginary: The Human and the Price of Dehumanized Labor*. University of Minnesota Press, 2018.

Richards, Neil M., and William D. Smart. "How Should the Law Think About Robots?" In *Robot Law*, edited by Ryan Calo, A. Michael Froomkin, and Ian Kerr, 3–22. Edward Elgar, 2016.

Richardson, James F. *Urban Police in the United States*. Kennikat Press, 1974.

Richardson, Kathleen. *An Anthropology of Robots and AI: Annihilation Anxiety and Machines*. Routledge, 2015.

Richardson, Kathleen. The Asymmetrical "Relationship" Parallels Between Prostitution and the Development of Sex Robots." *Computers and Society*, 45(3): 290–293, 2016.

Riek, Laurel, and Don Howard. "A Code of Ethics for the Human–Robot Interaction Profession." In *Proceedings of We Robot*, 2014.

Rifkin, Jeremy. *The End of Work: The Decline of the Global Labor Force and the Dawn of the Post-Market Era*. Putnam's Sons, 1995.

Rinehart Kochel, Tammy. "Constructing Hot Spots Policing: Unexamined Consequences for Disadvantaged Populations and for Police Legitimacy." *Criminal Justice Policy Review*, 22(3): 350–374, 2011.

Ritchie, Andrea J. *Invisible No More: Police Violence Against Black Women and Women of color*. Beacon Press, 2017.

Roberts, Dorothy. "Race." In *The 1619 project: A New American Origin Story*, edited by Nikole Hannah-Jones. 45–62, Random House, 2021.

Robertson, Jennifer. "Gendering Humanoid Robots: Robo-Sexism in Japan." *Body & Society*, 16(2): 1–36, 2010.

Robertson, Jennifer. "Robo Sapiens Japanicus: Humanoid Robots and the Posthuman Family." *Critical Asian Studies*, 39(3): 369–398, 2007.

Robinette, Paul, Wenchen Li, Robert Allen, Ayanna M. Howard, and Alan R. Wagner. "Overtrust of Robots in Emergency Evacuation Scenarios." In *Proceedings of the 11th ACM/IEEE International Conference on Human–Robot Interaction (HRI)*, 101–108, IEEE, 2016.

Rogers, Caitlin. "The Aftermath of Police Blowing Up a Maine Man's Home with a Bomb Robot." *Bangor Daily News*, August 28, 2018.

Romanenko, Taras, and Polina Shcherbinina. "Robot vs Worker." *Technology and Language*, 3(1):17–28, 2022.

Rosemont, Henry, and Donald J. Munro. "On Reappraising Ancient Chinese Philosophy: The Concept of Man in Early China." *Philosophy, East and West*, 21(2): 203–217, 1971.

Rosemont, Henry Jr., and Roger T. Ames. *Confucian Role Ethics: A Moral Vision for the 21st Century?* V&R Academic, 2016.

Rosenthal-von der Pütten, Astrid M., Nicole C. Krämer, Laura Hoffmann, Sabrina Sobieraj, and Sabrina C. Eimler. "An Experimental Study on Emotional Reactions Towards a Robot. *International Journal of Social Robotics*, 5: 17–34, 2013.

Rowbotham, Judith, Kim Stevenson, Samantha Pegg. "The Beginnings of Crime Intelligence: 1800–1860." In *Crime News in Modern Britain: Press Reporting and Responsibility, 1820–2010*, 14–38. Springer, 2013.

Rowe, Niamh. "'It's Destroyed Me Completely': Kenyan Moderators Decry Toll of Training of AI Models." *The Guardian*, August 2, 2023.

Ryland, M. "Hypervisibility: How Scrutiny and Surveillance Makes You Watched, But Not Seen." *The body is not an apology*, 2013. https://www.tumblr.com/thebodyisnotanapology/57763238146/hypervisibility-how-scrutiny-and-surveillance

Šabanović, Selma. "Robots in Society, Society in Robots: Mutual Shaping of Society and Technology as a Framework for Social Robot Design." *International Journal of Social Robotics*, 2(4): 439–450, 2010.

Saggese, Alessia, Mario Vento, and Vincenzo Vigilante. "MIVIABot: A Cognitive Robot for Smart Museum." In *Proceedings of the International Conference on Computer Analysis of Images and Patterns*, 15–25, Springer, 2019.

Samora, Julian, Joe Bernal, and Albert Pena. *Gunpowder Justice: A Reassessment of the Texas Rangers*. University of Notre Dame Press, 1979.

Sandygulova, Anara, and Gregory M. P. O'Hare. Age-Related Differences in Children's Associations and Preferences for a Robot's Gender." In *Companion of the ACM/IEEE International Conference on Human–Robot Interaction*, 235–236, ACM, 2018.

Sandygulova, Anara, Mauro Dragone, and Gregory M. P. O'Hare. "Real-Time Adaptive Child–Robot Interaction: Age and Gender Determination of Children Based on 3D Body Metrics." In *Proceedings of the 23rd IEEE International Conference on Robot and Human Interactive Communication*, 826–831, IEEE, 2014.

Santiago, John. "Confucian Ethics in the Analects as Virtue Ethics." In *Philosophical Ideas and Artistic Pursuits in the Traditions of Asia and the West: An NEH Faculty Humanities Workshop*, 2008.

Sanusi, Ismaila Temitayo, and Sunday Adewale Olaleye. "An Insight into Cultural Competence and Ethics in K-12 Artificial Intelligence Education." In *2022 IEEE Global Engineering Education Conference (EDUCON)*, 790–794. IEEE, 2022.

Sarathy, Vasanth, Matthias Scheutz, Yoed N. Kenett, Mowafak Allaham, Joseph L. Austerweil, and Bertram F. Malle. "Mental Representations and Computational Modeling of Context-Specific Human Norm Systems. In *Proceedings of the Annual Meeting of the Cognitive Science Society (CogSci)*, 1035–1040, Cognitive Science Society, 2017.

Saunders, Trevor J., and T. A. Sinclair, transl. *Aristotle: The Politics*. Penguin Classics, 1981.

Schaut, Scott. *Robots of Westinghouse, 1924–Today*. Mansfield Memorial Museum, 2006.

Schelly, Bill. *Otto Binder: The Life and Work of a Comic Book and Science Fiction Visionary*. North Atlantic Books, 2016.

Scheuerman, Morgan Klaus, Alex Hanna, and Emily Denton. "Do Datasets Have Politics? Disciplinary Values in Computer Vision Dataset Development." In *Proceedings of the ACM on Human–Computer Interaction*, 5(CSCW2), 1–37, ACM, 2021.

Scheuerman, Morgan Klaus, Jacob M. Paul, and Jed R. Brubaker. "How Computers See Gender: An Evaluation of Gender Classification in Commercial Facial Analysis

Services." In *Proceedings of the ACM on Human–Computer Interaction*, 3(CSCW), 1–33, ACM, 2019.

Scheuerman, Morgan Klaus, Kandrea Wade, Caitlin Lustig, and Jed R. Brubaker. "How We've Taught Algorithms to See Identity: Constructing Race and Gender in Image Databases for Facial Analysis." In *Proceedings of the ACM on Human–Computer Interaction*, 4(CSCW1), 1–35, ACM, 2020.

Scheuerman, Morgan Klaus, Madeleine Pape, and Alex Hanna. "Auto-Essentialization: Gender in Automated Facial Analysis as Extended Colonial Project." *Big Data & Society*, 8(2), 1–15, 2021.

Scheutz, Matthias. "The Inherent Dangers of Unidirectional Emotional Bonds Between Humans and Social Robots." In *Robot Ethics: The Ethical and Social Implications of Robotics*, edited by Patrick Linn, Keith Abney, and George A. Bekey, 205–221. MIT Press, 2011.

Schmitt, Christopher. "Plea Bargaining Favors Whites, as Blacks, Hispanics Pay Price." *San Jose Mercury News*, December 8, 1991.

Schneier, Bruce, and B. Schneier. *Beyond Fear: Thinking Sensibly About Security in an Uncertain World*. Springer, 2003.

Scientific American. "Automaton Extraordinary." *Scientific American*, 3(34):272, 1848.

Scorzin, Pamela C. "> More human than human <. Digital dolls on social media." *Denkste: Puppe-multidisziplinäre zeitschrift für mensch-puppen-diskurse*, 4(1): 157–166, 2021.

Scott, Ridley. "Blade Runner," 1982.

Seaborn, Katie, and Alexa Frank. "What Pronouns for Pepper? A Critical Review of Gender/ing in Research." In *CHI Conference on Human Factors in Computing Systems*, 1–15, ACM, 2022.

Seaborn, Katie, and Peter Pennefather. "'Neither 'Hear' nor 'Their': Interrogating Gender Neutrality in Robots." In *Companion of the ACM/IEEE International Conference on Human–Robot Interaction (Extended Abstracts)*, 1030–1034, IEEE, 2022.

Seaborn, Katie, Peter Pennefather, and Haruki Kotani. "Exploring Gender-Expansive Categorization Options for Robots." In *Extended Abstracts of the CHI Conference on Human Factors in Computing Systems*, 1–6, ACM, 2022.

Seigel, Micol. *Violence Work: State Power and the Limits of Police*. Duke University Press, 2018.

Sepkowitz, Kent A., and Leon Eisenberg. "Occupational Deaths Among Healthcare Workers." *Emerging Infectious Diseases*, 11(7): 1003, 2005.

Settles, Isis H., NiCole T. Buchanan, and Kristie Dotson. "Scrutinized but Not Recognized: (In)visibility and Hypervisibility Experiences of Faculty of Color." *Journal of Vocational Behavior*, 113: 62–74, 2019.

Shen, Solace, Petr Slovak, and Malte F Jung. "'Stop. I See a Conflict Happening.' A Robot Mediator for Young Children's Interpersonal Conflict Resolution." In *Proceedings of the ACM/IEEE International Conference on Human–Robot Interaction*, 69–77, ACM, 2018.

Shevlin, Henry. "How Could We Know When a Robot Was a Moral Patient?" *Cambridge Quarterly of Healthcare Ethics*, 30(3): 459–471, 2021.

Shiner, Amy. "Police Helicopter Surveillance." *Arizona Law Review*, 15(1): 145–171, 1973.

Short, Elaine, Katelyn Swift-Spong, Jillian Greczek, Aditi Ramachandran, Alexandru Litoiu, Elena Corina Grigore, David Feil-Seifer, et al. "How to Train Your Dragonbot: Socially Assistive Robots for Teaching Children About Nutrition Through Play." In *Proceedings of the 23rd IEEE International Conference on Robot and Human Interactive Communication*, 924–929, IEEE, 2014.

Sidner, Sara, and Mallory Simon. "How Robot, Explosives Took out Dallas Sniper in Unprecedented Way." *CNN*, July 12, 2016.

Siebert, Wilbur H. "The Underground Railroad in Massachusetts." *The New England Quarterly*, 9(3): 447–467, 1936.

Sierra-Arévalo, Michael. "American Policing and the Danger Imperative." *Law & Society Review*, 55(1): 70–103, 2021.

Sierra-Arévalo, Michael. "Reward and 'Real' Police Work." In *The Ethics of Policing: New Perspectives on Law Enforcement*, edited by Ben Jones and Eduardo Mendieta, 66–90. New York University Press, 2021.

Simon, Bernd, and David L. Hamilton. "Self-Stereotyping and Social Context: The Effects of Relative In-Group Size and In-Group Status." *Journal of Personality and Social Psychology*, 66(4): 699–711, 1994.

Simons, Patricia. "(Check) Mating the Grand Masters: The Gendered, Sexualized Politics of Chess in Renaissance Italy." *Oxford Art Journal*, 16(1): 59–74, 1993.

Simpson, Ruth, and Patricia Lewis. "An Investigation of Silence and a Scrutiny of Transparency: Re-Examining Gender in Organization Literature Through the Concepts of Voice and Visibility." *Human Relations*, 58(10): 1253–1275, 2005.

Sinclair, Kat. "The Feminised Robot: Labour and Harawayan Afterlives." In *Artificial Intelligence and Its Discontents: Critiques from the Social Sciences and Humanities*, edited by Ariane Hanemaayer, 43–60. Springer, 2022.

Slevin, Colleen, and Thomas Peipert. "Prosecutors Investigating Police Who Handcuffed Black Girls." *Washington Post*, August 7, 2020.

Slobogin, Christopher. "Testilying: Police Perjury and What To Do About It." *University of Colorado Law Review*, 67: 1037–1060, 1996.

Smedley, Audrey. "Antecedents of the Racial Worldview." In *Race and Racilization: Essential Readings*, edited by Tania Das Gupta, Carl E. James, Roger C. A. Maaka, Grace-Edward Galabuzi, and Chris Andersen, 31–44. Canadian Scholars Press, 2007.

Smedley, Audrey, and Brian D. Smedley. "Race as Biology is Fiction, Racism as a Social Problem is Real: Anthropological and Historical Perspectives on the Social Construction of Race." *American Psychologist*, 60(1): 16–26, 2005.

Smedley, Audrey, and Brian D. Smedley. *Race in North America: Origin and Evolution of a Worldview*. Westview Press, 2012.

Smialek, Eric. "Who Needs to Calm Down? Taylor Swift and Rainbow Capitalism." *Contemporary Music Review*, 40(1): 99–119, 2021.

Smith, Neil. *The New Urban Frontier: Gentrification and the Revanchist City*. Psychology Press, 1996.

Smitherman, Geneva. "Word from the Hood: The Lexicon of African-American Vernacular English." In *African-American English*, 223–248. Routledge, 2021.

Solomon, Cynthia, Brian Harvey, Ken Kahn, Henry Lieberman, Mark L. Miller, Margaret Minsky, Artemis Papert, and Brian Silverman. "History of Logo." In *Proceedings of the ACM on Programming Languages*, 4(HOPL): 1–66, ACM, 2020.

Song-Nichols, Kallyn, and Andrew Young. "Gendered Robots Can Change Children's Gender Stereotyping." In *Proceedings of the Annual Meeting of the Cognitive Science Society (CogSci)*, 2480–2485, Cognitive Science Society, 2020.

Songalia, Ryan. "The Activist Who Streamed a Bizarre Standoff with the NYPD Will Keep Fighting." *VICE*, August 11, 2020.

Søraa, Roger Andre. "Mechanical Genders: How do Humans Gender Robots? *Gender, Technology and Development*, 21: 99–115, 2017.

Soss, Joe, and Vesla Weaver. Police Are Our Government: Politics, Political Science, and the Policing of Race–Class Subjugated Communities." *Annual Review of Political Science*, 20: 565–591, 2017.

Sparrow, Robert. "Do Robots Have Race?: Race, Social Construction, and HRI." *IEEE Robotics & Automation Magazine*, 27(3): 144–150, 2019.

Sparrow, Robert. "Robotics Has a Race Problem." *Science, Technology, & Human Values*, 45(3): 538–560, 2020.

Spearman, Charles. "'General Intelligence' Objectively Determined and Measured." *American Journal of Psychology*, 15(2): 201–293, 1904.

Spearman, Charles. "The Heredity of Abilities." *The Eugenics Review*, 6(3): 219–237, 1914.

Spearman, Charles. "The Measurement of Intelligence." *The Eugenics Review*, 6(4): 577–578, 1915.

Speri, Alice. "The FBI Has Quietly Investigated White Supremacist Infiltration of Law Enforcement." *The Intercept*, January 31, 2017.

Standage, Tom. *The Turk: The Life and Times of the Famous Eighteenth-Century Chess-Playing Machine*. Walker, 2002.

Star, Susan Leigh, and Anselm Strauss. "Layers of Silence, Arenas of Voice: The Ecology of Visible and Invisible Work." In *Proceedings of the ACM on Human–Computer Interaction*, (CSCW), 8: 9–30, ACM, 1999.

Stead, Valerie. "Learning to Deploy (In)Visibility: An Examination of Women Leaders' Lived Experiences." *Management Learning*, 44(1): 63–79, 2013.

Steed, Ryan, and Aylin Caliskan. "Image Representations Learned with Unsupervised Pre-Training Contain Human-Like Biases." In *Proceedings of the ACM conference on Fairness, Accountability, and Transparency (FAccT)*, 701–713, ACM, 2021.

Steele, Catherine Knight. *Digital Black Feminism*. New York University Press, 2021.

Stevens, Kate. "Police, Pringle Robotics Collaborate for Red White & Boom Safety." *Central Illinois Proud*, June 30, 2023.

Stolp-Smith, Michael, and Tom Williams. "More than Binary: Transgender and Nonbinary Perspectives on Human Robot Interaction." In *Proceedings of the ACM/IEEE International Conference on Human–Robot Interaction*, 697–705, ACM, 2024.

Stone, Wesley L. "The History of Robotics." In *Robotics and Automation Handbook*, edited by Thomas R. Kurfess, 8–19. CRC Press, 2018.

Stoughton, Seth. "Law Enforcement's "Warrior" Problem." *Harvard Law Review Forum*, 128(6): 225–234, 2014.

Strait, Megan, Ana Sánchez Ramos, Virginia Contreras, and Noemi Garcia. "Robots Racialized in the Likeness of Marginalized Social Identities Are Subject to Greater Dehumanization Than Those Racialized as White." In *Proceedings of the 27th IEEE International Conference on Robot and Human Interactive Communication (RO-MAN)*, 452–457, IEEE, 2018.

Strauss, Claudia. "The Imaginary." *Anthropological Theory*, 6(3): 322–344, 2006.

Strengers, Yolande, and Jenny Kennedy. *The Smart Wife: Why Siri, Alexa, and Other Smart Home Devices Need a Feminist Reboot*. MIT Press, 2021.

Su, Norman Makoto, Amanda Lazar, Jeffrey Bardzell, and Shaowen Bardzell. "Of Dolls and Men: Anticipating Sexual Intimacy with Robots." *ACM Transactions on Computer–Human Interaction (TOCHI)*, 26(3): 1–35, 2019.

Suchman, Lucy. *Human–Machine Reconfigurations: Plans and Situated Actions*. Cambridge University Press, 2007.

Suchman, Lucy. "Making Work Visible." *Communications of the ACM*, 38(9): 56–64, 1995.

Suchman, Lucy A. *Plans and Situated Actions: The Problem of Human–Machine Communication*. Cambridge University Press, 1987.

Sue, Derald Wing, Christina M. Capodilupo, Gina C. Torino, Jennifer M. Bucceri, Aisha Holder, Kevin L. Nadal, and Marta Esquilin. "Racial Microaggressions in Everyday Life: Implications for Clinical Practice." *American Psychologist*, 62(4): 271–286, 2007.

Sue, Derald Wing, Christina M. Capodilupo, Kevin L. Nadal, and Gina C. Torino. "Racial Microaggressions and the Power to Define Reality." *American Psychologist*, 63(4): 277–279, 2008.

Sunstein, Cass R. *Why Nudge? The Politics of Libertarian Paternalism*. Yale University Press, 2014.

Swann, Thomas. "Information, Cybernetics and the Second Industrial Revolution." *Ephemera*, 17(2): 457–465, 2017.

Tan, Haodan, Fuchang Yang, Nava Teja Tummalapalli, Chetan Bhatia, and Kaustubh Barde. "Raven: A Street Robot to Address Homelessness." In *Proceedings of the Companion of the 2017 ACM/IEEE International Conference on Human–Robot Interaction*, 397–398, ACM, 2017.

Tanksley, Tiera. "Employing an Abolitionist, Critical Race Pedagogy in CS: Centering the Voices, Experiences and Technological Innovations of Black Youth." *Journal of Computer Science Integration*, 6(1): 1–16 2023.

Tate, Katherine, James Lance Taylor, and Mark Q Sawyer. *Something's in the Air: Race, Crime, and the Legalization of Marijuana*. Routledge, 2013.

Tatman, Rachael, and Conner Kasten. "Effects of Talker Dialect, Gender and Race on Accuracy of Bing Speech and YouTube Automatic Captions." In *Proceedings of Interspeech*, 934–938, Curran Associates, 2017.

Teitelbaum, Sheldon. "Scientists Say Asimov Put the Stars in Their Eyes." *Los Angeles Times*, April 8, 1992.

Tennent, Hamish, Solace Shen, and Malte Jung. "MicBot: A Peripheral Robotic Object to Shape Conversational Dynamics and Team Performance." In *Proceedings of the 14th ACM/IEEE International Conference on Human–Robot Interaction (HRI)*, IEEE, 133–142, 2019.

Tewolde, Amanuel Isak. "'Becoming Black': Racial Formation of Eritrean Migrants in Israel." *African Diaspora*, 13(1–2): 183–203, 2021.

The Racine Advocate. "A Singular Automaton." *The Racine Advocate*, May 10, 1848.

Theobald, Robert. *Free Men and Free Markets*. Anchor Books, 1965.

Thevenot, Brian, and Erwin Seba. "Dallas Sniper Plotted Bigger Assault, Taunted Officers: Police Chief." *Reuters*, July 11, 2016.

Tiger, Rebecca. *Judging Addicts: Drug Courts and Coercion in the Justice System*. New York University Press, 2013.

Tighe, Scott, and William Brown. "The Militarization of Law Enforcement: Bypassing the Posse Comitatus Act." *Justice Policy Journal*, 12(2): 1–39, 2015.

Todo, Takayuki, and Takanari Miisho. "Interactions on Eyeballs of Humanoid-Robots." In *Proceedings of the 2nd International Conference on Human–Agent Interaction*, 281–283, ACM, 2014.

Tompkins, D. A. "The Cotton Industry." *Publications of the American Economic Association*, 5(1): 144–153, 1904.

Torre, Ilaria, and Sébastien Le Maguer. "Should Robots Have Accents?" In *Proceedings of the 29th IEEE International Conference on Robot and Human Interactive Communication (RO-MAN)*, 208–214, IEEE, 2020.

Touretzky, David, Fred Martin, Deborah Seehorn, Cynthia Breazeal, and Tess Posner. "Special Session: AI for K-12 Guidelines Initiative." In *Proceedings of the 50th ACM Technical Symposium on Computer Science Education*, 492–493, ACM, 2019.

Tran, Nhan, Trevor Grant, Thao Phung, Leanne Hirshfield, Christopher Wickens, and Tom Williams. "Now Look Here! ↓ Mixed Reality Improves Robot Communication Without Cognitive Overload." In *Proceedings of the International Conference on Virtual, Augmented, and Mixed Reality (VAMR), held as part of the International Conference on Human–Computer Interaction (HCI)*, Springer, 2023.

Trenchard, John, and Thomas Gordon. *The Independent Whig*. Vol. 1, 7th ed. J. Peele, 1743.

Tronto, Joan. *Moral Boundaries: A Political Argument for an Ethic of Care*. Routledge, 2020.

Tronto, Joan C. "Creating Caring Institutions: Politics, Plurality, and Purpose." In *Ethics and Social Welfare*, 4(2): 158–171, 2010.

Turbil, Cristiano. "Memory, Heredity and Machines: From Darwinism to Lamarckism." In Samuel Butler's *Erewhon*. *Journal of Victorian Culture*, 25(1): 31–46, 2020.

Turing, Alan M. "Computing Machinery and Intelligence." *Mind*, 49, 433–460, 1950.

Turing, Alan Mathison. "On Computable Numbers, with an Application to the Entscheidungs Problem." *Proceedings of the London Mathematical Society*, s2–42: 230–265, 1937.

Turkle, Sherry. *Alone Together: Why We Expect More from Technology and Less from Each Other*. Hachette UK, 2017.

Uchida, Craig D. "The Development of the American Police: An Historical Overview." *Critical Issues in Policing: Contemporary Readings*, 3: 19–35, 1993.

United States v. Brignoni-Ponce. 422 U.S. 873, 1975, https://supreme.justia.com/cases/federal/us/422/873/

Van Wynsberghe, Aimee. "Service Robots, Care Ethics, and Design." *Ethics and Information Technology*, 18(4): 311–321, 2016.

Van Wynsberghe, Aimee, and Scott Robbins. "Critiquing the Reasons for Making Artificial Moral Agents." *Science and Engineering Ethics*, 25(3): 719–735, 2019.

Vaughn, Alyssa. "After Saying Sex with Minors is Not Always Sexual Assault, MIT Scientist Resigns." *Boston Magazine*, September 18, 2019.

Verbeek, Peter-Paul. "Materializing Morality: Design Ethics and Technological Mediation." *Science, Technology, & Human Values*, 31(3): 361–380, 2006.

Verbeek, Peter-Paul. *Moralizing Technology: Understanding and Designing the Morality of Things*. University of Chicago Press, 2011.

Verran, Helen. "Re-Imagining Land Ownership in Australia." *Postcolonial Studies: Culture, Politics, Economy*, 1(2): 237–254, 1998.

Villaluz, Kathleen. "The World's First Robot Police Officer Begins Patrolling in Dubai." *Interesting Engineering*, May 30, 2017.

Vincent, J. "Google Assistant Adds Pretty Please Functionality, Updated Lists, and Other Features." *The Verge*, November 29, 2018.

Vincent, James. "NYPD Used Facial Recognition to Track Down Black Lives Matter Activist." *The Verge*, August 18, 2020.

Vitale, Alex S. *The End of Policing*. Verso Books, 2017.

von Windisch, Karl Gottlieb. *Inanimate Reason, or a Circumstantial Account of That Astonishing Piece of Mechanism, M. de Kempelen's Chess-Player; Now Exhibiting at No. 8 Saville-Row, Burlington Gardens*. Bladon, 1784.

Waddell, Kaveh. "Hey Siri, Is That You? Apple's New Voices Resonate with Some Black iPhone Users." *Consumer Reports*, April 23, 2021.

Wade, Richard C. *Slavery in the Cities: The South 1820–1860*. Oxford University Press, 1967.

Wagner, Ben. "Ethics as an Escape from Regulation. From 'Ethics-Washing' to Ethics-Shopping?" In *Being Profiled: Cogitas ergo sum: 10 Years of Profiling the European Citizen*, edited by Emre Bayamlioglu, Irina Baraliuc, Liisa Albertha Wilhelmina Janssens, and Mireille Hildebrandt, 84–89. Amsterdam University Press, 2018.

Wagner, Cosima. "'The Japanese Way of Robotics': Interacting 'Naturally' with Robots as a National Character?" In *Proceedings of the IEEE International Conference on Robot and Human Interactive Communication (RO-MAN)*, 510–515. IEEE, 2009.

Wald, Johanna, and Daniel J. Losen. "Defining and Redirecting a School-to-Prison Pipeline." *New Directions for Youth Development*, 2003(99): 9–15, 2003.

Walia, Harsha. *Border and Rule: Global Migration, Capitalism, and the Rise of Racist Nationalism*. Haymarket Books, 2021.

Walk, Hunter. "Amazon Echo is Magical. It's Also Turning My Kid into an Asshole." *Hunter Walk*, April 6, 2016.

Walker, Mark. "A Moral Paradox in the Creation of Artificial Intelligence: Mary Poppins 3000s of the World Unite!" In *Proceedings of the AAAI Workshop on Human Implications of Human–Robot Interaction*, 23–28, ACM, 2006.

Walker, Michael, Thao Phung, Tathagata Chakraborti, Tom Williams, and Daniel Szafir. "Virtual, Augmented, and Mixed Reality for Human–Robot Interaction: A Survey and Virtual Design Element Taxonomy." *ACM Transactions on Human–Robot Interaction (T-HRI)*, 12(4): 1–39, 2023.

Wall, Tyler. "Unmanning the Police Manhunt: Vertical Security as Pacification." *Socialist Studies/Études Socialistes*, 9(2): 32–56, 2013.

Walton, Gregory M., Mary C. Murphy, and Ann Marie Ryan. "Stereotype Threat in Organizations: Implications for Equity and Performance." *Annual Review of Organizational Psychology and Organizational Behavior*, 2(1): 523–550, 2015.

Walzer, Michael. "Soldiers and Police." In *The Ethics of Policing: New Perspectives on Law Enforcement*, edited by Ben Jones and Eduardo Mendieta, 93–106. New York University Press, 2021.

Wang, Dan. "Data on Police Officers Killed Since 1961," 2015. https://danwang.co/statistics-on-police-fatalities/

Ward, Tom. "Dubai Wants Robots to Make Up 25% of its Police Force by 2030." *Futurism*, May 23, 2017.

Weaver, Vesla. "Policing Narratives in the Black Counterpublic." In *The ethics of policing: New perspectives on law enforcement*, edited by Ben Jones and Eduardo Mendieta, 149–178. New York University Press, 2021.

Wei-Ming, Tu. "Self-Cultivation as Education Embodying Humanity." In *Proceedings of the Twentieth World Congress of Philosophy*. Vol. 3, 27–39, Bowling Green State Univ Philosophy, 1999.

Weisman, Kara. "Extraordinary Entities: Insights into Folk Ontology from Studies of Lay People's Beliefs About Robots." In *Proceedings of the Annual Meeting of the Cognitive Science Society*, 3493–3499, Cognitive Science Society, 2022.

Weisswange, Thomas H., Hifza Javed, Manuel Dietrich, Tuan Vu Pham, Maria Teresa Parreira, Michael Sack, and Nawid Jamali. "What Could a Social Mediator Robot Do?

Lessons from Real-World Mediation Scenarios." In *Proceedings of the ICRA Workshop "Towards a Balanced Cyberphysical Society: A Focus on Group Social Dynamics,"* 2023.

Weizenbaum, Joseph. "Eliza—A Computer Program for the Study of Natural Language Communication Between Man and Machine." *Communications of the ACM*, 9(1): 36–45, 1966.

Wen, Ruchen, Alyssa Hanson, Zhao Han, and Tom Williams. Fresh Start: Encouraging Politeness in Wakeword-Driven Human–Robot Interaction." In *Proceedings of the ACM/IEEE International Conference on Human–Robot Interaction*, 112–121, ACM, 2023.

Wen, Ruchen, Boyoung Kim, Elizabeth Phillips, Qin Zhu, and Tom Williams. "Comparing Norm-Based and Role-Based Strategies for Robot Communication of Role-Grounded Moral Norms." *ACM Transactions on Human–Robot Interaction (T-HRI)* 12(3): 1–25, (2023).

Wen, Ruchen, Brandon Barton, Sebastian Fauré, and Tom Williams. "Unpretty Please: Ostensibly Polite Wakewords Discourage Politeness in Both Robot-Directed and Human-Directed Communication." In *Proceedings of the ACM International Conference on Multimodal Interaction*, 181–190, ACM, 2022.

Wen, Ruchen, Zhao Han, and Tom Williams. "Teacher, Teammate, Subordinate, Friend: Generating Norm Violation Responses Grounded in Role-Based Relational Norms." In *Proceedings of the 17th ACM/IEEE International Conference on Human–Robot Interaction (HRI)*, 353–362, IEEE, 2022.

West, Mark, Rebecca Kraut, and Han Ei Chew. "I'd Blush if I Could: Closing Gender Divides in Digital Skills Through Education." Technical report. UNESCO, 2019.

Western, Bruce, and Becky Pettit. "Incarceration and Social Inequality." *Daedalus*, 139(3): 8–19, 2010.

Whedon, Joss. "Dollhouse, Mutant Enemy Productions," 2009.

While, Aidan H., Simon Marvin, and Mateja Kovacic. "Urban Robotic Experimentation: San Francisco, Tokyo and Dubai." *Urban Studies*, 58(4): 769–786, 2021.

Wiener, Norbert. *Cybernetics: Control and communication in the animal and the machine*. Hermann, 1948.

Wiener, Norbert. *I Am a Mathematician: The Later Life of a Prodigy*. MIT Press, 1964.

Wiener, Norbert. *The Human Use of Human Beings: Cybernetics and Society*. Da Capo Press, 1988.

Wikipedia. "Community Insights 2021 Report." https://meta.wikimedia.org/wiki/Community Insights/Community Insights 2021 Report, 2021.

Wilkerson, Isabel. *Caste: The Origins of Our Discontents*. Random House, 2020.

Williams, Jay G. "On Reading a Confucian Classic: The Rhetoric of the Lun Yu." *Journal of Chinese Religions*, 19(1): 105–111, 1991.

Williams, Kristian. *Our Enemies in Blue: Police and Power in America*. AK Press, 2015.

Williams, Logan D. A. "Data Racism, Machine Learning and Contextualizing Design." In *Because Technology Discriminates: Anti-Racist Counter-Expertise*, 127–156. Springer, 2023.

Williams, Nathaniel. *Gears and God: Technocratic Fiction, Faith, and Empire in Mark Twain's America*. University of Alabama Press, 2018.

Williams, Nathaniel. "War Machines and Child Geniuses: American Edisonades." In *The Cambridge History of Science Fiction*, edited by Gerry Canavan and Eric Carl Link, 86–100. Cambridge University Press, 2019.

Williams, Randi. How to Train Your Robot: Project-Based AI and Ethics Education for Middle School Classrooms. In *Proceedings of the 52nd ACM Technical Symposium on Computer Science Education*, 1382–1382, ACM, 2021.

Williams, Tom. "No Justice, No Robots." 2020. https://nojusticenorobots.github.io/.

Williams, Tom. "The Eye of the Robot Beholder: Ethical Risks of Representation, Recognition, and Reasoning Over Identity Characteristics in Human–Robot Interaction." In *Companion of the ACM/IEEE International Conference on Human–Robot Interaction (alt.HRI), 1–10*. ACM, 2023.

Williams, Tom. "Understanding Roboticists' Power Through Matrix Guided Technology Power Analysis." In *Companion of the 2024 ACM/IEEE International Conference on Human–Robot Interaction (alt.HRI)*, 46–56, ACM, 2024.

Williams, Tom, Cynthia Matuszek, Kristiina Jokinen, Raj Korpan, James Pustejovsky, and Brian Scassellati. "Voice in the Machine: Ethical Considerations for Language-Capable Robots." *Communications of the ACM (CACM)*, 66(8): 20–23, 2023.

Williams, Tom, Cynthia Matuszek, Ross Mead, and Nick DePalma. "Scarecrows in Oz: The Use of Large Language Models in HRI." *ACM Transactions on Human–Robot Interaction (T-HRI)*, 13(1): 1–11, 2023.

Williams, Tom, Daniel Grollman, Mingyuan Han, Ryan Blake Jackson, Jane Lockshin, Ruchen Wen, Zachary Nahman, and Qin Zhu. "Excuse Me, Robot": Impact of Polite Robot Wakewords on Human–Robot Politeness." In *Proceedings of the International Conference on Social Robotics*, 404–415, Springer, 2020.

Williams, Tom, and Kerstin Haring. "No Justice, No Robots: From the Dispositions of Policing to an Abolitionist Robotics." In *Proceedings of the AAAI/ACM Conference on Artificial Intelligence, Ethics, and Society (AIES)*, 566–575, ACM, 2023.

Williams, Tom, Qin Zhu, Ruchen Wen, and Ewart J. de Visser. "The Confucian Matador: Three Defenses Against the Mechanical Bull." In *Companion of the ACM/IEEE International Conference on Human–Robot Interaction (alt.HRI)*, 25–33, ACM, 2020.

Williams, Tom, and Ruchen Wen. "Human Capabilities as Guiding Lights for the Field of AI-HRI: Insights from Engineering Education." In *Proceedings of the AAAI Fall Symposium on Artificial Intelligence for Human–Robot Interaction (AI-HRI)*, AAAI, 2021.

Williamson, Ben. "Calculating Children in the Dataveillance School: Personal and Learning Analytics." In *Surveillance Futures*, 50–66. Routledge, 2016.

Winkle, Katie. "Robots from Nowhere: A Case Study in Speculative Sociotechnical Design and Design Fiction for Human-Robot Interaction." In *Companion of the ACM/IEEE International Conference on Human–Robot Interaction (alt.HRI)*, IEEE, 2025.

Winkle, Katie, Donald McMillan, Maria Arnelid, Katherine Harrison, Madeline Balaam, Ericka Johnson, and Iolanda Leite. "Feminist Human–Robot Interaction: Disentangling Power, Principles and Practice for Better, More Ethical HRI." In *Proceedings of the ACM/IEEE International Conference on Human–Robot Interaction (HRI)*, 72–82, ACM, 2023.

Winkle, Katie, Emmanuel Senft, and Séverin Lemaignan. "LEADOR: A Method for End-to-End Participatory Design of Autonomous Social Robots." *Frontiers in Robotics and AI*, 8:1–18, 2021.

Winkle, Katie, Gaspar Isaac Melsión, Donald McMillan, and Iolanda Leite. "Boosting Robot Credibility and Challenging Gender Norms in Responding to Abusive Behaviour: A Case for Feminist Robots." In *Companion of the ACM/IEEE International Conference on Human–Robot Interaction (alt.HRI)*, 29–37, ACM, 2021.

Winkle, Katie, Ryan Blake Jackson, Gaspar Isaac Melsión, Dražen Brščić, Iolanda Leite, and Tom Williams. "Norm-Breaking Responses to Sexist Abuse: A Cross-Cultural Human Robot Interaction Study." In *Proceedings of the 17th ACM/IEEE International Conference on Human–Robot Interaction (HRI)*, 120–129, IEEE, 2022.

Winkle, Katie, Séverin Lemaignan, Praminda Caleb-Solly, Paul Bremner, Ailie J. Turton, and Ute Leonards. "In-Situ Learning from a Domain Expert for Real World Socially Assistive Robot Deployment." In *Proceedings of Robotics: Science and Systems*, RSS Foundation, 2020.

Winkle, Katie, Séverin Lemaignan, Praminda Caleb-Solly, Ute Leonards, Ailie Turton, and Paul Bremner. "Effective Persuasion Strategies for Socially Assistive Robots." In *Proceedings of the 14th ACM/IEEE International Conference on Human–Robot Interaction (HRI)*, 277–285, IEEE, 2019.

Withers, Laura. "How Bearcats Became Toys: The 1033 Program and its Effect on the Right to Protest." *George Washington Law Review*, 84(3): 812–841, 2016.

Witz, Anne and Mike Savage. "The Gender of Organizations." *The Sociological Review*, 39: 3–62, 1991.

Wong, David. "Chinese Ethics." In *The Stanford Encyclopedia of Philosophy*, edited by Edward N. Zalta and Uri Nodelman. Metaphysics Research Lab, Stanford University, Fall 2023. https://plato.stanford.edu/archives/fall2023/entries/ethics-chinese/

Woods, Heather Suzanne. "Asking More of Siri and Alexa: Feminine Persona in Service of Surveillance Capitalism." *Critical Studies in Media Communication*, 35(4): 334–349, 2018.

Woolf, Nicky. "Did Police's Militarised Response to a Small Town Pumpkin Patch Riot Just Make It Worse?" *The Guardian*, October 22, 2014.

Wooten, Melissa E., and Enobong H. Branch. "Defining Appropriate Labor: Race, Gender, and Idealization of Black Women in Domestic Service. *Race, Gender & Class*, 19(3/4): 292–308, 2012.

Worrall, John L. "Addicted to the Drug War: The Role of Civil Asset Forfeiture as a Budgetary Necessity in Contemporary Law Enforcement." *Journal of Criminal Justice*, 29(3): 171–187, 2001.

Wright, Alan. *Policing: An Introduction to Concepts and Practice*. Routledge, 2002.

Wright, James. *Robots Won't Save Japan: An Ethnography of Eldercare Automation*. Cornell University Press, 2023.

Wrigley, Julia. "Feminists and Domestic Workers." *Feminist Studies*, 17(2): 317–329, 1991.

Wu, Yunhan, Daniel Rough, Anna Bleakley, Justin Edwards, Orla Cooney, Philip R Doyle, Leigh Clark, and Benjamin R Cowan. "See What I'm Saying? Comparing Intelligent Personal Assistant Use for Native and Non-Native Language Speakers." In *Proceedings of the 22nd International Conference on Human–Computer Interaction with Mobile Devices and Services*, 1–9, ACM, 2020.

Yancey-Bragg, N'dea. "Aurora Police Apologize After Drawing Guns on Black Family, Mistaking Vehicle as Stolen." *USA Today*, August 4, 2020.

Yang, Jessica L., Yolanda Anyon, Malina Pauline, Katherine E. Wiley, Donna Cash, Barbara J. Downing, Eldridge Greer, Ellen Kelty, Thomas Lee Morgan, and Lisa Pisciotta. "'We Have to Educate Every Single Student, not Just the Ones That Look Like Us': Support Service Providers' Beliefs About the Root Causes of the School-to-Prison Pipeline for Youth of Color." *Equity & Excellence in Education*, 51(3–4): 316–331, 2018.

Ye, Xin and Lionel Robert. "Human Security Robot Interaction and Anthropomorphism: An Examination of Pepper, RAMSEE, and Knightscope Robots." In *Proceedings of the IEEE International Conference on Robot & Human Interactive Communication (RO-MAN)*, 982–987, IEEE, 2023.

Young, Deborah. "Unnecessary Evil: Police Lying in Interrogations." *Connecticut Law Review*, 28: 425, 1995.

Yun, Sangseok, Jongju Shin, Daijin Kim, Chang Gu Kim, Munsang Kim, and Mun-Taek Choi. "Engkey: Tele-Education Robot." In *Proceedings of the International Conference on Social Robotics*, 142–152, Springer, 2011.

Zenderland, Leila. *Measuring Minds: Henry Herbert Goddard and the Origins of American Intelligence Testing*. Cambridge University Press, 2001.

Zhang, Zhengyou. "Microsoft Kinect Sensor and Its Effect." *IEEE Multimedia*, 19(2): 4–10, 2012.

Zhao, Jieyu, Yichao Zhou, Zeyu Li, Wei Wang, and Kai-Wei Chang. "Learning Gender-Neutral Word Embeddings." In *Proceedings of the Conference on Empirical Methods in Natural Language Processing (EMNLP)*, 4847–4853, ACL, 2018.

Zhu, Qin. "Engineering Ethics Education, Ethical Leadership, and Confucian Ethics. *International Journal of Ethics Education*, 3(2): 169–179, 2018.

Zhu, Qin, Tom Williams, Blake Jackson, and Ruchen Wen. "Blame-Laden Moral Rebukes and the Morally Competent Robot: A Confucian Ethical Perspective." *Science and Engineering Ethics*, 26: 2511–2526, 2020.

Zhu, Yifei, Colin Brush, and Tom Williams. "Designing Augmented Reality Robot Guidance Interactions Through the Metaphors of Re-Embodiment and Telepresence." In *Proceedings of the IEEE International Conference on Robot and Human Interactive Communication (RO-MAN)*, 29–36, IEEE, 2024.

Zhu, Yifei, Ruchen Wen, and Tom Williams. "Robots for Social Justice (R4SJ): Toward a More Equitable Practice of Human–Robot Interaction." In *Proceedings of the ACM/IEEE International Conference on Human–Robot Interaction (HRI)*, 850–859, IEEE, 2024.

Zimring, Franklin. "When Police Do Not Need to Kill." In *The Ethics of Policing: New Perspectives on Law Enforcement*, edited by Ben Jones and Eduardo Mendieta, 107–119. New York University Press, 2021.

Zuboff, Shoshana. *The Age of Surveillance Capitalism: The Fight for a Human Future at the New Frontier of Power*. Profile Books, 2019.

Zuckerman, Ethan. "On Me, and the Media Lab." 2019. https://ethanzuckerman.com/2019/08/20/on-me-and-the-media-lab/.

Zuev, Dennis, and Gary Bratchford. "The Citizen Drone: Protest, Sousveillance and Droneviewing." *Visual Studies*, 35(5): 442–456, 2020.

Index

ACM, 202
Albuquerque Police Department, 196
Alexander, Michelle, 134, 177, 184, 201
Algorithms, 25, 123, 138–139, 141–142, 144, 160, 184
  and bias, 146
  and bureaucracy, 17, 186
  and categorization, 147–149, 154, 162, 164, 167
  and datasets, 59
  and decision-making, 50, 62, 145
  and discrimination, 145
  fraud-detection, 185
  and perception, 167
  and racialization, 157, 161
  recognition, 118, 163, 181
  and systems, 115, 151, 186
Amazon, 11, 14, 58–60, 62, 142, 148, 195
  Alexa, 11, 57, 76, 140
  Astro, 194, 207
  Ring, 181
Anti-racism, 46
Artificial intelligence (AI), 4, 26, 62, 67, 91, 148, 164
  "artificial," 62, 129
  and Cortana, 76
  and cybernetics, 49–51
  and Dartmouth AI conference, 50
  and ethics, 81
  and machine morality, 115
  models, 59, 146
  origins of, 50, 52
  pioneers of, 51–52, 59, 205
  and robots, 53, 84, 129, 213
  systems, 90, 94, 186, 221
  and technologies, 207
  and Whiteness, 50, 73
Artificial Intelligence Lightweight Android (AILA), 1–2, 17, 66, 70, 72–74, 91, 109, 133
Asimov, Isaac, 3, 6, 45–47, 49, 51, 111, 113, 117, 120, 132, 135
Automation, 46–48, 58, 180, 185
Axe, David, 195

Bayley, David, 190
Behaviors, 9–10, 21, 81, 121, 124, 132, 138, 156–158, 165–166, 184, 192
  codes of, 119
  emotional, 83
  human, 11, 13, 30, 54, 137, 173
  linguistic, 78
  and norms, 115, 120, 134, 159
  overt, 11
  robot, 12, 19, 87, 114, 119, 122–127, 130, 148, 204
  rules of, 117–118
  social, 7
  standards of, 149
  surplus, 84
  user, 14

Biases, 21, 140, 143
  algorithmic, 145–146
  anti-feminine, 94
  and facial recognition, 145
  gender, 81–82, 87, 166, 217
  misogynist, 90
  and patriarchy, 84, 109
  and race, 51, 82, 187
  and robots, 80, 85, 87, 91, 103, 167
  and shooter bias test, 82–83
  and White patriarchy, 69, 110
Big Tech, 14, 22, 60, 85, 98, 104, 129, 145
Black feminism, 19, 24, 131, 135, 218–219
Black Feminist Thought, 15, 17, 58
Black Lives Matter, 146–147
Black Panther Party, 189
Black people, 4, 52, 77, 119, 121–122, 143–144, 175–176, 180, 208
  and anti-Blackness, 83
  and Black Codes, 30, 38, 174
  and criminality, 58, 183–184, 186, 212
  and culture, 79
  and emancipation, 42, 170, 174
  enslaved, 4–6, 26, 29, 32–33, 37, 39–42, 47, 49, 60, 157
  and hypervisibility, 147, 156
  and invisibility, 147
  labor of, 6, 62
  and masculinity, 202
  and men, 49, 82, 130, 177, 179
  and police violence, 187, 196, 200, 205
  and racism, 89, 152, 154, 201
  and robots, 36, 92, 98–99, 103
  and slave uprisings, 43
  and speech recognition systems, 140
  and surveillance, 181
  and women, 54–56, 59, 64–65, 142, 151, 165, 178, 182
  and workers, 47–48
Black studies, 20, 155–156, 220

Čapek, Karel, 3–4, 6, 43–44, 49
Color-blind racism, 41, 149, 165
Colorism, 84
Communications of the ACM, 23
Computer chips, 1, 70
Computer graphics, 95
Computer science, 23–24, 48–49, 93, 105, 125, 144–145, 151–152, 163, 180, 183, 205
Computer vision systems, 137, 142–144, 147, 149–152, 154, 158, 160–163–164, 188, 199, 215
Computing, 23–24, 49, 99, 107, 156, 180, 182–183, 185
Confucian robot ethics, 218–219
Confucian role ethics, 124–128, 131, 134, 219
Crime, 34
  and criminal intent, 202
  and criminal records, 182
  drug, 177, 209
  fear of, 172
  fighting, 176–177, 182, 190, 193–194
  and high-crime neighborhoods, 184
  nonviolent, 144
  prevention, 173, 175–176, 190, 209
  protection from, 166
  race-neutral, 174
  rates, 178, 190, 198
  risk, 183
  sexual, 187
  violent, 185
Criminal justice system, 201

Danaher, John, 53
Darling, Kate, 100
Dartmouth Conference on Artificial Intelligence, 49–51
Darwin, Charles, 42
Data Feminism, 23, 128
Design, 4, 7, 23, 40, 90, 130, 144
  aesthetics, 109
  biases, 110

counterfactual design fiction, 104–105, 111
cross-stereotypic design, 91–92, 94
and dark design pattern, 14
gendered, 85–86
justice, 108, 111, 129, 134, 168, 218, 220
justice philosophy, 129, 134, 168
participatory, 107–108, 128, 219
and social identity design cues, 70–76, 80, 91–92, 98–102, 104, 133, 215–216, 218
and *The Steam Man of the Prairies*, 44
technology, 57, 77, 157, 220
trends, 3
and user-oriented design, 108

Education, 25, 142, 179, 184, 195, 221
and assessments, 141
computing, 99
and educational robots, 65, 119–120
engineering, 106
Engineering, 1–2, 23, 43, 49, 92
audio, 101
and Black engineers, 99
and machines, 62
patents, 40, 105
and pop culture, 71–72
and robotics, 26, 34, 41, 90
and robots, 36, 38, 45, 68, 99, 108
and social justice, 128, 210–211, 218, 220
solutions, 106, 220
English-language, 64
and institutions, 51, 64
and opportunities, 48
and technology, 118
Ethnicity, 16, 29, 49, 64, 101, 110, 120, 148, 150, 153–155
and heterogeneity, 79
and hierarchies, 88
and homogeneity, 67
Ethnic studies, 20

Face recognition technology, 54, 141–147, 160, 163, 166, 181, 188, 194, 200–201, 207
Facial expressions, 95
Facial structure, 39–40, 75
Femininity, 1, 53–54, 78, 91, 98, 167, 190
and features, 100
and feminine-presenting faces, 141–142
and labor, 56, 58, 85–86
and morphology, 70
and names, 76, 84
racialized, 65
and robots, 13, 16, 41–42, 55, 66, 71–75, 81–82, 86, 90, 93–94, 100, 110
and sexualization, 109
and stereotypes, 56
and voice assistants, 57, 84
and voices, 77, 101
White, 64

Gender, 6, 142, 147
and AILA, 72
and artificial intelligence, 50
biases, 80, 166
binary, 18, 151–152
blindness, 165
and class, 43–44, 50–51
cues, 70, 91, 101
equality, 92
fluidity, 97
heteropatriarchy, 72
identities, 68, 81, 104, 152, 161–163, 165
neutrality, 100–101
and oppression, 15–17
and race, 22, 26, 44, 51, 87, 138, 148–149, 156–158, 160, 164, 167, 199, 217
relations, 43
and robotics, 53, 69, 74–76, 78, 80, 94, 110, 148–149, 166

Gender (cont.)
  and sexism, 82
  and sex robotics, 86
  and sexuality, 21, 73
  and social identity design cues, 98, 100
  stereotypes, 93
  studies, 81
  violence, 151
  and Whiteness, 73, 205
Gender recognition, 152, 166

Heterosexuality, 57, 65–66, 100, 109, 149, 163–164
Howard, Ayanna, 8, 83
Human intelligence, 51
Human morality, 124–125
Humanoid robots, 1–2, 5, 42–43, 67–68, 73, 158
  and abuse, 13
  design of, 41, 53, 74, 76, 109–110
  features of, 100
  Japanese, 65
  and roboticists, 15
  as servants, 26
  and speech, 66
  White, 140
Hypersexuality, 71–72, 86

Intelligence, 84
  government, 182
  and masculinity, 91
Intelligence testing, 51
International Space Station, 1, 70

Juvenile justice system, 120

Kansas City Police Department, 202–203

Language capability, 108, 110, 200
Language-capable robots, 17, 20, 63, 77–78, 88, 91, 110, 114, 139, 141, 216
Language-interactive systems, 87
Language model design, 90
Large language models, 59–60, 79, 87–90
Law, 5, 20, 172, 182, 220
  and Black people, 30
  enforcement, 119, 146, 160, 166, 170–171, 174–177, 181, 185–186, 190–191, 194, 207, 212
  immigration, 193
  plantation, 29
  and racial categories, 153
Los Angeles Police Department, 183
Lowell (MA), 31–33, 35
  and mills, 33, 55
Lyon, David, 186

Machine morality, 113, 115–118, 120–122, 127, 129, 132–133, 135, 217
Mechanical animals, 34
Mechanical arms, 46
Mechanical bodies, 3, 5–6, 36, 42–44, 108
Mechanical devices, 34
Mechanical slaves, 37, 40, 45, 47–49
MicBot, 10
Mills, Charles, 5, 121–122
Moral action, 127
Moral agents, 9, 13, 132
Moral buffering, 198
Moral cognition, 115, 117–118
Moral communication, 115–116, 124–126
Moral competence, 133
Moral decision-making, 126
Moral ecosystems, 12
Moral harms, 13
Moral influence, 218
Moral norms, 9–10, 12, 21, 116, 120–121, 158
Moral philosophy, 6, 115, 121, 124, 128–132, 220

Moral principles, 130–134, 137, 219
Moral psychology, 20, 118, 220
Moral reasoning, 126
Moral self-cultivation, 127
Moral selves, 123
Moral standards, 113–114, 120

Native Americans, 5, 28, 36, 38, 153, 171, 176, 196
Natural language, 79, 88–90, 102, 110
New Jim Cobots, 7, 22, 115, 117, 120, 122, 124, 133–134, 138, 141, 147, 159, 163, 167, 188, 215, 217
New York Police Department, 147

Oppression, 16, 20, 47, 179, 208, 218
  and colonization, 176
  cycles of, 201
  gender, 15, 17
  of humans, 114, 137
  justification of, 85, 88
  patriarchal, 159
  race-based, 172, 175, 184
  reinforcement of, 18, 114, 118–119, 153
  and robots, 211
  structuring of, 18, 169, 212
  subversion of, 19, 215
  systems of, 88–89

People of color, 141, 162, 179, 181, 199
  communities of color, 176, 187
  and death rates, 90
  and enslavement, 178, 201
  and face recognition, 145
  and imprisonment, 143
  and invisibility, 146
  and labor, 64, 69
  men of color, 183
  and patriarchy, 157
  and robots, 46, 59, 84, 90, 104
  students of color, 119
  and surveillance, 194
  and violence, 141, 185, 211–212
  and War on Drugs, 178
  women of color, 24, 55–56, 59, 63–65, 144, 217
Police
  and Amazon, 14
  and Black men, 143
  cameras, 183
  and communities of color, 176, 179
  and "crime fighting," 175–177, 182, 185
  departments, 173–179, 203, 210
  drones, 192, 194
  Dubai, 166
  and face recognition technology, 144, 146
  and killings, 144
  militarization, 190–192, 203, 212
  reform, 176
  and robots, 22, 163, 168–170, 188–190, 192, 195–198, 200–202, 204–213, 217
  and slave patrols, 38, 171–172, 174–176
  and Spot robot, 33
  surveillance, 180–181, 183–184, 186–188, 192–194, 204
  and technology, 180
  violence, 107, 145, 147, 169–170, 195–199, 202, 206
Police officers, 204
  in Dubai, 109
  White, 18, 133
Police stations, 6, 169
Policing, 175–177, 188, 197, 200, 208–213
  antidrug, 189
  community, 190
  evolution of, 19, 22, 146, 168, 170, 179–182, 186, 201
  and facial recognition, 142
  model of, 173–174
  and over-policing, 187

Policing (cont.)
and participatory design, 108
predictive, 183–185, 202, 207
psychology of, 190
racialized, 174, 199
and robotics, 18, 145, 191, 194–195
and surveillance, 192, 204
system of, 30, 171–172, 193
and violence, 192, 206

Race, 79, 122, 153, 170
and America, 155
and artificial intelligence, 50, 148
and biological explanations, 88
categories of, 6, 15, 19, 157, 159
and class, 21, 29–30, 41, 51, 103, 175, 188
and gender, 70, 100–101, 110, 138, 148–149, 152, 156–157, 160, 162–167, 199
nature of, 31, 84
neutrality, 141, 174–175, 177–178
and oppression, 16, 44, 175
origins of, 26–27, 154
and politics, 52
and roboticists, 17
and robotics, 21, 24, 147, 152, 164
science, 121, 135
and science fiction, 45
and slavery, 39
and spatial dynamics, 161, 163
systems of, 22, 34, 179, 205, 217
and technology, 31, 44, 77, 200
and violence, 213
and White supremacy, 150
Racial justice, 146
Racism, 27, 37, 41, 48, 51–52, 87–90, 134, 165, 175, 190, 202
and Black people, 58
and categorization, 138, 167
and colorism, 84
and logics, 151, 156
and minstrel shows, 44
and policing, 186
and robots, 47, 49, 83
scientific, 150
and sexism, 86
and slavery, 47
and values, 120
Robot design, 18, 54, 63, 66, 71–72, 84, 104, 111, 113, 120, 122–123, 128, 132, 161, 166–167, 213, 217
and aesthetics, 69, 137
agender, 98
and anthropomorphism, 100, 194
and Blackness, 41, 54
and ethics, 117, 126, 219
and diversity, 92, 168, 220
and feminism, 128
and flexibility, 97
gendered, 80, 91, 165
humanoid, 110, 194
Japanese, 75
and language capability, 102
and masculinity, 94
and patriarchy, 21, 65, 109
and policing, 212
and popular culture, 68
and power, 17, 22
principles of, 106
and racial ontology, 152
and SAE, 79
sex, 53
social, 103, 107, 164, 208
and social identity design cues, 74, 80
status quo of, 20, 70, 103, 138
and trauma, 106
and Whiteness, 87
and White women, 65
Roboticists, 23, 53, 68, 92, 102, 109, 128–129, 170, 191–192, 221
and abolitionist robotics, 211
and algorithms, 159–160
and care ethics, 131
and cultural power, 21, 67, 69, 91
and data, 195

and design, 75, 107, 166
and disciplinary power, 111, 114, 117–120, 124, 134–135, 137–138, 167–168
and efficiency, 123
and engineering processes, 106
and ethics, 123, 126–127, 132, 147–149
and gender, 74, 76, 164–165
and humanoid robots, 110, 158
and human relationships, 132
Indigenous, 79
industrial, 46
and interpersonal power, 89
and Japan, 66
and language-capable robots, 91
and masculinity, 77, 197
and moral cognition, 126
and morality, 113, 115–116, 120, 124, 134
and obedience, 123
and oppression, 215
and patriarchy, 18, 24, 108, 110, 133, 166, 211–213, 218
and policing, 19, 201–202, 204, 207–208, 211–212
and power, 7, 9, 15–20, 24–26, 103, 159, 205–206, 216–219
and race, 76
and robotic slaves, 49
and sexism, 84
social, 13
and social identity design cues, 79–80, 99–100, 104
and social justice, 220
and stereotypes, 85
and structural power, 22, 169
and White supremacy, 21, 22, 90, 108, 147

Science fiction, 2–4, 26, 31, 33, 44
and Edward Ellis, 36
evolution of, 52
and Isaac Asimov, 45–46, 113, 121
and robots, 39, 67–68, 108, 216
Sex, 70, 73, 91, 151, 205
characteristics, 157
Sexism, 11, 27, 80–90, 92–94, 119, 175
Sex robots, 53–54, 71–72, 74, 86
Sexual abuse, 185, 187
Sexual desires, 5, 57, 69
Sexual dominance, 5, 41, 43, 86
Sexual harassment, 78
Sexual identity, 104
Sexuality, 19, 21, 73
Sexualization, 75, 78, 84, 91, 109
Sexual orientation, 16
Sexual stereotypes, 56, 88
Sexual violence, 86, 129, 152, 171, 187
Skin color, 39–40, 75–76, 83, 142–143, 157, 165
Slave patrols, 30, 38, 170–175
Social agents, 4, 9, 13, 39
Social behaviors, 7
Social beings, 3
Social cobots, 120
Social cognition, 117
Social contexts, 15
Social control, 205
Social dominance, 5
Social harms, 13
Social identity trait perception, 22, 78, 84, 138, 147, 156, 159, 161, 164, 166–167, 215, 217
Social imaginaries, 38, 46
Social inequality, 201
Social injustice, 91
Social institutions, 16
Social justice, 22, 105–106, 123, 133, 207, 211, 218, 220
frameworks, 128
work, 24
Social media, 7, 200
Social norms, 9–10, 21
Social power structures, 17

Social relations, 30, 49, 116, 124, 210–211
Social responsibility, 131
Social roboticists, 92
Social robots, 4, 8, 14, 22, 59, 76, 95, 102–103, 107, 119, 130–131, 138–139, 143, 164, 194–195, 208–209
Social role theory, 85
Social root causes, 19–20, 109, 134, 167, 213
Social science, 23
Social status, 42, 125
Social stratification, 29, 51, 70, 75–76
Social systems, 26, 30–31, 34, 67, 152, 154–155, 163
Speech recognition, 118, 139–141, 160
The Steam Man of the Prairies, 4–5, 36–37, 39–45, 52, 108, 134, 196, 200
Stereotypes, 17, 42, 55, 69, 80, 91, 99–101, 217
  gendered, 81–82
  patriarchal, 110
  racist, 27, 73
  reinforcement of, 102–103, 108–109
  and robots, 84–85
  sexist, 85, 87–88, 90, 93
  subversion of, 94
  unfeminine, 56
  capitalism, 135, 149
  gendered, 22, 138, 158, 168
  police, 180–181, 183, 186–187, 192–195, 202, 204, 208
  racialized, 181, 195
  and robotics, 119, 137, 201
  state, 180
  systems, 14, 55
  technologies, 137, 147, 156–158, 160, 163–164, 166–167, 182, 188, 190
  and violence, 192, 202–203, 208
Surveillance, 16, 146, 161, 171, 184, 200

Turk, 3–4, 59–60, 62–63, 148, 224n8

Universal law, 6

Violence, 22, 40, 88, 171, 188, 200, 211–213
  administrative, 185–186
  data, 146
  epistemic, 120
  gender, 151
  graphic, 71
  police, 107, 145, 147, 169–170, 180, 188, 196–199, 202, 206, 208–210
  racial, 175–176, 186–187, 195, 203
  sexual, 86, 129, 152, 171, 187
  and surveillance, 119, 192, 202
  tendency toward, 28
  workers, 206
Visual recognition, 161, 165
Voice assistant design, 93

Wagner, Alan, 8
Wakeword-driven design, 11
White Americans, 5, 7, 26, 58, 64, 76–77, 122, 176–177, 205
White patriarchy, 70, 85, 89–90, 110, 135, 138, 157, 164, 166–167, 170, 208–209
Whiteness, 38
  and artificial intelligence (AI), 50, 73, 76, 109, 133, 140, 157
  neutrality of, 72
  privileges of, 24
  and robots, 79, 87
  and voice assistants, 77
  and biases, 84
  and norms, 81, 115, 117, 120, 133–134
  and order, 59
  and power structures, 7, 15, 20–22, 24, 26–27, 49, 53, 55, 58, 64–66, 68, 87, 103, 108, 114, 159, 166, 200, 210–211, 213, 217

and robot imaginary, 2, 6–7, 27, 46, 67–69, 71–72, 80, 91, 100, 104, 108–109, 111, 113, 123, 132, 159, 163, 201, 215–216
Williams, Robert, 143–144
Winkle, Katie, 8, 13, 92–93, 105–107, 128–129, 166
Workers, 6, 32, 59, 62
and automation, 58
Black, 47–48, 174
care, 66–67
of color, 141
enslaved, 174
exploitation of, 60, 129, 148
labor of, 68
and Lowell's mills, 33
needs of, 105
revolts of, 43
and teleoperation, 65
violence, 206

Publisher contact:
The MIT Press
Massachusetts Institute of Technology
77 Massachusetts Avenue, Cambridge, MA 02139
mitpress.mit.edu

EU Authorised Representative:
Easy Access System Europe, Mustamäe tee 50,
10621 Tallinn, Estonia
gpsr.requests@easproject.com

Printed by Integrated Books International,
United States of America